The McGraw-Hill
Telecommunications
Factbook

Other McGraw-Hill Communications Books of Interest

The McGraw-Hill Telecommunications Factbook

Joseph A. Pecar

Roger J. O'Connor

David A. Garbin

CPE - customer premises equipment

McGraw-Hill, Inc.
New York St. Louis San Francisco Auckland Bogotá
Caracas Lisbon London Madrid Mexico City Milan
Montreal New Delhi San Juan Singapore
Sydney Tokyo Toronto

Library of Congress Cataloging-in-Publication Data

Pecar, Joseph A.
    The McGraw-Hill telecommunications factbook / by Joseph A. Pecar,
Roger J. O'Connor, and David A. Garbin.

        p.   cm.
    Includes index.
    ISBN 0-07-049183-6 (pbk.)
    1. Telecommunication. I. O'Connor, Roger J. II. Garbin, David
A. III. Title. IV. Title: Telecommunications factbook.
    TK5101.P334 1992
    621.382—dc20   92-24340
                                                                          CIP

    6 7 8 9 0   DOC/DOC   9 9 8 7 6

ISBN 0-07-049183-6

*The sponsoring editor for this book was Neil Levine, the editor was April D. Nolan,
and the director of production was Katherine G. Brown.
This book was set in Century Schoolbook. It was composed by
TAB-MCGRAW-HILL.*

Printed and bound by R. R. Donnelley & Sons Company.

Contents

Part 3   Data Telecommunications Fundamentals

Part 4   Data Services

Introduction

The worldwide telecommunications industry, currently growing at 10% to 15% per year, reached $500 billion in 1990. As the established leader for new telecommunications technologies, products, and services, the United States accounts for nearly one half of this global market. Because businesses expend a large share of the dollars that create the market, business people have a vital need for knowledge that permits them to select the array of services providing the greatest benefit and competitive advantage at the lowest cost.

To succeed in today's workplace, people must be able to exchange information quickly and accurately. Job content is therefore increasingly influenced by telecommunications. Required business management skills now include knowledge of available telecommunications services, their application to changing organizational needs, the ability to work with technical professionals, and the expertise to acquire and use telecommunications services efficiently and cost-effectively.

Over the past decade, U.S. business telecommunications usage has grown at an unprecedented rate. At the same time, sweeping regulatory changes and the rapid development of new technologies have created so many options that merely remaining abreast of developments has become a major managerial challenge.

Even for telecommunications professionals, keeping up-to-date on the plethora of public and private offerings for the delivery of voice, data, image, video, and other telecommunications services is now a formidable task. Moreover, it is difficult for engineers to fully appreciate day-to-day business needs in order to support them with telecommunications applications.

By the same token, it is difficult for business people to articulate their needs in terms that telecommunications professionals can understand—and that suppliers can address with accurate, competitive proposals. What is required is a "knowledge bridge" between business needs and the growing spectrum of telecommunications offerings. *The McGraw-Hill Telecommunications Factbook* is designed to fulfill that requirement.

Objectives

The principal objective of this book is to provide a comprehensive introduction and insightful perspectives into modern telecommunications services and their underlying technologies. A second objective—no less important than the first—is to employ a presentation style easily understood by government and commercial telecommunications planners, managers, users, and professionals who do not have the time to sift through multiple publications, complex formulae, and mathematics only to be forced to draw their own conclusions regarding technology, performance, and market alternatives.

All important telecommunications services and technologies are treated, but the quantity of information is limited to that needed for a complete understanding. In addition, rather than just treating topics individually, expert interpretations provide a valuable grasp of "bottom line" relationships among emerging services, technologies, and industry standards.

Simplicity of presentation style does not sacrifice the ability to familiarize readers with industry terminology and essential concepts—which is often the case with introductory material. To accomplish this, we systematically present basic definitions as part of explanations of larger concepts. This equips the reader not only with terminology, but also with rationale for real-world applications, a tremendous advantage for thorough understanding and memory retention.

Plan of the Text

Although many of its topics are the subject of individually published textbooks, our book's material has been carefully selected so that the reader does not have to deal with more information than is necessary to achieve the learning objectives.

Under this approach, new material is placed into the context of material already presented. Telephony and the historical development of voice networks is treated first, since the vast majority of U.S. network traffic is still voice, with the addition of other information such as data, image, and video. It is from today's voice networks that tomorrow's integrated information networks will evolve.

Part 1, Telecommunications Fundamentals, begins with terminology and background material that can be covered in several hours. Because government legislation and regulation has had such a profound impact on U.S. telecommunications, a historical review of the structure it has imposed is presented as a foundation for succeeding technical material.

Part 2 describes voice services, Parts 3 and 4 treat data telecommunications fundamentals and data services, respectively, and Part 5 discusses integrated services. In Part 6, other telecommunications technologies are explored, and the final Part 7 provides conclusions and outlooks on the future. All parts and chapters of the book emphasize available telecommunications services, and corresponding business applications.

Of course, writing about any rapidly developing technical subject is much like aiming at a moving target. While much of its content addresses telecommunications principles and terms of reference that are relatively unchanging, the book's modular structure is designed to be augmented by new editions covering future developments in U.S. telecommunications.

Telecommunications Fundamentals

1

Definitions, Terminology, & Background

This chapter introduces telecommunications by defining its terms. Next it describes the current telecommunications structure in the United States, which has been shaped by regulatory initiatives and judicial decisions. Telephony and the historical development of voice networks are emphasized, since the vast majority of traffic carried on today's networks is still voice. Moreover, from today's voice networks, tomorrow's integrated voice/data networks will evolve.

The general background and terminology provided in this chapter serve as context for succeeding chapters. Topics introduced here are revisited later in greater detail, describing operational and technical characteristics of telecommunications technologies and services.

Chapter 2 discusses concepts, techniques, and device capabilities essential to a basic understanding of telecommunications. Later chapters treat voice, data and integrated services, and the telecommunication systems, networks, hardware, and software components that deliver them.

To make this book easy to use, discussions focus on telecommunications services that support business applications. We first provide the reader with a working knowledge of telecommunications services and technologies, and then match them to business applications. Later sections furnish business telecommunications planners and users with guidance needed to develop *technical specifications* and *request for proposal* (RFP) packages, as well as to negotiate contracts with suppliers. In short, this book provides the means to apply advanced telecommunications capabilities to modern business needs, enhancing efficiency, profitability and competitive advantage.

Telecommunications Defined

What is telecommunications? The word is derived from the Greek *tele*, "far off," and the Latin *communicare*, "to share." *Communications* is the process of representing, transferring, interpreting, or processing information (data) among persons, places, or machines. The process implies a sender, a receiver, and a transmission medium over which the information flows. It is important that the meaning assigned to the data is recoverable without degradation.

More specifically, telecommunications is any process that enables one or more users (persons or machines) to pass to one or more other users information of any nature delivered in any usable form—by wire, radio, visual, or other electrical, electromagnetic, optical, acoustic, or mechanical means. A telecommunications service is a specified set of information transfer capabilities delivered to a group of users by a telecommunications system. In this book, telecommunications services are treated in terms of voice services (Chapters 8 through 11), data services (Chapters 12 through 15), and integrated services (Chapters 16 and 17). Services not covered under these headings are discussed in Chapter 18.

At the highest level, business applications are unique aggregations of telecommunications services that satisfy particular enterprise needs, e.g., medical/health care, hospitality, airline, retailing, etc. Most lower-level business applications, such as station-to-station calling within a business premises, are enterprise-independent. All businesses use them.

It is necessary to distinguish between telecommunications *services* and the telecommunications systems/networks/equipment/components (*facilities*) by which the services are delivered, since one of the major decisions users must make is the extent to which they will satisfy their telecommunications needs via privately owned facilities versus obtaining them from service providers, such as telephone companies, that retain ownership of the facilities.

This distinction might appear trivial, but economic trade-offs among these options are extremely complex and are the subject of continuing controversy. Distinctions between options are often subtle. One example involves *virtual private network services*, provided by public telephone companies. Here the provider retains ownership but offers services that are intended to be indistinguishable from those obtained when privately owned or dedicated-leased facilities are used. Arrangements described below include obtaining services using shared public facilities, leased but dedicated private facilities, leased virtual private facilities, and privately owned facilities. Services delivered over facilities available to the general public are usually provided under tariffs approved by regulatory agencies.

System, Network, & Component Descriptions

The following network definitions and descriptions are derived from those provided by AT&T in Bibliography Reference 9.

A *telecommunications network* is a system of interconnected facilities designed to carry the traffic that results from a variety of telecommunications services. The network has two different but related aspects. In terms of its physical components, it is a facilities network. In terms of the variety of telecommunications services that it provides, it can support many traffic networks, each representing a particular interconnection of facilities. (The distinction will become more evident later.)

A network consists of nodes and links. *Nodes* represent switching offices, facility junction points, or both, and *links* are transmission facilities. *Traffic* is the flow of information within the network, among nodes, over links.

Three characteristics influence the nature of a network. First, traffic must be carried over large geographic areas. Second, traffic is generated at virtually any time, although the duration of each call may be short. Third, the ability to exchange information (i.e., connections) must be available with relatively short delays.

Figure 1.1 illustrates key aspects of a telecommunications network. Part A of Fig. 1.1 shows a situation in which no switching is used and pairs of telephones at four *end points* (business locations) are directly connected by transmission links. This example illustrates that a single user would need three telephones and three transmission links to satisfy his or her needs. Such a network is expensive since it requires the greatest number of telephones and transmission links. The design is improved by including a switch at each end point location, as shown in Part B of Fig. 1.1, allowing each telephone to access any one of three transmission links. Part C shows the most efficient design with switching at a central point, which minimizes the number of switches and transmission paths.

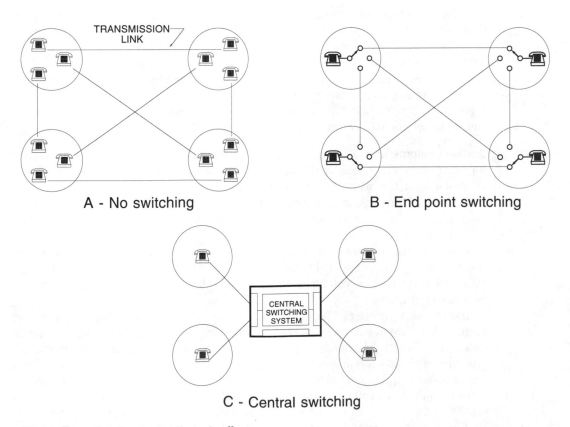

Fig. 1.1 Transmission and switching trade-offs.

In a network with no switches, if there are n locations, the number of independent transmission paths required is $n \times (n-1)$ divided by 2, or 6 for this example. The number of required telephones is $n \times (n-1)$, or 12 for this case. With central switching, the number of telephones and transmission paths required is reduced to n. This most economical network design is achieved at a performance price: the number of simultaneous connections through the network is limited. While the network in Fig. 1.1A, supports six simultaneous connections, that of Fig. 1.1C supports only two simultaneous connections.

A limited capacity for simultaneous calls is not normally a problem in well-engineered networks since the simultaneous use by all or even many users is unlikely. The simplified example shown in Fig. 1.1 illustrates one of the more important considerations in network design—the trade-off between network resource sizing and the probability that calls cannot be completed. This performance parameter is referred to as *grade of service* (GOS) and will be discussed in Chapters 4 and 7.

The components of a facilities network are divided into three categories: switching, transmission facilities, and station equipment.

Switching systems. Switching systems interconnect transmission facilities at various locations and route traffic through a network.

Transmission facilities. Transmission facilities provide the communication paths that carry user and network control information between nodes in a network. In general, transmission facilities consist of a medium (e.g., the atmosphere, copper wire cable, fiber-optic cable) and various types of electronic equipment located at points along the medium. This equipment amplifies or regenerates signals, provides termination functions at points where transmission facilities connect to switching systems, and can combine many separate sets of call information into a single "multiplexed" signal to enhance transmission efficiency.

Station equipment. Station equipment is generally located on the user's premises. Its function is to transmit and receive user information (traffic), and to exchange control information with the network to place calls and access services from the network. This information is conveyed in the form of electrical signals. Station equipment is one of several varieties of what is known as *customer premises equipment* (CPE).

Switching and transmission systems provide for *signaling*, a major network function that is rapidly advancing in standards and techniques. Signaling is the transmission of information to establish, monitor, or release connections and to control network operations.

In many cases there is a hierarchy or a series of facilities networks used to complete calls between station equipment. For example, long-distance calls usually involve two local telephone company networks, one long-distance network, and customer premises equipment (comprising inside wiring, station, and other customer-owned equipment). Thus, an end-user-to-end-user facility network includes station equipment, transmission facilities, and switching systems. However, station equipment and other CPE is normally owned by the users, so it is not included in either local or long-distance telephone company network segments.

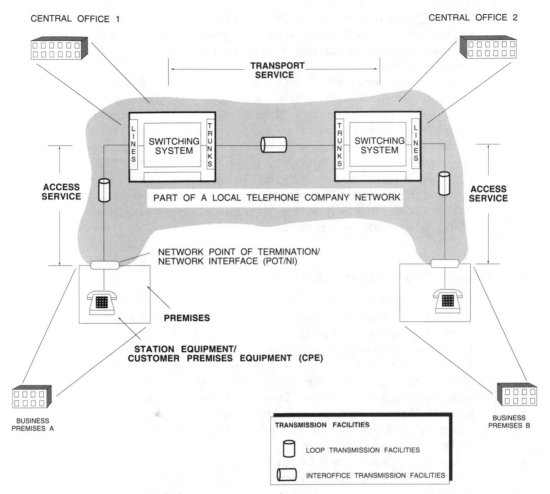

Fig. 1.2 Components of a local telephone company network.

Figure 1.2 illustrates the components a local telephone company must have to support local calling. The network shown comprises two switching *central offices* (COs), a term referring to a telephone company building in which network equipment is installed. The number of central-office switches in local telephone companies depends on the size of the area served, but in this example two switches are sufficient to illustrate the local network operation.

Loop transmission facilities connect switching systems to customer station equipment located in various business premises throughout the served area. A *local loop* is a transmission path between a user/customer's premises and a central office. The most common form of loop, a pair of wires, is called a *line*. The connection between the telephone company's network and CPE is formally called the network *point of termination* (POT) or, alternatively, the *network interface* (NI).

Interoffice transmission facilities connect the telephone company's switching systems. In a network, a *trunk* is a communication path connecting two switching systems, used to establish end-to-end connections between loops to customer station equipment.

As described above, station equipment is furnished by the user to obtain network services from the local telephone company, i.e., completing calls to and from any other user's station equipment. More specifically, the network services needed to complete a call include originating *access service*, to connect his station equipment to the local telephone company's central-office switch serving his area; *transport service*, to route the call through the network to the central office serving the called party; and connection to the called station equipment via terminating access service, again via the local loop. This segmentation of telecommunications network services into access and transport services applies to all network configurations and will be used (and expanded upon) throughout this book.

Telecommunications Structure in the U.S.

In the United States, federal and state governments are authorized to intervene in the telecommunications marketplace when free enterprise is deemed inadequate to ensure the economical supply and distribution of products and services. Telecommunications regulation at the federal level involves the Communications Act of 1934 (as amended), administered by rulings of the Federal Communications Commission (FCC). State regulatory bodies, such as *public utility commissions* (PUCs), handle state-level regulation. The most influential force in U.S. telecommunications is, however, the U.S. District Court's *Modification of Final Judgment* (MFJ), which concluded the U.S. Justice Department's antitrust suit against AT&T by modifying an earlier (1956) consent decree.

Regulatory impact has been so profound that an historical review of the structure it has imposed on telecommunications in the United States is a prerequisite to further discussions.

History

The American Telephone and Telegraph (AT&T) company dominated both local and long-distance markets until 1982, by which time it had reached $155 billion in assets and over 1 million employees. AT&T, the parent company of an entity known as the Bell System, served over 144 million telephones through 24 Bell-operating telephone companies. The remaining 36 million telephones were served by some 1450 independently operating telephone companies.

Since its incorporation in 1885, AT&T had been the subject of recurrent Department of Justice antitrust actions. Following burgeoning growth in the 1950s and 1960s, AT&T's market dominance prompted the 1975 Justice Department suit. After seven years in the courts, AT&T finally accepted a restructuring agreement (the MFJ) which was approved by U.S. District Court Judge Harold Greene in August of 1983, and became effective January 1, 1984.

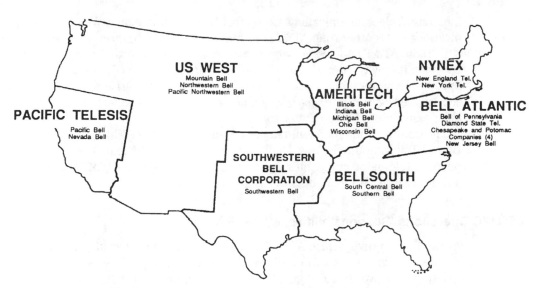

Fig. 1.3 The 22 Bell-Operating Companies (BOCs) organized into seven regional Bell companies (RBHCs) that were created as a result of MFJ.

The breakup or *divestiture* of AT&T resulted in the creation of 22 *Bell-operating companies* (BOCs) organized into seven *regional Bell holding companies* (RBHCs), also called *regional holding companies* (RHCs), as shown in Fig. 1.3. In addition to divestiture of the BOCs, the MFJ called for the creation of BOC service areas, called *Local Access and Transport Areas* (LATAs), to distinguish local from long-distance calling markets.

A LATA is a geographic area within each BOC's territory that has been established in accordance with the provisions of the MFJ, which defines the area in which a BOC may offer its telecommunications services. In 1989, there were 198 LATAs in the United States.

An exchange carrier (or *local exchange carrier*, LEC) is any company, BOC or independent, which provides *intra-LATA* telecommunications within its franchised area. Intra-LATA is a term used to describe services, revenues, functions, etc., that relate to telecommunications originating and terminating within a single LATA. *Independent telephone companies* (ITCs), also called *non-Bell exchange carriers* (NBECs), continue to provide telecommunications services within their franchised areas. In this book, the term LEC will be used when referring to either BOC or independent local exchange carriers.

Inter-LATA is a term used to describe services, revenues, functions, etc., that relate to telecommunications originating in one LATA and terminating outside that LATA. An *interexchange carrier* (IXC) is a company which provides telecommunications services between LATAs.

The major IXCs include AT&T, MCI, US Sprint, Allnet, ITT, Cable & Wireless, ATC/Microtel, and members of the National Telecommunications Network (NTN). NTN members include Advanced Telecommunications Corp., Consolidated Networks

Inc., Litel Telecommunications Corp., RCI Long Distance, and Williams Telecommunications. IXCs other than AT&T are referred to as *other common carriers* (OCCs).

In 1985, AT&T had a 82% market share, MCI 7%, US Sprint 4%, and all others 7%. By 1992, it is estimated the AT&T will have dropped to 69% with MCI having 12%, Sprint 8% and all others 11%. Altogether, if one counts carriers, resellers, rebillers, service aggregators, subaggregators, and agents, there are over 1500 long distance businesses in the United States today.

The dominant independent telephone companies are GTE (including Contel) and United Telecommunications. Due to mergers and acquisitions, the number of independent telephone companies had dropped to 1371 by the end of 1988, according to a United States Telephone Association report.

LEC/IXC Operations, Responsibilities, & Restrictions

At the time of divestiture, AT&T in its 1983 annual report maintained that the Bell System had given the United States the biggest and the best telecommunications system in the world and that it had achieved the goal of making telephone service available to everyone in the country at an affordable price. Most people still would not challenge that claim. Indeed, a post-divestiture concern is whether competition within a market once dominated by a monopoly can maintain quality, integrity, and survivability (from a national security and emergency preparedness perspective) of the nation's telecommunications at predivestiture levels.

This question involves the technical ability of the divested Bell operating companies (BOCs) and IXC facilities to conduct business operations that result in effective competition, without undermining quality, integrity, and survivability. That is, if the Modification of Final Judgment (MFJ) results in a technical or business environment within which only AT&T can provide quality IXC service or succeed financially, or if independent equipment manufacturers are denied market share due to predatory practices of the BOCs, then the great efforts associated with the MFJ, the 1968 Carterfone decision, and Computer Inquiry I/II/III, will have been for naught. (Bibliography Reference 11 treats in detail the legislative, judicial, and regulatory impact on U.S. telecommunications.)

Without attempting to predict the outcome of such issues, this book describes LEC/IXC operations, responsibilities, and restrictions specified in the MFJ and in subsequent judicial and FCC rulings.

The best way to describe post-MFJ telecommunications facilities, service responsibilities and restrictions is to trace an inter-LATA call via LEC and IXC *Public Switched Telephone Networks* (PSTNs). The PSTN includes those portions of LEC and IXC networks that provide public switched telephone services. Figure 1.4 will be used to analyze an inter-LATA call and to introduce additional terms arising from the MFJ.

For this example, a calling party in LATA x places a call to a receiving party in LATA y. The calling party is provided with dial tone via PSTN access services from his LEC. (Types of access services and their signaling, protocol, and operating characteristics are described in Chapter 9.) LEC PSTN access services are used for both intra-LATA and inter-LATA calls, with the central office (CO) or *end office* (EO) switches interpreting the digits dialed by the calling party and routing the call accordingly.

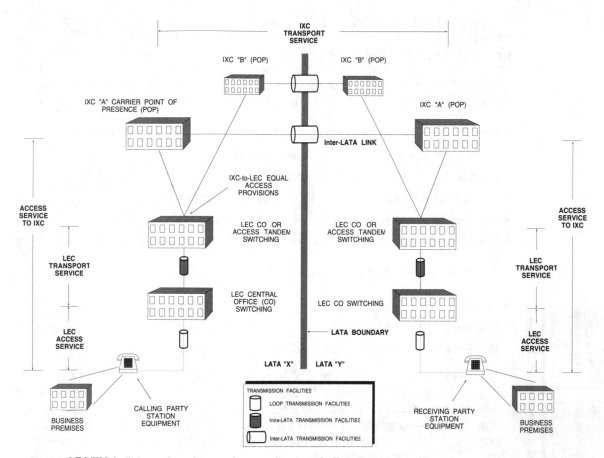

Fig. 1.4 LEC/IXC facilities and services used to complete inter-LATA calls.

An EO is a BOC or an ITC switching system within a LATA where loops to customer stations are terminated for purposes of interconnection with each other and to trunks. (Note that CO and EO can be used interchangeably.)

For our example inter-LATA call, the LEC routes the call through its intra-LATA network from the originating EO to one of its *access tandems,* which connects to trunks terminating at an IXC *point-of-presence* (POP). An access tandem (AT) is a BOC (i.e., LEC) switching system that provides a traffic concentration and distribution function for inter-LATA traffic originating or terminating within a LATA. The AT provides the IXC with access to more than one EO within a LATA, although more than one AT may be required to provide access to all EOs within a LATA. In a LATA served by more than one IXC, the IXC to be used is selected (on a permanent or a call-by-call basis) by the calling customer.

A POP is a physical location within a LATA at which an IXC establishes itself for the purpose of obtaining LATA access and from which the LEC provides access services to its intra-LATA networks. An IXC may have more than one POP within a LATA, and the POP may support public and private, switched and non-switched services.

Once the call is accepted by the IXC, the IXC provides transport service through its network to its POP in LATA y, for our example. From that point the call is completed via the LATA y LEC network to the called party's station equipment, in a fashion similar to that described for the call originating LATA x.

The MFJ restricts the lines of business that the BOCs and the AT&T may engage in. For example, under the terms of the MFJ, the 22 divested BOCs are allowed to:

- Offer local exchange service within specified geographical operating areas
- Provide equal access to BOC facilities for all IXCs
- Provide Yellow Pages directory publications
- Sell but not manufacture customer premises equipment
- Provide cellular mobile communications services
- Provide voice storage retrieval, voice messaging, and electronic mail services

Under MFJ terms, BOCs may not:

- Provide interexchange services
- Provide information services
- Provide any product or service that is not tariffed or otherwise regulated

The provisions of the MFJ are constantly under review and revision. In October 1991, for example, the Supreme Court cleared the way for the seven regional Bell holding companies to enter into the electronic information services business by rejecting an emergency request from the American Newspaper Publishers Association to rescind a recent FCC ruling that permits BOCs to offer video programming services (known as *video dial tone services*).

The MFJ imposes no obligation on independent telephone companies, nor does it restrict the lines of business in which they can engage or the types of service they may provide. However, GTE is bound by a consent decree similar to but not identical to the MFJ.

There is ambiguity in the MFJ reference to "exchange" services. Prior to divestiture, *exchange services* commonly meant any services delivered by central office switches. The MFJ did not, however, prohibit BOCs from providing intra-LATA transport services between their owned CO switches (exchanges). Rather, BOCs were prohibited from offering inter-LATA services, (confusingly described by the MFJ as "interexchange" services).

Inter-LATA carriers are described by the MFJ as providing interexchange services, when in fact, IXCs are prohibited from owning the facilities (EOs, ATs, or transmission) used for intra-LATA access or transport.

LEC/IXC Tariffs

IXCs other than AT&T were originally regulated by the FCC. In the Competitive Carrier proceeding (Docket 70-252), the FCC decided to forbear from regulating OCCs.

As the dominant IXC, AT&T is still regulated and required to file *tariffs* for its services. A tariff fixes the allowed rate (price) for a specific telecommunications service, equipment, or facility, and constitutes a contract between the user and the telecommunications carrier.

Tariffed services and rates are established by a formal process in which carriers submit filings for government regulatory review, possible amendment, and approval. Tariffs, therefore, contain the most complete and precise descriptions of carrier offerings. Further, tariffed offerings cannot be dropped or changed without government approval. LEC tariffs are under the jurisdiction of state authorities. Chapters 7 and 10 describe tariffs and network design and engineering techniques used to compare and evaluate the extent to which tariffed carrier offerings satisfy business applications.

IXC Access to LEC Networks

One of the more significant directives of the MFJ ordered all BOCs to provide IXCs with "equal access" to BOC networks. Switched IXC access to BOC networks means that, on a call-by-call basis, calls are routed through the LEC PSTN to the IXC POP. Because equal-access service is based on dialing, signaling, routing, and transmission plans that require particular EO capabilities, access services other than equal access continue to be offered to IXCs for customers served by nonconforming EOs.

LEC access services are characterized by *line-side access* or *trunk-side access* at the EO switching system. *Feature Group A* access to the LEC networks is through a two-wire, line-side connection to the EO. This service exhibits the poorest quality, and, because it does not support *automatic number identification* (ANI), users must first dial an IXC local number and then, after receiving a second dial tone or a recorded message, enter up to 14 more digits (for caller identification and billing purposes) in addition to the dialed long-distance number.

ANI is the capability of an EO switching system to automatically identify the calling station. ANI is usually used for LAMA or CAMA. *Local automatic message accounting* (LAMA), is a process using equipment in an end office for automatically recording billing data for local message rate calls and for station-to-station toll calls. *Centralized automatic message accounting* (CAMA) is an arrangement that provides detailed billing information at a centralized location other than the end office.

Feature Group A access can result in local message unit charges in addition to the IXC long-distance charges. It is no longer a principal IXC access method but remains useful for making long-distance calls on the road. *Feature Group B access* is provided via higher-quality four-wire trunk-side connections and usually avoids local message unit charges. Like Feature Group A dialing, Feature Group B permits users to access an IXC by dialing 950-0/1xxx, where xxx is a *carrier identification code* (CIC) assigned by Bell Communications Research (Bellcore). *Bellcore* is an organization (jointly owned by the seven regional holding companies) that conducts research, establishes industry-wide standards, acts as a centralized technical resource, and provides management services to the RBHCs.

Feature Group C, an access arrangement available only to AT&T, offers "1 Plus" dialing, using four-wire trunk-side connections and ANI at most locations. Feature Group C was largely in place prior to divestiture. Direct-access line trunks run directly from EOs to AT&T POPs, further enhancing quality and speed of service. As upgrades to equal access are completed, EOs and AT&T convert from Feature Group C to Feature Group D.

Feature Group D represents the long-term solution for equal access and, except for the fact that calls are generally routed through the LEC PSTNs rather than directly from EOs, offers service comparable to Feature Group C.

As indicated earlier, compliance with "equal access" has required conversion and upgrade of local EO facilities which, at this point, is nearly complete. Also, as IXC competition began in the early 1980s, not all OCCs had the technical capability to take advantage of trunk-side access.

To equalize the competitive positions among new IXCs and AT&T following divestiture, the BOCs were ordered to charge lower origination and termination charges for access arrangements providing lower-quality service. These charges are paid by IXCs to LECs and constitute a large percentage of the long-distance charges that IXCs charge their customers. For example, in 1989, the amount of these charges paid by AT&T to LECs was roughly half of its revenue. As equal access conversion is completed and as customers demand higher-quality service, the large user cost differences among IXCs attributable to "unequal" LEC access will tend to disappear. Figure 1.5 illustrates the range of access capabilities available to IXCs.

One of the consequences of equal access and the use of access tandem switches as points of traffic concentration and distribution for all IXCs serving a LATA, is that

FEATURE GROUP	QUALITY	AUTOMATIC NUMBER IDENTIFICATION (ANI)	DIALING	COMMENTS
A	TWO - WIRE LINE-SIDE ACCESS	NO	UP TO 21 DIGITS	•POOREST QUALITY •MAY INVOLVE LOCAL "MESSAGE UNIT" CHARGE •NOT COMMON ANYMORE
B	FOUR - WIRE TRUNK-SIDE ACCESS	PARTIAL	UP TO 21 DIGITS	•HIGH QUALITY •NO MESSAGE UNIT CHARGES •950-0/1XXX ACCESS CODE DIALING
C	FOUR - WIRE TRUNK-SIDE ACCESS	YES	"ONE PLUS" OR "DIAL 1"	•HIGH QUALITY •AT&T ONLY •GENERALLY DIRECT LEC "EO" TO AT&T "POP" ROUTING •MUST CONVERT TO FEATURE GROUP D WHEN LECS CONVERT TO EQUAL ACCESS
D	FOUR - WIRE TRUNK-SIDE ACCESS	YES	"ONE PLUS" OR "DIAL 1"	•REPRESENTS THE EQUAL ACCESS DIALING, SIGNALLING, ROUTING, AND TRANSMISSION ARRANGEMENT

Fig. 1.5 LATA access services (feature groups) provided by LECs to IXCs.

the ATs can represent *single points of failure* for all long distance service, even if a customer elects to use multiple IXCs. The consequences of this reliability vulnerability were graphically illustrated several years ago in the Hinsdale, Illinois, fire, which interrupted long-distance service completely for a large number of users in the Chicago area.

This chapter has introduced telecommunications networks and services, using terminology and providing definitions currently in use by the industry. To this point, voice networks and services have dominated the discussion—an appropriate beginning, as noted earlier, since the vast majority of traffic carried is still voice, with the recent addition of other types of information such as data and facsimile.

Although significant differences exist between voice and data communications, economic and performance benefits have already produced common voice and data designs for transmission systems and intrafacility wiring. Later chapters introduce the full range of modem voice, data and integrated services and highlight the similarities and distinctions between voice and data facilities.

2

Basic Concepts, Techniques, & Devices

Analog Electrical Signals

In telecommunications networks, information is transferred in the form of *signals*. A signal is usually a time-dependent value attached to an energy-propagating phenomenon used to convey information. For example, an audio (sound) signal is one in which the information is characterized by loudness and pitch. Until the early 1960s, the PSTN evolved as an *analog* network. The meaning of the term "analog" in networks is illustrated in Fig. 2.1. Figure 2.1A is a pictorial representation of a voice sound wave, i.e., how the compression and expansion of air would look if sketched on paper as a function of time. A microphone or *transducer* (the telephone handset in the figure) intercepting the sound wave converts differences in acoustic pressure to "analogous" differences in electrical signal amplitudes, as shown in Fig. 2.1B. Once converted to electrical format, these signals can be transported through networks to loudspeakers (other transducers), which convert the electrical signals back to sound waves.

Time sketches or representations such as those shown in Fig. 2.1A and B are referred to as *waveforms*.

An *analog electrical signal* is a continuous signal that varies in some direct correlation with an impressed phenomenon, stimulus, or event that bears intelligence. Sound waves and their electrical analogs are characterized by loudness (*amplitude*) and *pitch*. Analog signals can assume any of an infinite number of amplitude values or states within a specified range, in accordance with (analogous to) an impressed stimulus. Pitch refers to how many times per second the signal swings between high and low amplitudes.

Simple sound waves and signals can be made up of only a single tone, like a single note on a piano, as shown in Fig. 2.1C. In this case the waveform consists of

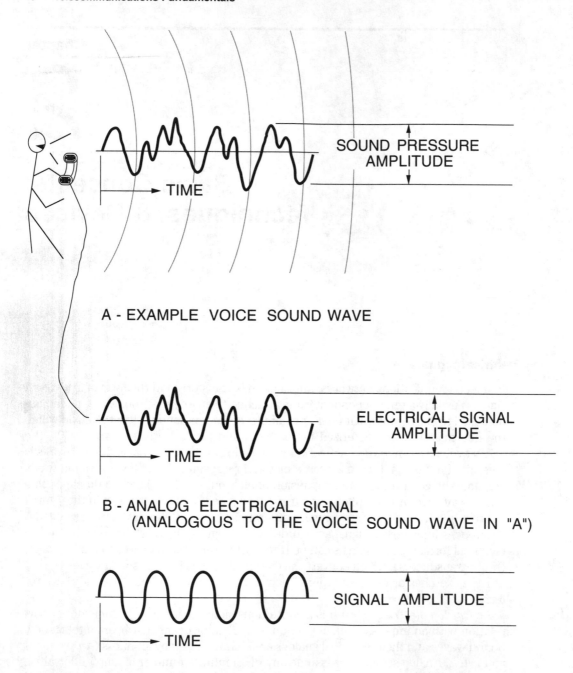

A - EXAMPLE VOICE SOUND WAVE

B - ANALOG ELECTRICAL SIGNAL
     (ANALOGOUS TO THE VOICE SOUND WAVE IN "A")

C - SINGLE FREQUENCY ANALOG ELECTRICAL SIGNAL

Fig. 2.1 Analog voice acoustic and electrical signals.

repeating identical *cycles* and is said to be of a *single frequency*, measured as the number of cycles that occur in one second of time. In communications, frequency was traditionally expressed in cycles per second (CPS), but is now expressed in *hertz* (Hz), still equal to one cycle per second. Thus, 1000 cycles per second is equal to 1000 hertz, or a *kilohertz* (kHz).

Complex waveforms, such as those representing voice, are made up of combinations of many different single-frequency signals, each with a potentially different amplitude. *Bandwidth* is a range of frequencies, usually specified as the number of hertz of the band or the upper and lower limiting frequencies. Since transmission costs are directly related to bandwidth, analog signals used to provide commercially acceptable quality (*toll quality*) for telephone communications are normally limited to the range of frequencies in which most of the voice energy occurs—that is, the range between 200 Hz and 3.5 kHz.

Digital Electrical Signals

A *digital signal* is an electrical signal in which information is carried in a limited number of different (two or more) discrete states. The most fundamental and widely used form of digital signals is *binary*, in which one amplitude condition represents a binary digit 1, and another amplitude condition represents a binary digit 0. Thus a binary digit, or *bit*, is one of the members of a set of two in a numeration system that has two—and only two—possible different values or states.

Figure 2.2 shows an example time profile of a binary signal waveform (also known as a *bitstream*), and illustrates the two-level, bit structure. The signal corresponds to information generated by a *data terminal equipment* (DTE), any device that can act as a *data source* (transmitter), a *data sink* (receiver), or both.

Most DTEs format information into 8-bit bytes. (A *byte* is an 8-bit quantity of information also generally referred to as an *octet* or *character*.) The number of different bit patterns possible in a byte is 256. Assignment of byte patterns (e.g., various 8-bit sequences of 1s and 0s) to represent specific alphanumeric characters, punctuation marks, control signals, or other signs and symbols enables transmission of information, e.g., the English language, via a process called *binary coding*. The assignment of bit patterns to English letters and other symbols must be agreed upon by both the data source and sink DTEs.

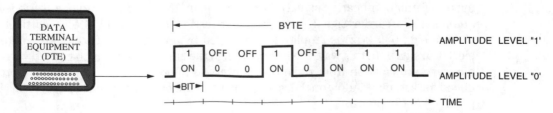

Fig. 2.2 Digital signal example.

The *American Standard Code for Information Interchange* (ASCII), for example, uses 8-bit bytes (7-bit bytes for data encoding and an eighth *parity* bit for error detection). The parity bit is set at the source DTE so that the total number of binary 1s is odd (*odd parity*) or even (*even parity*). The sink DTE checks each byte and can thereby detect single bit per byte errors.

Analog Signal-to-Digital Signal Conversion: Encoding and Decoding

Virtually any analog signal can be converted to a digital signal using a *codec*. A codec (contraction of *co*der and *dec*oder) is a device that transforms (encodes) analog signals into digital signals for transmission through a network in digital format and decodes received digital signals, transforming them back to analog signals. The motivation for analog-to-digital conversion is that digital transmission reduces costs and improves communications quality. The analog-to-digital codec process is illustrated in Fig. 2.3.

Analog signal-to-digital signal (A-to-D) conversion involves sampling, quantizing, and digitizing. Figure 2.3A shows the process of sampling the amplitude of the analog signal at regular intervals. Figure 2.3B illustrates *quantizing*, the replacement of the actual amplitude of the samples with the nearest value from a finite set of specific amplitudes. The samples are then digitized completing the encoding process.

In binary coding, a binary digital signal is generated with sequences of bits that represent the quantized values of the analog signal amplitude samples. Figure 2.3C shows an example binary number code and the relationship to the decimal equivalents for the sample amplitudes shown in Fig. 2.3B. As with the decimal numbering system, the least significant digit is farthest to the right, i.e., bit 1.

In binary format, the number 00000001 is equivalent to the decimal system number 1. The binary equivalent to the decimal number 2 is 00000010. The binary equivalent to the decimal number 4 is 00000100. As can be seen, each movement of digits to the left corresponds to a multiplication by a factor of 2 rather than 10, as in the decimal system. Thus the binary number system is said to be of *base 2* whereas the decimal system is said to be of *base 10*.

A-to-D conversion is essential to *pulse code modulation* (PCM). Modulation is the process of varying certain parameters of a carrier signal—i.e., a signal suitable for modulation by an information signal—by means of another signal (the modulating or information-bearing signal). For instance, in AM broadcast radio, a *radio frequency* (RF) carrier signal (at a frequency assigned by the FCC to a particular station, e.g., 630 kHz) is amplitude modulated by an analog voice or music audio electrical signal. In this way, the information in the audio signal can be carried via radio wave propagation from radio transmitters to radio receivers where it is reproduced for listening. Additional description of generic carrier transmission systems is provided in Chapter 3.

In our PCM example, the analog modulation signal is sampled, and the sample is quantized and digitized into a defined number of equal duration *binary-coded pulses* (bits) representing the quantized amplitude samples of the analog signal. At

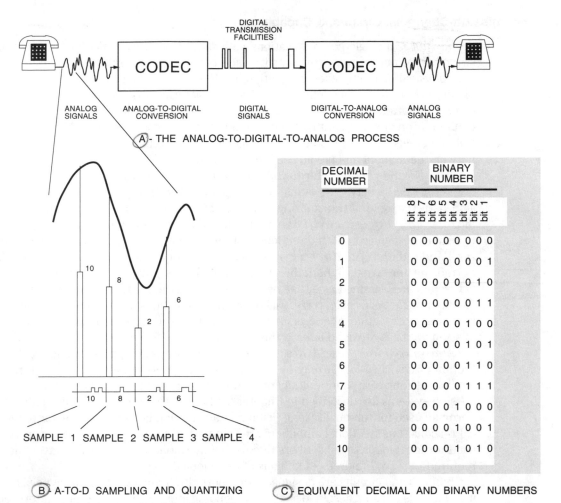

DIGITAL TRANSMISSION FACILITIES

CODEC — CODEC

ANALOG SIGNALS

ANALOG-TO-DIGITAL CONVERSION

DIGITAL SIGNALS

DIGITAL-TO-ANALOG CONVERSION

ANALOG SIGNALS

Ⓐ - THE ANALOG-TO-DIGITAL-TO-ANALOG PROCESS

SAMPLE 1   SAMPLE 2   SAMPLE 3   SAMPLE 4

DECIMAL NUMBER	BINARY NUMBER (bit 8 bit 7 bit 6 bit 5 bit 4 bit 3 bit 2 bit 1)
0	0 0 0 0 0 0 0 0
1	0 0 0 0 0 0 0 1
2	0 0 0 0 0 0 1 0
3	0 0 0 0 0 0 1 1
4	0 0 0 0 0 1 0 0
5	0 0 0 0 0 1 0 1
6	0 0 0 0 0 1 1 0
7	0 0 0 0 0 1 1 1
8	0 0 0 0 1 0 0 0
9	0 0 0 0 1 0 0 1
10	0 0 0 0 1 0 1 0

Ⓑ A-TO-D SAMPLING AND QUANTIZING   Ⓒ EQUIVALENT DECIMAL AND BINARY NUMBERS

Fig. 2.3 Analog signal-to-digital signal (A-to-D) conversion.

a receiving point in a telecommunications system, the process is reversed and the original analog signal reconstructed.

Binary-coded numbers of a fixed length (i.e., the quantity of bits—word length—used to represent binary numbers) can only represent an analog signal amplitude with a limited degree of precision. This means that no finite PCM process can ever be totally without quantizing noise, the difference between the converted binary value and the actual analog signal's amplitude.

For voice applications, quantizing noise is controllable, and toll-quality voice can be produced for speech signals limited to less than 4 kHz bandwidth, if a sampling rate of 8000 samples per second and 8 bits per sample (256 quantizing levels) are used. The impact of quantizing noise is further controlled in networks by preventing more than two or three successive PCM encoding/decoding processes from occurring on end-to-end connections.

Transmission Channels, Circuits, & Capabilities

A *channel* is a single communications path in a transmission medium connecting two or more points in a network, with each path being separated by some means; e.g., physical or multiplexed separation, such as frequency or time division multiplexing. (Multiplexing is defined and discussed later in this section.) *Channel* and *circuit* are often used interchangeably; however, circuit can also describe a physical configuration of equipment that provides a network transmission capability for multiple channels. The characteristics of channels and circuits are determined by the network equipment and media used to support them.

Channels and circuits refer to *unidirectional* (one-way) paths or *bidirectional* (two-way) paths between communicating points. A *simplex circuit* is a transmission path (capable of transmitting signals in one direction only, e.g., broadcast radio). A *half duplex circuit* is a bidirectional transmission path (capable of transmitting signals in both directions, but only in one direction at a time, e.g., Citizen Band radio). A *full duplex circuit* is a bidirectional transmission path (capable of transmitting signals in both directions simultaneously, e.g., the telephone).

Circuits are classified as either two-wire or four-wire, regardless of whether they use fiber or metallic cable, terrestrial or satellite radio links, infrared or other optical transmission, in atmospheric or free space media. (A description of various transmission media is provided later in this chapter.)

A *four-wire circuit* uses two sets of one-way transmission paths, one for each direction of transmission. It may be two pairs (four wires) of metallic conductors or equivalent four-wire as in multichannel transmission systems. (Multichannel transmission systems are described in Chapter 3.) Four-wire circuits, or equivalents, are normally used for toll-quality long-distance transmission facilities, where, for technical reasons, unidirectional amplifiers are required. *Two-wire circuits* are normally used for local loops to subscribers due to the enormous wire and cable investment required to serve hundreds of millions of telephones.

Channels and circuits are designed to support either analog or digital signals. Digital channels must be designed so that the sending and receiving terminals interpret, in an identical way, each transmitted bit of information. For example, the receiving terminal must know whether the first bit of a byte received corresponds to the most significant bit or the least significant bit. Two mechanisms have been designed to provide this type of "synchronization."

In *asynchronous transmission*, each byte or character is marked with distinctive START and STOP bits so that each of the bits between the START and STOP bits can be interpreted identically by the sending and receiving terminals. In *synchronous transmission*, a means is provided to match up sending and receiving terminals so that for a continuous stream of bytes, characters, or words, the significance of each bit is agreed to on a continuing basis. Synchronous transmission systems eliminate the need to include START and STOP bits in each transmitted byte and, hence, are significantly more efficient.

Transmission Capacity & Quality

Within an analog network, bandwidth is the signal capacity characteristic that specifies the rate at which information can be exchanged between two points. A *voice-*

band channel is defined as having a 4 kHz bandwidth—although, as noted earlier, in voice-grade analog channels, the speech signal is typically limited to a range of frequencies from 200 Hz to 3.5 kHz. The additional bandwidth allows for a guard band on either side of the speech signal to lessen interference between channels in some multichannel transmission systems. Analog circuits are available with larger bandwidths to support either multiple voiceband channels or special services such as broadcast program or video material.

The capacity characteristic associated with digital signals is *channel rate* or *bit rate*, that is, the number of bits (or bytes) per second that a channel or circuit will support. For example, a transmission facility that can support data exchange at the rate of 1 megabit per second (1 Mbps or 1,000,000 bits per second), delivers the same quantity of information, i.e., *throughput*, as a 1 kilobit per second (kbps or 1,000 bits per second) facility, but in only $\frac{1}{1000}$ of the time.

Greater bandwidth and higher bit rates use more network resources and, consequently, cost more. However, telecommunication service charges are also proportional to the length of time required to complete a voice call, or in the case of data communications the time required to complete an information exchange transaction. Unfortunately, the cost trade-offs between bandwidth, bit rate, and call holding times are neither linear nor simple. As a result, matching business requirements with telecommunication service options is not straightforward, but involves the use of sophisticated modeling and estimation techniques.

Voice coding schemes that can, without noticeably degrading quality, pack more voice channels into a digital channel of a given bit rate capacity, will generally be more economical.

Telecommunications signals are subject to *transmission impairments*, or degradation caused by practical limitations of channels (e.g., signal-level loss or attenuation, echo, various types of signal distortion, etc.), or interference induced from outside the channel (such as power-line hum or interference from heavy electrical machinery).

The measurement of transmission impairments is an important aspect of predicting whether or not telecommunications systems will sustain the business applications they are intended to support. Signal-to-noise ratio, percent distortion, frequency response, and echo are measurements that relate to impairments most noticeable to users in analog voice systems.

Bit error rate (BER) is the ratio of the number of bits received with errors to the total number of bits transmitted. BER and the average number of error-free seconds are the dominant impairment measurements for digital channels. A bit error or sequence of errors in channels supporting digitized voice may only be perceived as a "pop" or burst of noise to the listener. Thus, for digital voice applications, a BER of one part in one thousand (10^{-3}) yields acceptable results. However, with data transmission relating to financial or other information, even single errors can be catastrophic. It is for this reason that error-detection and correction schemes have been designed to result in virtually error-free data information exchange.

The ability to specify impairment limits and to test procedures to assure compliance, as a condition of new-system acceptance, is essential for developing requests for proposals and contractual documents for telecommunications systems and services. This subject will be treated in more detail in Chapters 4 and 11. For now, it is sufficient to understand that one reason for transitioning from analog to digital facilities is that analog signals are more susceptible to transmission impairments than are digital signals. Moreover, analog transmission facilities require more expensive repeaters and amplifiers, more precise tuning and adjustment, and greater levels of maintenance.

Similar cost savings and performance improvements result from the use of digital switching in lieu of older electromechanical and even electronic analog designs. While 20 years ago, the per channel cost of PCM conversion (the process that makes possible conversion of network implementation from analog to digital) was high, today it is accomplished with inexpensive, integrated circuits. Therefore, the trend in modern telecommunications systems is to minimize use of analog facilities. Note that the transition to digital switching and transmission facilities within the public switched telephone network is being accomplished in a manner transparent to users and either preserves or improves the integrity and performance of voiceband channels while lowering costs.

Although the conversion to digital technologies was undertaken to improve voice service economic and technical performance, these technologies also facilitate emerging data and other nonvoice services, and the sharing of facilities by those services.

Transmitting Digital Signals Over Voice Networks

Modems (*mo*dulator/*dem*odulators) are devices that convert digital signals generated by data terminal equipment (DTEs) to analog signal formats suitable for transmission through the extensive, worldwide connectivity of public and private, switched (dial-up) and nonswitched telephone voice networks. Modems are designed to overcome analog network limitations and to support data communications over virtually any channel capable of delivering ordinary telephone service.

Figure 2.4 illustrates the use of modems in public and private voice networks. The pair of modems used on any given connection incorporates compatible modulation/demodulation designs. Some manufacturers use proprietary designs to achieve higher speeds and to reduce errors caused by network impairments. Figure 2.4 identifies some of the past and current Bell System and CCITT (International Consultative Committee for Telegraphy and Telephony, an international telecommunications standards-setting group) standards supported by modem manufacturers. Appendix A identifies and discusses U.S. and international standards-setting groups.

Today's modems not only support a maximum rate in accordance with an international standard, e.g., V.22 bis (*bis* means the second iteration of the standard) at 2400 bps, but also work with modems that support only lower-rate standards. Modems can be procured as stand-alone single-channel devices, as plug-in cards for personal computers, or in rack-mounted, multichannel configurations suitable for *modem pooling*. Reliable V.22 bis stand-alone modems can be purchased for $100–150.

Interface between stand-alone modems and DTEs is almost universally via the 25-pin connector specified in the Electronic Industry Association's (EIA) RS-232-C or EIA-232-D standards. The telephone network interface is via RJ-11C or other standard modular telephone plug and jack connections.

"Smart" modems, used in conjunction with communications software installed in personal computers or other DTEs, execute a wide range of performance-improving and convenience functions. Examples include error control, automatic answering,

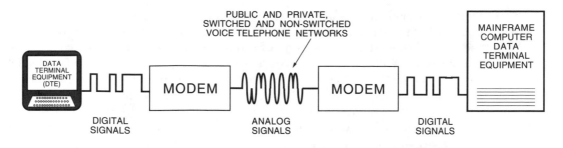

BELL STANDARD	CCITT STANDARD	SPEED FULL DUPLEX	PAGES PER HOUR @25.6 Kbits/PAGE
103	V.21	300 bps	42
212A	V.22	1200 bps	168
224	V.22bis	2400 bps	336
	V.32	9600 bps	1344

Fig. 2.4 Data communications through voice networks using modems.

called-party number storage and retrieval, auto-dialing, modem/terminal/network diagnostics, and data compression. ·

Data-compression techniques remove redundancy in transmitted bit patterns to produce effective transmission rate enhancements of 20% to 200%. Over the same network channel, a modem designed to send and receive data at 1200 bps without data compression may be capable of supporting 2400 bps with data compression.

Investing in higher-speed modems and data compression means that using channels with identical costs, one can drastically improve the rate at which information can be exchanged. The impact of modem speed and data compression on the number of pages that can be transmitted per hour over identical circuits is shown in Fig. 2.4. For users with large data communications requirements, the figure indicates that for dial-up lines through the PSTN, or for private services where bill-back charges are based on the length of the call, investments in high-speed modems and data-compression techniques can quickly be recovered.

Alternative Voice Processing/Coding Techniques

The voice coding PCM technique described above produces 64 kbps digital signals—i.e., 8,000 samples per second times 8 bits per sample equals 64,000 bits per second. The specific implementation used in North America and Japan has been designated as μ-*law* (μ=255) *PCM*, and uses a process called *companding* to enhance the signal-to-quantizing noise, particularly for analog signals at small amplitudes.

In Europe, the European Conference of Posts and Telecommunications (CEPT) has standardized on a 64 kbps PCM technique that uses a slightly different companding method and has been designated as *A-law PCM*. Both techniques result in high-quality speech and are widely used in telecommunications systems throughout the world. Experience has proven PCM voice and data (modem) traffic quality superior to that of most analog techniques. This is essential to the growth of digital networks, since neither LEC nor IXC facilities are designed to detect whether voice or data traffic is being carried, precluding special treatment for data traffic.

From a private-line user's point of view, economics often dictates the use of lower than 64 kbps-bit rate voice processing techniques. Low-bit rate or *narrowband* voice processing permits more voice channels to be transmitted over digital channels of given bit rate or throughput capacity. Narrowband voice is accomplished by using microcomputer or specialized hardware-based processing power to remove redundancies inherent in speech signals, thereby producing digitized voice signals at less than 64 kbps.

One such technique that has achieved ANSI (American National Standards Institute) standard status is *adaptive differential PCM* (ADPCM). ADPCM requires only 32 kbps to digitize analog voice signals and produces voice quality that is not perceptibly lower than that of 64 kbps PCM digitization.

The widespread application of ADPCM within public networks has yet to occur, as it requires special administration because of link compatibility, signaling provisions, data limitations, restrictions on the number of tandem coding/decoding points, and other problems. However, the use of ADPCM in private networks is growing rapidly due to the great potential for cost savings and the fact that private network operators always have control over their own traffic and do not need to be responsive to random public-traffic demands.

A simple example illustrates the potential for savings using ADPCM. AT&T's most recent tariff for a 2500 mile private, dedicated digital circuit, capable of supporting 24 channels of voice digitized at the 64 kbps rate (DS1 service), was $20,755 per month, or about $865 per month per voice channel.

Using ADPCM, the same AT&T circuit can support 44 channels of digitalized voice, reserving four channels for signaling. This means that the costs can be reduced to $472 per voice channel per month once the cost for ADPCM equipment has been amortized. A user with a need for 44 channels can save $17,290 per month after a payback period of two to three months. For very large users, including the government, with $100 million and larger annual voice service costs, the potential for savings is very significant.

Digital speech interpolation (DSI) can result in further savings when used in conjunction with either PCM 64 kbps or ADPCM 32 kbps codecs. DSI takes advantage of the fact that in normal conversation, "quiet periods" permit nearly doubling the number of voice signals that can be accommodated if the equipment serving a large number of channels allocates channels only when a talker is active. DSI is a proven technology. An analog form called *time assigned speech interpolation* (TASI), has been used for years to enhance the efficiency of transoceanic communications.

Other voice processing techniques have been developed that produce intelligible, but not toll-quality, voice at rates as low as 2400 bps. For example, the National Security Agency has developed and produced several hundred thousand STU-III secure telephones that protect the highest levels of classified government information.

The STU-IIIs include modems so that they can be used over virtually any channel capable of supporting ordinary voice service.

The STU-IIIs use a narrowband voice processing algorithm known as *linear predictive coding* (LPC) and produce digitized voice at 2400 bps. Newer models use a technique known as *code excited linear predictive coding* (CELP), which greatly increases voice quality and produces digitized voice at 4800 bps. LPC and CELP could be used in nonsecure environments to achieve cost savings beyond what is possible with ADPCM in applications where good intelligibility without toll quality is sufficient.

Although the $2000 cost-per-unit for STU-IIIs is nearly an order of magnitude less than its $20,000 predecessors, it still represents a sizable cost differential relative to nonsecure telephone equipment. For nonsecure applications, a well-designed system would not require replacement of all station equipment but rather would only require STU-IIIs for trunks, a reduction by a factor of 10 or more relative to the number of stations.

The need to protect sensitive technological and economic information (versus military secrets, as distinguished in the 1984 Computer Security Act) might generate a large requirement for secure telephones in both nonmilitary government agencies and in industry. Should this occur, there might be an opportunity to achieve information security and cost savings simultaneously.

Figure 2.5 contrasts voice processing or coding techniques to show that reduced transmission bit rates are achieved by greater amounts of computer processing power embedded in narrowband voice processing equipment. In the future, it can be expected that the levels of voice quality achieved at lower rates will improve.

VOICE CODING/PROCESSING TECHNIQUE	BIT RATE (Kbps)	COMPLEXITY (MIPS)*	VOICE QUALITY
PCM/ (Waveform Transformation)	64	0.01	HIGH, TOLL
ADPCM/ (Waveform Transformation)	32	0.1	HIGH, TOLL
CELP/ (Analysis/Synthesis)	4.8	1.2	INTERMEDIATE
LPC/ (Analysis/Synthesis)	2.4	1	LOW

* MILLION INSTRUCTIONS PER SECOND

Fig. 2.5. Comparison of voice coding/processing techniques.

3

Transmission Systems

Building on the definitions and basic concepts presented in Chapters 1 and 2, the next two chapters treat transmission and circuit switching, which along with station equipment constitute the principal components of telecommunications networks. As noted in Chapter 1, transmission facilities provide the communication paths that carry user and network control information between nodes in a network. In general, transmission facilities consist of a medium and various types of electronic equipment located at points along the transmission route. This equipment amplifies or regenerates signals, provides termination functions at points where transmission facilities connect to switching systems, and provides the means to combine many separate sets of call information into a single "multiplexed" signal to enhance transmission efficiency.

Transmission Media

A *transmission medium* is any material substance or "free space," (i.e., a vacuum) that can be used for the propagation of suitable signals, usually in the form of electromagnetic (including lightwaves), or acoustic waves, from one point to another; unguided in the case of free space or gaseous media, or guided by a boundary of material substance. *Guided media*, including paired metallic wire cable, coaxial cable, and fiber optic cable, constrain electromagnetic or acoustical waves within boundaries established by their physical construction. *Unguided media* is that in which boundary effects between free space and material substances are absent. The free space medium may include a gas or vapor. Unguided media including the atmosphere and outer space support terrestrial and satellite radio and optical transmission. This section provides brief descriptions of and defines terms associated with prevalent media. Later sections depict how each medium relates to and supports various telecommunications services.

Guided media

For many decades, the most common medium supporting voice applications within residential and business premises (i.e., within the local loop) has been copper *unshielded twisted pair* (UTP). UTP, illustrated in Fig. 3.1, is basically two wood-pulp or plastic-insulated copper wires (conductors), twisted together into a pair. The twists, or *lays*, are varied in length to reduce the potential for signal interference between pairs. Wire sizes range from 26 to 19 AWG (American Wire Gauge, i.e., 0.016 to 0.036 inch in diameter), and are manufactured in cables consisting of 2 to 3600 pairs. A *cable* is a group of metallic conductors or optical fibers that are bound together, with a protective sheath, a strength member and insulation between individual conductors/fibers, and contained within a jacket for the entire group.

In electrical circuits, a *conductor* is any material that readily permits a flow of electrons (electrical current). In twisted pair, the electrical signal wave propagates from the sending end to the receiving end in the *dielectric* material (insulation) between the two conductors. Due to the finite conductivity of copper, the medium for guided wave transmission is fundamentally *dispersive*. With dispersion, complex signals are distorted because the various frequency components which make up the signal have different propagation characteristics and paths.

Dispersion limits the upper bit rate that a medium can support by distorting signal waveforms to the extent that transitions from one information state to another cannot be reliably detected by receiving equipment (e.g., logical 1 to logical 0 value changes).

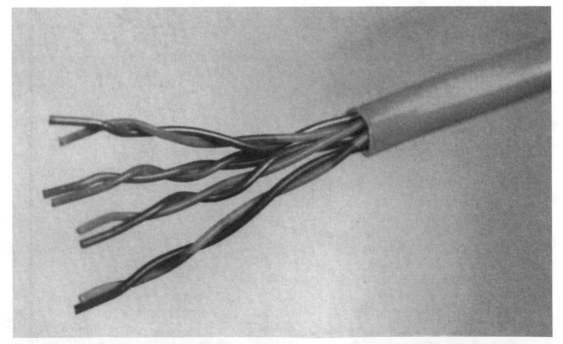

Fig. 3.1 Unshield twisted pair (UTP) cable.

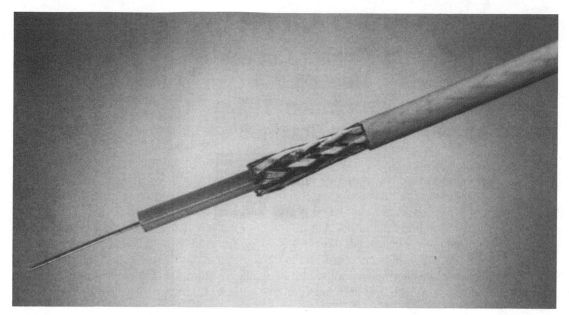

Fig. 3.2 Coaxial cable.

Shielded twisted pair (STP) cable is similar to UTP, but the twisted pairs are surrounded by an additional metallic sheath before being clad with an insulating jacket.

Coaxial cable, shown in Fig. 3.2, consists of an insulated central copper or aluminum conductor surrounded by an outer metallic sheath which is clad with an insulating jacket. The outer sheath consists of copper tubing or braid. Coaxial cable with solid metallic outer sheaths reduces leakage of signals relative to braid-type designs. Because of its strength characteristics, cable-television distribution systems normally use aluminum coaxial cable with solid outer sheaths to minimize *radio frequency interference* (RFI) with aircraft navigation and other life safety systems.

Cables and other media differ in the following ways:

- Bandwidth capabilities (for example the number of voice conversations that can be supported per circuit)

- Susceptibility to electrical interference from other communications circuits or from unrelated electrical machinery or natural sources such as lightning

- The ability to handle either analog or digital signals

- Cost

With UTP, one pair is commonly used for each voice conversation. However, with multiplexing schemes to be described later, UTP has been adapted to support 24 voice grade analog signals per pair. Coaxial cable, on the other hand, exhibits useful bandwidths in the hundreds of MHz (1 MHz=1,000,000 hertz or cycles per second=one *megahertz*), and certain cables are able to transmit several thousand voice channels. Because it is considerably more expensive per foot than UTP, applications for coaxial cable predominantly require large bandwidths such as multiple channel voice, cable television, image transfer, and high-speed data networks.

Until the Institute of Electrical and Electronic Engineers (IEEE) established its 10BaseT specification variant to its IEEE 802.3 local area network (LAN) standard, UTP had not played a major role in the support of digital signals or data communications. Now UTP cabling that complies with 10BaseT can support error-free transmission at rates up to 10 Mbps over distances up to 300 feet. This accomplishment represents a breakthrough for *premises wiring* in that a single type of low cost cabling and modular connectors can now support both voice and data applications. Research at several companies is directed at ways to permit UTP to be used for premises transmission at rates up to 100 Mbps. Chapter 6 describes modern approaches and standards applying to premises wiring systems.

Optical fibers are composed of concentric cylinders made of dielectric materials (i.e., nonmetallic materials that do not conduct electricity). At the center is a core comprising the glass or plastic strand or fiber in which the lightwave travels. Cladding surrounds the core and is itself enclosed in a light-absorbing jacket that prevents interference among multifiber cables. Figure 3.3 illustrates multifiber cable that can be purchased with between two and 144 fibers.

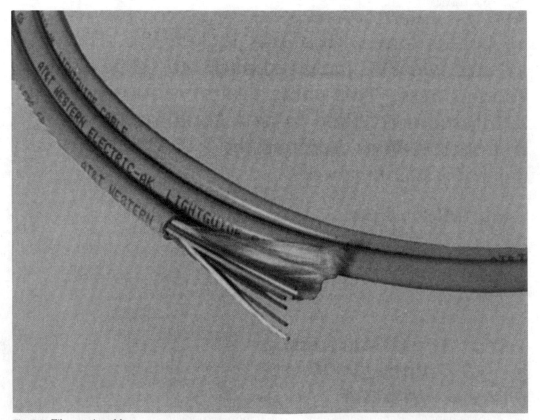

Fig. 3.3 Fiber optic cable.

The optical fiber medium requires that electrical signals be converted to light signals for transmission through hair-thin strands of glass or plastic to light-sensitive receivers, where light signals are converted back to electrical signals. Some present and future technologies might eventually permit cost-competitive signal processing, filtering, switching, and multiplexing to occur in the optical-versus-electrical signal domain.

Optical fibers are either *single mode* or *multimode*. Multimode fibers, with much wider cores, allow the electromagnetic wave (lightwave) to enter at various angles, and reflect off core-cladding boundaries as light propagates from transmitter to receiver. Single-mode fibers have sufficiently small core diameters that the lightwaves are constrained to travel in only one transverse path from transmitter to receiver. This requires the utmost in angular alignment of light-emitting devices at points where light enters the fiber, and it results in higher transmitter/termination costs than multimode fiber systems.

From a technical performance point of view, single mode fiber exhibits bandwidths of up to 100,000 MHz while multimode bandwidth is in the range of 1,000 to 2,000 MHz (1,000 MHz=one billion hertz=one *gigahertz*=1 GHz).

The principal advantages of lightwave transmission using fiber optic cable (also referred to as *lightguide*) are ultra-wide bandwidth, small size and weight, low *attenuation* relative to comparable metallic media (attenuation is signal level or amplitude loss per foot of length), and virtual immunity to interference from electrical machinery and man-made or natural atmospheric electrical disturbances. Disadvantages include the added cost for electro-optical transmitters and detectors, higher termination costs (largely manpower for making physical connections and splices), and overall higher installed cost in short distance applications such as premises wiring systems.

Advantages outweigh the disadvantages, however, and fiber optic cable is rapidly becoming the transmission medium of choice in applications such as high capacity multichannel metropolitan and wide area interoffice trunks, transoceanic cables, and high-speed data communications. In addition, decreasing costs have resulted in fiber optic penetration into the realm of premises wiring. This is occurring in enterprises that require wide bandwidth communications among campus buildings, and/or directly to desktops for specialized applications, such as image file transfer. Reflecting this development, AT&T now offers a composite cable that includes eight metallic twisted pairs (UTP) and two fiber strands within a common outer sheath. Such composite or *hybrid* cable is intended for information outlets within offices serving desktop voice and data terminals.

Unguided Media

Unguided media, that is the atmosphere and outer space, is used in terrestrial microwave radio transmission systems, satellite, mobile telephone, and personal communications systems. *Terrestrial microwave radio* transmission systems consist of at least two radio transmitter/receivers (transceivers) connected to high-gain antennas (directional antennas that concentrate electromagnetic or radiowave energy in narrow beams) focused in pairs on each other, as illustrated in Fig. 3.4. The operation is *point-to-point*; that is, communications can be established between only two

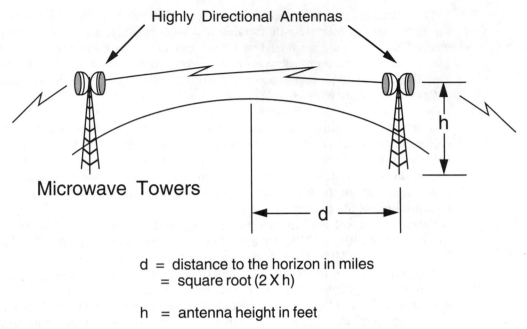

Fig. 3.4 Terrestrial microwave.

installations. This is contrasted to *point-to-multipoint* systems, such as broadcast radio or citizen band radios.

In long-distance carrier applications, terrestrial microwave is an alternative to guided metallic or fiber optic cable transmission media. For this application, antennas are normally mounted on towers and require an unobstructed or *line-of-sight* path between the antennas, which typically can be separated by up to 30 miles. Strings of intermediate or relay towers, each with at least two antennas and repeater/transceivers to detect and amplify signals, interconnect switching centers in different metropolitan locations. By 1980, AT&T had installed 500,000 miles of terrestrial microwave facilities. While not much growth in mileage has occurred since then, significant progress has been made in conversion from analog to digital microwave transmission technologies.

The term *microwave* is often taken, somewhat arbitrarily, to mean frequencies above 890 MHz. The telecommunications industry use of "microwave" radio propagation usually refers to operating frequencies above 2.56 GHz. The Federal Communications Commission (FCC) has assigned frequencies in the 4-, 6-, and 11-GHz bands for long-haul telecommunications common-carrier use. Microwave radio, operating in the FCC-allocated 18- and 23-GHz bands, is now a popular short-haul transmission medium within private networks. Private microwave terrestrial applications include high-capacity transmission between privately owned switches and direct multichannel access to exchange carrier networks.

In the latter application, private microwave transmission is used to connect CPE directly to carrier facilities in lieu of LEC-provided alternatives. *Digital termina-*

tion service (DTS) provided by some carriers permits operators of private networks to use digital microwave equipment to access carrier networks. Private microwave equipment constitutes a fast-growing industry segment that is expected to reach $200 million by 1993. Depending upon capacity and other capabilities, private point-to- point microwave sites range in cost from $15,000 to $150,000.

One advantage of terrestrial microwave is that it obviates the need to acquire right-of-ways (except for towers) or to bury cable or construct aerial facilities. In areas where the cost of right-of-ways is very high, or, as is true in some rural areas where cable installation costs might be prohibitive, significant savings can be realized with microwave. However, as a transmission medium, the earth's atmosphere creates problems not encountered with other media. For example, trees, obstructions, heavy ground fog, rain, and very cold air over warm terrain can cause significant attenuation or signal power loss. Decades of experience have led to conservative tower spacing, space and frequency diversity (alternate physical paths and frequencies), and other engineering practices to offset these difficulties and achieve reliable operation. Another problem associated with microwave transmission is the possible inability to obtain operating licenses due to dense usage and the associated frequency congestion in a number of metropolitan areas.

Commercial satellite communications entails microwave radio, line-of-sight propagation from a transmitting earth terminal (usually ground-based, but potentially ship- or airborne) through free space (the atmosphere and outer space) media to a satellite, and back again to earthbound receiving terminals. In essence, satellites are equivalent to orbiting microwave repeaters.

Terminals, sometimes termed *earth stations*, consist of antennas and electronics necessary to:

- Interface satellite equipment with terrestrial systems
- Modulate and demodulate radio frequency (RF) carrier signals with multiple (multiplexed) voice and data signals
- Transmit and receive RF carrier signals to and from satellites
- Otherwise establish, support and control communications among earth terminals

Figure 3.5 illustrates the satellite network topology.

Commercial satellites operate in three frequency bands: C band (4/6 GHz), Ku band (11/14 GHz), and Ka band (20/30 GHz). In each case, the lower-frequency band is used for *downlink* (satellite-to-earth terminal) transmission and the higher-frequency band for *uplink* transmission. For high-capacity C-band communications, antennas are typically parabolic in shape and 10 meters (about 33 feet) in diameter. Ku- and Ka-band antennas can be as small as 1.5 and 1.0 meters (5 and 3 feet), respectively.

At present, there are approximately 150 satellites in *geosynchronous orbit* (a circular orbit 22,300 miles above the equator). Satellites in geosynchronous orbits appear stationary from any point on earth. In these orbits, normally only three satellites are required to provide nearly global earth coverage (the exceptions are the mostly uninhabited polar areas). Even though radio signals travel at the speed of

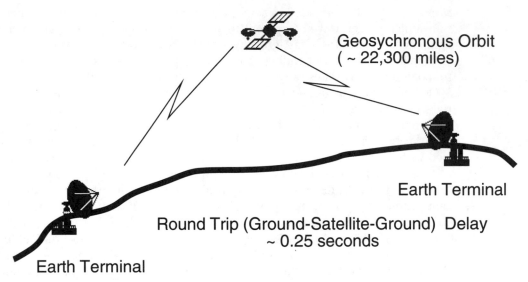

Geosychronous Orbit
(~ 22,300 miles)

Earth Terminal

Round Trip (Ground-Satellite-Ground) Delay
~ 0.25 seconds

Earth Terminal

Fig. 3.5 Satellite network topology.

light, the time required for them to travel from the earth terminal to the satellite is approximately 0.12 second. For a single "hop"—that is, a connection involving only one uplink and one downlink—the round-trip delay is about a quarter of a second. This is noticeable and sometimes annoying during voice conversations. It should also be noted that the long, round-trip transit time demands special techniques to permit efficient data communications.

The *very small aperture terminal* (VSAT) earth terminal market has grown to nearly $1 billion in the past decade. This C- or Ku-band technology uses small antennas (1.5 to 6 feet in diameter) and is intended to support the low-capacity requirements (9.6 to 56 kbps data rates) of individual motel/hotel or retail outlet sites. Small sites use two-hop links and transmit to *public* hub stations, which relay signals to other corporate locations, as shown in Fig. 3.6. Installed costs per terminal are in the less than $10,000 range, and VSAT services are generally significantly less than the tariff associated with equivalent, concatenated LEC intra-LATA and IXC inter-LATA services.

Other portions of this book describe how media are incorporated into transmission systems, and how they support specific voice and data telecommunications services.

Voice Frequency & Carrier Transmission Systems

Two broad categories of transmission systems are *voice frequency* (VF) transmission systems and *carrier* transmission systems. VF transmission supports the ubiquitous local loops connecting business and residential customer premises with serving central offices. The median loop length is 1.7 miles, with 95% of all loops less than 5.2 miles. VF transmission is also used for short interoffice connections. The medium most often used for VF transmission is copper twisted pair, multipair cable.

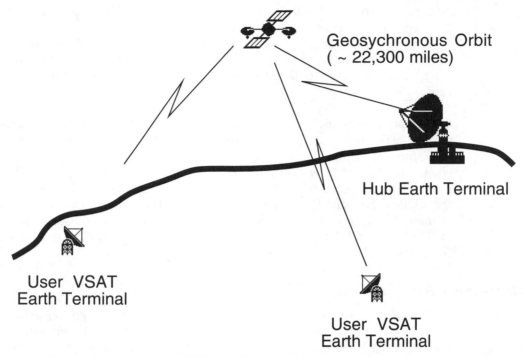

Geosychronous Orbit
(~ 22,300 miles)

Hub Earth Terminal

User VSAT
Earth Terminal

User VSAT
Earth Terminal

Fig. 3.6 Very small aperture terminal (VSAT) satellite network topology.

In local loops, VF transmission includes LEC main, branch, distribution, and drop cables, collectively known as *outside plant*, with inside wiring completing connections to on-premises telephones and other CPE devices.

In 1980, the predivestiture Bell System investment in transmission facilities was $60 billion. Sixty percent of this investment was in local-loop plant, with the interoffice 40% being split 26% for metropolitan areas, 5% rural, and 9% long distance. In the future, fiber optic cable can be expected to replace copper twisted pair cable in the local loop environment as it proves economical. Since today's copper cable provides power for telephones, with the advent of fiber, some alternative mechanism will have to be devised if telephone operation independent of other electrical power availability is to be maintained. It is likely that normal ac convenience outlets will power tomorrow's fiber-connected telephones.

A *carrier system* transmits multiple channels of information by processing and converting them to a form suitable for the transmission medium used. Many information channels can be carried by one broadband carrier system. Broadband carrier systems are classified as either analog carrier systems or digital carrier systems.

Analog carrier systems. Analog carrier systems use repeaters that correct for medium impairment characteristics, and produce at their outputs linear scaled versions of the input signals. Analog carrier systems can carry speech, data, video and supervisory signals, although they are best suited for speech signals. Analog carrier systems operating over multipair cable, "N-carrier," coaxial cable, "L carrier," and radio systems are rapidly being replaced by digital facilities.

An economic comparison of analog versus digital carrier facilities must include whether or not the interconnected switching systems are analog or digital. With the current transition to digital switching, increasing emphasis will be on digital carrier systems.

Digital carrier systems.   Digital carrier systems are designed to transmit digital signals, using regenerative versus linear repeaters, and time division multiplexing. *Multiplexing* is a technique that enables a number of communications channels to be combined into a single broadband signal and transmitted over a single circuit. At the receiving terminal, demultiplexing of the broadband signal separates and recovers the original channels. The primary purpose of multiplexing is to make efficient use of transmission capacity to achieve a low per-channel cost. Two basic multiplexing methods used in telecommunications systems are *frequency division multiplexing* (FDM) and *time division multiplexing* (TDM).

Multiplexing Systems

FDM divides the frequency bandwidth (spectrum) of a broadband transmission circuit into many subbands, each capable of supporting a single, full-time communications channel on a noninterfering basis with other multiplexed channels. FDM multiplexing is suitable for use with analog carrier transmission systems. Standard AM and FM broadcast radio is an example of FDM where different stations occupy FCC assigned portions of the standard broadcast band. Cable TV is another example where different stations are assigned frequency bands on a single-cable medium, and are selected by appropriate frequency conversion equipment using either standalone converter boxes or cable-ready TV set tuners.

The quality of signals that undergo a frequency division multiplexing and demultiplexing process is dependent upon the design and precision of analog componentry used in FDM equipment. Because TDM multiplex equipment incorporates largely digital techniques, it possesses the "digital" versus "analog" advantage over FDM multiplexing equipment.

In TDM, a transmission facility is shared in time rather than frequency. Figure 3.7 illustrates how this is accomplished using D-type TDM channel bank equipment. *D-type channel bank* is terminal equipment used for combining (multiplexing) individual channel signals on a time division basis. D-type channel banks provide interfaces for n analog signal inputs. Each input signal is directed to a codec for conversion to PCM samples. Channel bank equipment and a repeatered digital line together comprise a digital carrier system.

For the example in Fig. 3.7, the codecs use the 64 kbps North American standard previously described. Each individual codec produces an 8-bit sample at the rate of 8,000 samples per second (8 kHz), or one 8-bit sample every 125 microseconds. (A *microsecond* is 1-millionth of a second.)

Each of these samples is interleaved in time sequence into a single high speed digital signal, as shown on the figure, beginning with one signal designated as channel

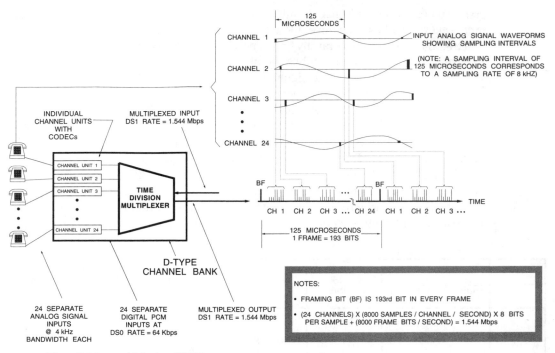

Fig. 3.7 Time division multiplexing (TDM).

1, and continuing through channel 24. The process is then repeated for successive 8-bit samples from each channel. Following the 24th channel's sample a *framing* bit (*bit BF*) is added by the transmitting equipment to permit the receiving equipment to derive frame synchronization and thus identify the time correlation between 8-bit information segments and each of the input analog channels.

In a TDM system, a *frame* is a sequence of time slots, each containing a sample from one of the channels served by the multiplex system. The frame is repeated at the sampling rate, and each channel occupies the same sequence position in successive frames. The TDM process can be described as assigning a *time slot* in each frame on a high-speed TDM *bus* (also referred to as a *highway*), to each of *n* input channel signals. Here *bus* is defined as one or more conductors (or some medium) that connect a related group of devices.

In our example, the frame rate is 8 kHz, the sampling and frame interval is 125 microseconds, and each frame consists of 24 time slots. The TDM bus bit rate (and the TDM output bit rate) is 1.544 Mbps, which is the product of 24 times 64 kbps for each PCM input signal, plus frame bits occurring at 8 kbps (24×64,000=1.536 Mbps+0.008 Mbps=1.544 Mbps).

The example TDM process concentrated 24 voice channels into one equivalent four-wire, full duplex circuit. Transmission facilities of this type are referred to as *T1-type digital carrier* (or simply *T carrier*) facilities. Other transmission facilities have capacities greater than 1.544 Mbps. For example, the AT&T FT3 lightwave fiber transmission system supports 44.736 Mbps, or 672 individual 64 kbps, PCM channels. To permit higher levels of multiplexing concentration, a multilevel TDM digital signal hierarchy has been developed.

The *DS1 level* in the hierarchy corresponds to the 1.544 Mbps TDM signal already described. Although not formally a member of the hierarchy, the *DS0 level* refers to the individual time slot digital signals at channel rates of 64 kbps. Two DS1 signals are digitally multiplexed to produce a *DS1C level* signal containing 48 DS0 channels and they require a transmission facility that supports 3.088 Mbps. Four DS1 signals comprise a *DS2 level* signal containing 96 DS0 channels, requiring a 6.312 Mbps transmission facility. A *DS3 level* signal results from the digital multiplexing of 7 DS2 signals, supports 672 DS0 channels and requires a 44.736 Mbps transmission facility. Finally, a *DS4 level* signal supports 6 DS3 level signals (4032 DS0 level signals) and requires a 274.176 Mbps transmission facility.

The DS designation refers to the signal level hierarchy and is independent of the type of carrier facility, except of course for compatibility requirements. For example,

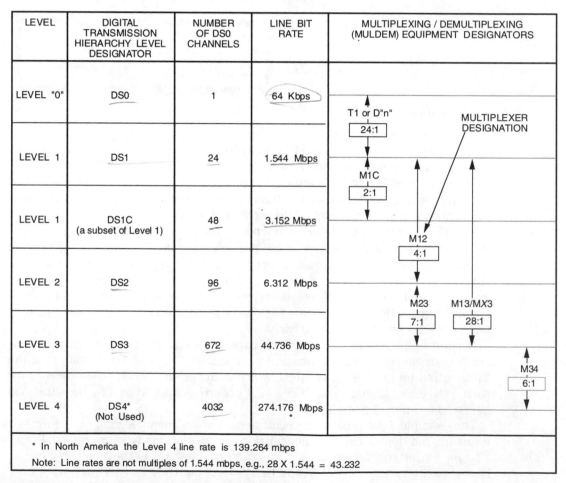

LEVEL	DIGITAL TRANSMISSION HIERARCHY LEVEL DESIGNATOR	NUMBER OF DS0 CHANNELS	LINE BIT RATE	MULTIPLEXING / DEMULTIPLEXING (MULDEM) EQUIPMENT DESIGNATORS
LEVEL "0"	DS0	1	64 Kbps	
LEVEL 1	DS1	24	1.544 Mbps	
LEVEL 1	DS1C (a subset of Level 1)	48	3.152 Mbps	
LEVEL 2	DS2	96	6.312 Mbps	
LEVEL 3	DS3	672	44.736 Mbps	
LEVEL 4	DS4* (Not Used)	4032	274.176 Mbps *	

* In North America the Level 4 line rate is 139.264 mbps

Note: Line rates are not multiples of 1.544 mbps, e.g., 28 X 1.544 = 43.232

Fig. 3.8 The U.S. time division multiplexing (TDM) digital signal hierarchy.

T1-types carry DS1 signals and FT3 lightwave systems carry DS3 signals. Figure 3.8 summarizes the digital signal hierarchy in the United States. Associated capacities and multiplexer designators defined originally by AT&T, are today established by ANSI (American National Standards Institute). Note that in North America the level 4 line rate is 139.264 Mbps rather than the rate originally defined.

In a network, a *clock* is a device that generates a signal that provides a timing reference. Clocks are used to control functions such as setting sampling intervals, establishing signal rates (bps), and timing of the duration of signal elements, such as bit intervals. In a completely synchronous TDM network, all participating nodes must use timing provided by a single master network clock. Because this is not the case in today's TDM networks, note in the digital signal hierarchy that the line rates at level 2 and above are higher than the sum of constituent member rates. The added "overhead" bits are necessary to maintain synchronization in an inherently asynchronous TDM operation resulting from less-than-perfect network timing devices spread throughout networks. (Future networks based on a synchronous digital hierarchy are discussed in Chapter 17.)

In addition to the multiplexing functions just described, the channel units in channel bank equipment provide proper voice frequency and signaling interfaces for central office trunk, loop, or other assigned circuits. Thus, from the analog inputs to a multiplexer, to the analog outputs from a remote demultiplexer, the TDM equipment is essentially transparent to both the user information and the signaling associated with various forms of telephone service. (Signaling associated with voice services is discussed in Chapter 5.)

T1 Carrier Systems

Historically, the T1 types of digital carrier systems have played a major role in the transition from analog to digital facilities. One of the fastest growing LEC and IXC facility segments, T carrier has become commonplace serving large private networks, and, due to declining costs and wider service options, is increasingly found in moderate sized networks, given that over 120 million miles span the North American continent. Introduced in 1962 within the old Bell System, T1 carrier supports 24 full-duplex voice channels using just two pairs of unshielded twisted pair (UTP) 19 AWG cable, constituting a low cost and reliable alternative to single, analog channel-per-twisted-pair voice frequency or other analog frequency division multiplexing (FDM) carrier systems. Because of its importance, this section expands on the preceding definitions to describe representative DS1 facilities, T1 carrier system signal structures and operational capabilities.

Figure 3.9 illustrates a typical installation which provides 24 analog voice grade channel end-to-end connectivity between two business premises, via LEC and/or IXC networks. Within the business premises, 24-channel DS1 signals are generated by customer premises equipment (CPE), for example, the D-type channel bank equipment described above or a T1 multiplexer. *Channel service units* (CSUs), at the least, and possibly *Data service units* (DSUs) are required to connect the CPE to the DS1 service. As shown in the figure, CSUs terminate telephone company digital circuits, and protect the network from harmful signals. In fact, CSUs or equipment incorporating

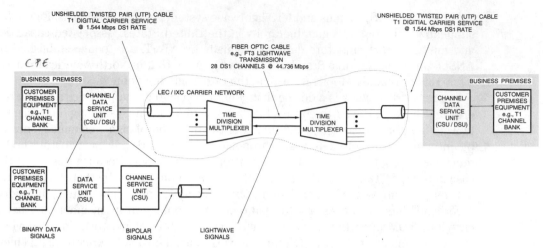

Fig. 3.9 DS1/T1 service connection arrangement example.

CSU functions must be designed in accordance with Part 68 of the FCC Rules and Regulations, entitled "Connection of Terminal Equipment to the Telephone Company."

Other CSU functions include line conditioning and equalization, error control, and the ability to respond to local and network "loop-back" circuit testing commands. In a telephone company, *line conditioning and equalization* is the spacing and operation of amplifiers so that gain provided by the amplifiers for each transmission frequency compensates for line signal loss at the same frequency.

DSUs, meanwhile, provide transmit and receive control logic, frame synchronization and timing recovery across T1 and other digital circuits (when these functions are not implemented in other CPE). DSUs also convert ordinary binary signals generated by CPE to special *bipolar signals*. Bipolar signals, described below, are designed specifically to facilitate transmission at 1.544 Mbps rates over UTP cable, a media originally intended for 3 kHz voiceband signals. As indicated in Fig. 3.9, the trend today is for manufacturers to combine CSU and DSU functions in a single device for interface with DS1, DS3 and the new digital services described in Chapters 14 and 16. Representative CSU/DSU manufacturers include ADC Kentrox, AT&T Paradyne, Coastcom, Timeplex, and Verilink. MCI, US Sprint, GTE Telephone Operations, U.S. West Communications Inc., WilTel Communications Systems, and other service providers supply equipment from various manufacturers in delivering digital services.

Figure 3.9 suggests that DS1 signals will undergo higher levels of multiplexing in long-haul end-to-end networks. An FT3 fiber optic transmission segment supporting 28 DS1 signals at a combined DS3 rate of 44.736 Mbps is the example in the figure. Note that binary, bipolar, and lightwave signal formats appear at various points in the end-to-end connection, but that these conversions remain transparent to user CPE.

DS1 bipolar signal format

As described above, T1 carrier systems carry 24 channels on two pairs of copper twisted-pair cable (one pair for each transmission direction). Signals are applied directly to cable pairs in bipolar format in which positive and negative pulses, always alternating, represent one binary signal state (for example the state corresponding to a binary bit value of 1). Under the *Alternate Mark Inversion* (AMI) convention, pulses correspond to binary 1s and the absence of pulses corresponds to binary 0 states. With *Alternate Space Inversion* (ASI), pulses correspond to binary 0s. Figure 3.10 illustrates the relationship between a binary pattern of 1s and 0s and signals in bipolar AMI format. The figure shows how 1s are transmitted, alternatively as positive or negative pulses, and 0s are conveyed by the absence of a pulse. The T1 line's bipolar signal can be transmitted approximately one mile over 19/22 AWG twisted pair before requiring a *repeater*. A repeater, which is normally dc-powered over the line, regenerates signals; that is, it detects 1s and 0s, recovers symbol timing, and generates reconstructed versions of the received signal for transmission down the next segment (span) of lines.

The bipolar format has significant advantages, particularly when used with twisted pair transmission media. One advantage is that single-bit errors can be easily detected at any point along the transmission path. For example, if at any instant in time, interference causes a pulse to be detected when in fact a no-pulse 0 condition is correct, a *bipolar violation* occurs. That is, instead of the polarity reversal (AMI) that should normally occur between two successive 1s, two successive pulses of the same polarity will be generated. Bipolar violations also occur when interference causes the annihilation of positive or negative pulses, causing a transmitted 1 state to be falsely detected as a 0. Detection of bipolar violations permits error correction and improves the quality of end-to-end transmission.

In a T1 carrier system, receivers and repeaters must synchronize to their incoming signals by detecting the timing between positive and/or negative pulses. Since transmitted 0s suppress the transmission of pulses, random binary information signals that generate long sequences of 0s degrade the ability of receivers to lock onto

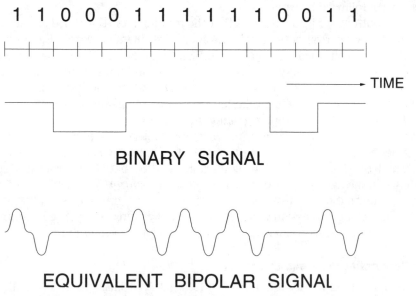

Fig. 3.10 Binary signal and bipolar signal format relationship.

and track the timing of incoming pulse signals. The original design in 1962 dictates that to maintain proper operation, on the average there must be at least one 1 in 15 bits, and at least three 1s in 24 bits.

Various *zero suppression coding* techniques have been implemented to prevent degradation due to long strings of 0s, a characteristic of nonvoice signals. Some early techniques limit DS0 channel capacity to 56 kbps. These techniques are acceptable for voice transmission but cannot support error-free 64 kbps data communications. The technique known as *bipolar eight zero substitution* (B8ZS) replaces a block of eight consecutive 0s with a code containing bipolar violations in the fourth and seventh bits. When eight 0s occur they are replaced with the B8ZS code before being multiplexed onto the T1 line. At the receiver, detection of the bipolar violations permits B8ZS codes to be converted back to 8-bit strings of 0s, allowing the full 64 kbps use of DS0 channels for data applications. Unfortunately, B8ZS is not compatible with early T1 facilities, and consequently carriers (particularly LECs) do not always have B8ZS facilities available. Telecommunications users must therefore be cautious when procuring digital transmission services for data applications.

T1 Superframe (SF) signal format

Over the years a number of T1 framing formats have been defined by AT&T. By far the most popular is the recent D3/Mode 3 D4 format for framing and channelization. Shown in Fig. 3.11, the D3/Mode 3 D4 bit stream is organized into superframes, each consisting of 12 frames. Note at the top of the figure that every other framing bit BF is determined and coded to produce patterns that are easily recognizable by receiving DSUs. BF(s) marking odd frames (i.e., BFt—or terminal framing bits) produce a sequence of alternating 1s and 0s, whereas BF(s) marking even-numbered frames (i.e., BFs—or signaling framing bits) produce a sequence of alternating groups of three 0s, followed by three 1s.

The bottom of Fig. 3.11 shows that the framing bits (the 193rd and last bit in each frame) are inserted between the 24th and 1st channel words. Each channel word consists of 8 bits (named B1 through B8). Channel words represent eight bits samples, taken at the rate of 8000 samples per second, and correspond to 24 different sources of voice or data information. In the D-type channel bank example described, the channel words represent digital versions of 24 analog voice signals.

For voice transmission, signaling information—or information exchanged between components of a telecommunications system to establish, monitor, or release connections—must be transmitted with the channel voice samples. This is accomplished by sharing the least significant bit (B8) between voice and signaling as shown in the figure, a process termed *robbed bit signaling* (RBS). The B8 bits carry voice information for five frames, followed by one frame of signaling information. This pattern of B8 assignment to voice and signaling is repeated during each successive group of six frames, as shown in the middle of the figure.

T1 Extended Superframe (ESF) signal formats

Widespread popularity and use of "T1 circuits"—that is, carrier-supplied DS1 service—within corporate networks makes their reliability crucial to business suc-

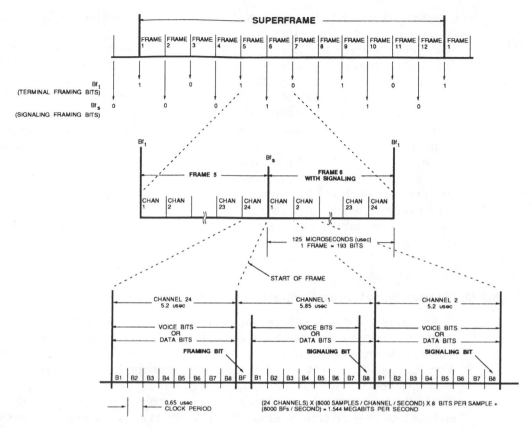

Fig. 3.11 DS1/T1 signal framing format.

cess. Yet until ESF was introduced, both users and carriers were often unaware of gradual deterioration of digital circuits until they failed catastrophically. ESF takes advantage of the fact that with modern technology, not every framing bit needs to be used for framing and synchronization. In ESF, the superframe is extended in length from 12 to 24 frames. Of the 24 framing bits in an extended superframe, six bits are used for framing (synchronization), six bits are used for error checking, and the remaining 12 bits are used for a 4 kbps *facility data link* (FDL), a communications link between channel service units (CSUs) and telephone company monitoring devices.

To permit error detection, the sending CSU station examines all of the 4,608 data bits within an extended superframe, and, using an algorithm, generates a unique *cyclic redundancy check* (CRC) code. This CRC code is transmitted using six of each group of 24 framing bits. Using received data bits, a remote station employs the same algorithm to recalculate the CRC code. The remote station detects the CRC code sent by the transmitting station, and compares it with the locally generated CRC code. Generally, if the two CRC codes match, there are no errors. This technique detects 98.4% of all possible bit error patterns. What's more, this information is stored and provides a historical record of performance over time so that degradation can be detected before total line outages occur.

The FDL can be used to exchange performance data among CSUs with ESF monitoring capabilities, and to remotely control CSU operational and test modes from central network management positions. For networks with multiple tandemed T1 bipolar UTP and fiber optic cable links (as in the Fig. 3.9 example), when combined with individual link performance indicators like the number of bipolar violations and various failure/degradation alarm signals, the end-to-end CRC data provides powerful predictive and diagnostic tools for T1 network management, control and maintenance. In fact the real motivation for and advantage of ESF, lies in startling increases in T1 uptime and availability. Reports indicate that once installed, ESF can result in 70% fewer incidents of T1 line outages, and 30% reductions in T1 circuit downtime per incident. Note that although T1 carrier originally connoted a specific capability, the term T1 has taken on a generic meaning in the industry and is used to describe all manner of transmission services and equipment operating at 1.544 Mbps.

Digital Multiplexer/Digital Cross-Connect System Equipment

Multiplexers and switches are fundamental, related building blocks in telecommunications systems. As will be explained in Chapter 4, today's digital circuit switches employ time division multiplexing techniques, and emerging technologies (such as Asynchronous Transfer Mode (ATM) discussed in Chapter 16) seamlessly combine both techniques. Currently, the simplest digital multiplexers are channel banks. Figure 3.12 illustrates the difference between simple channel bank multiplexers and the next level of sophistication, often referred to in the literature as *flexible* or *single aggregate* T1 multiplexers. Channel banks usually accommodate 24, 48, or 96 plug-in channel units and generate DS1 signals at 1.544 Mbps, which correspond to 24 DS0, 64 kbps signals.

Flexible T1 multiplexers support a variety of input signals ranging from sub-rate (less than DS0 rate) to multiple DS0 rate signals. *Networking multiplexers* (not shown in the figure) represent a more sophisticated level of T1 multiplexers that combines the capabilities previously described and permits simultaneous connection to multiple T1 transmission facilities for automatic or manual reconfiguration in response to circuit outages or changing traffic conditions.

The *digital cross-connect system* (DCS) is a new generation of switching/multiplexing equipment that permits per-channel DS0 (64 kbps) electronic cross-connection from one T1 transmission facility to another, directly from the constituent DS1 signals. Before DCS availability, if a DS0 signal arrived at a network switching node in one DS1 transmission facility and needed to be relayed to another DS1 transmission facility, an arrangement of back-to-back channel bank multiplexers and switches was necessary, as shown in the top right-hand portion of Fig. 3.13. That is, incoming composite 24-channel versions of the DS1 signals arriving on one DS1 transmission facility had to be demultiplexed into 24 separate DS0 signals, each connected to a switch, which in turn had to be connected to other channel banks for re-multiplexing, and finally connected to other DS1 transmission facilities.

In contrast, a DCS allows the 24 individual DS0 channels in a particular T1 line to be distributed among any of the other T1 lines connected to the DCS, as shown in the lower right portion of Fig. 3.13, with no requirement for channel banks or external switching. The cross-connect capability of DCS can be used to *groom* or segregate DS0 channels by type (e.g., voice, data, special services 4-wire/2-wire etc.), increasing the fill of T1 lines and enabling more efficient utilization of resources.

The compact Coastcom (a manufacturer of CPE T1 multiplex equipment) models shown in the figure are sized to support 8 and 16 T1 lines respectively, while some

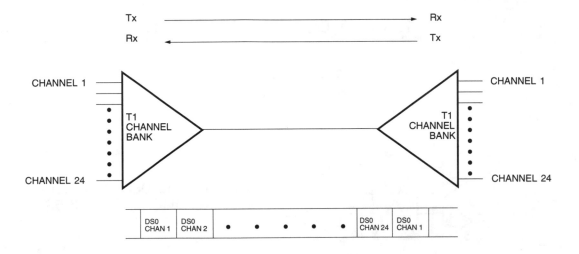

T1 CHANNEL BANK MULTIPLEXER

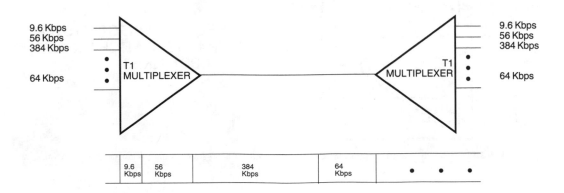

FLEXIBLE T1 MULTIPLEXER

Fig. 3.12 Time division multiplexer (TDM) configurations.

16 PORT DCS   8 PORT DCS

COASTCOM DXC
DCS PRODUCTS

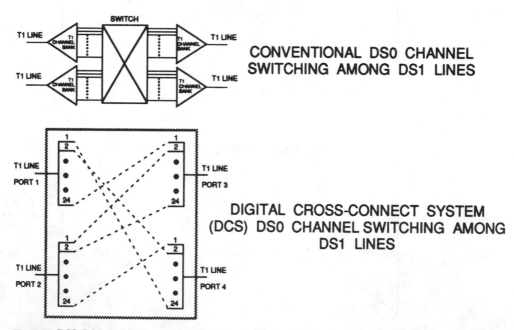

CONVENTIONAL DS0 CHANNEL
SWITCHING AMONG DS1 LINES

DIGITAL CROSS-CONNECT SYSTEM
(DCS) DS0 CHANNEL SWITCHING AMONG
DS1 LINES

Fig. 3.13 DCS, DS1, and DS0 grooming; DS0 concentration; DS0 segregation operations. (Photo courtesy of COASTCOM.)

AT&T models terminate up to 127 lines. Moreover, the connection paths within DCS are software-controllable, which facilitates network management, and eliminates the need for manual connections. While DCS configuration changes are not made on a call-by-call, or circuit-switched basis, a DCS does implement "channel switching"—

that is, the ability to reconfigure networks in response to outages, time-of-day traffic variations, or to accommodate growth and organizational changes.

Representative multiplexer manufacturers include AT&T Paradyne, Coastcom, Netrix, Timeplex, and others. DCS manufacturers include AT&T, Bytex Corp., Coastcom, Frederick Engineering Inc., Rockwell Network Transmission Systems, Tellabs, and others. AT&T refers to its proprietary DCS products as *digital access and cross-connect systems* (DACS).

Figure 3.14 illustrates the so-called "fractional T1" service that IXCs and some LECs are now able to offer using flexible multiplexing and DCS equipment. In the past, T1 service was available only in integral 24 channel DS0 increments. Business users with lesser requirements often had to resort to more expensive (on a per-channel basis) private line voice and private single channel data services. With fractional T1, users can obtain service in DS0 increments at per channel rates that are almost always less expensive than other alternatives. The lower portion of Fig. 3.14 illustrates cost comparisons of AT&T tariffs for 2500 mile circuits. Note that single DS0 fractional T1 (e.g., AT&T's ACCUNET Spectrum of Digital Services) service arrangements are less expensive than either a single private voice-grade line or AT&T's 56 kbps private-line Digital Data Service (DDS).

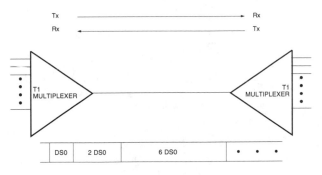

FRACTIONAL T1 EXAMPLE

AT&T MONTHLY IXC RATES (2500 MILES)		
SERVICE	COST	COST PER DS0 CHANNEL
ANALOG VOICE GRADE PRIVATE LINE	$1124.24	NA
56 Kbps DIGITAL DATA SERVICE	$5403.50	NA
FRACTIONAL T1 SERVICES		
64 Kbps ACCUNET SPECTRUM OF DIGITAL SERVICES	$1048.00	$1048.00
128 Kbps ACCUNET SPECTRUM OF DIGITAL SERVICES	$1997.00	$998.50
384 Kbps ACCUNET SPECTRUM OF DIGITAL SERVICES	$5966.00	$994.00
FULL T1 ACCUNET SPECTUM OF DIGITAL SERVICES	$20,775.00	$865.63

Fig. 3.14 Fractional T1 service examples.

At present, not all LECs offer fractional T1 service. Consequently, users might have to lease other more expensive LEC services to gain access to IXC networks and their fractional T1 service offerings. The cost justification for such instances must be established on an individual basis, often necessitating a thorough knowledge of carrier services and tariffs and computer-based network modelling and optimization tools. Applications of such tools and the elements of traffic engineering in telecommunications are treated in Chapter 7, and, where appropriate, throughout the remainder of this book.

Overall, digital multiplexing transmission services are key to the modernization of voice services, growth of video conferencing, and the development of metropolitan and wide-area data communications. In spite of the limitations of asynchronous and fixed bandwidth channelization operations, for the foreseeable future, TDM systems, originally designed for voice service, will continue to be adapted for voice, data, video, and integrated applications.

Chapter

4

Circuit Switching Systems

Up to now, we have covered telecommunications fundamentals dealing with the generation and exchange of information in electronic form over various transmission networks. The fundamentals of *circuit switching*, presented below, center around the need to share telecommunications resources. For example, if a separate telephone and transmission line were required to connect you with every other person in the company, you probably couldn't fit into your office. The earliest telephone systems used just such an arrangement, however, which quickly threatened to bury communities in telephone poles and wiring.

The basic concept of circuit switching was previously illustrated in Fig. 1.1 (see page 5), showing how it does away with the need for dedicated transmission media running between each pair of communicators. Instead, the idea of *resource sharing* is introduced, wherein each person can access transmission resources when they need to—"just-in-time" communications, so to speak.

Earliest switching arrangements revolved around telephone operators, who in the beginning were young men, often on roller skates, hurtling across central offices with plug-ended cords to connect telephone calls. These fellows tended to be rude to callers, however, and were soon replaced by better-mannered young ladies.

A major leap forward occurred with the advent of the *step-by-step* automatic telephone switch. Invented in 1889 by Almon T. Strowger, this switch was used, with only minor improvements, in telephone company COs up to the 1970s. An undertaker in wild and woolly Kansas City, MO of the mid-1880s, Mr. Strowger suspected his competitors of bribing central telephone operators to learn of the latest gunfights, and where the bodies could be found. His automatic switch was meant to foil their efforts. It did.

In modern packaging, circuit switching techniques are still used to connect a greater number of telephones than there are transmission paths between them. The "switching" is the method by which a particular transmission line is connected

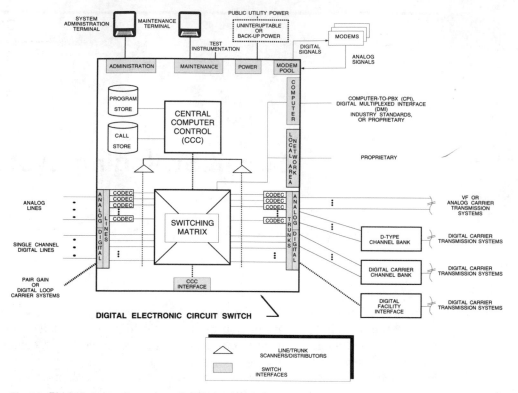

Fig. 4.1 Block diagram of a modern digital circuit switch.

between parties who wish to use it, disconnected following their use, and then re-connected between new parties.

Fundamentals

A switching system is an assembly of equipment arranged to establish connections between lines, between lines and trunks, or between trunks. Included are all kinds of related functions, such as monitoring the status of circuits, translating addresses to routing instructions, testing circuits for busy conditions, detecting and recording troubles, sensing and recording calling information, etc. Circuit switching establishes connections on demand and permits the exclusive use of those connections until they are released. Packet and message switching, primarily used in data communications networks, will be discussed in Chapter 12.

The earliest circuit switch designs employed plug boards or electromechanical switch contacts, and to a large extent were controlled by human operators. Today, digital circuit switches are essentially specialized minicomputers. Every digital circuit switch has three essential elements: switching matrices, a central control computer, and interfaces, which among other functions, provide termination points for input and output signals, as illustrated in Fig. 4.1.

Switch matrices

A *switch matrix* is the mechanism that provides electrical paths between input and output signal termination points. Modern matrices are electronic, using either time division or space division switching. A *time division switch* employs the TDM process described in chapter 3, in a *time-slot interchange* (TSI) arrangement. Figure 4.2 shows how TSI uses input and output TDM buses.

TSI can be thought of as having storage locations (buffers) associated with specific time slots in a TDM bit stream that corresponds to input signals, and with specific time slots in a TDM bit stream that correspond to output signals. As indicated in Fig. 4.2, the switching operation can change the order of the time slots by changing the order of the information from input and output buffers.

In space division, a physical, electrical (spatial) link is established through the switch matrix. Where older space division switches used electromechanical mechanisms with metallic contacts, modern space-division switches are implemented electronically using integrated circuits.

For small line/trunk sizes, switches can use time division alone. For large switches, combinations of time division and space division are required due to speed limitations over TDM buses. For example, to build a time division switch that would accommodate 100,000 connections, each representing a 64 kbps signal, would require a TDM bus speed of 100,000 times 64 kbps or 6.4 billion bits per second. Consequently, popular large switch designs today use concatenations of both time division and space division switch modules, e.g., time-space-space-time division (TSST) or even TSSST approaches. For these large switches the space division module is referred to as a *time multiplexed switch* (TMS). A TMS is defined as an element of a time division switching network that operates as a very high speed space division switch whose input-to-output paths can be changed for every time slot.

Although in Fig. 4.1 the switch matrix is shown as a single component, most manufacturers implement the matrix using TSI and TMS modules. Small-capacity

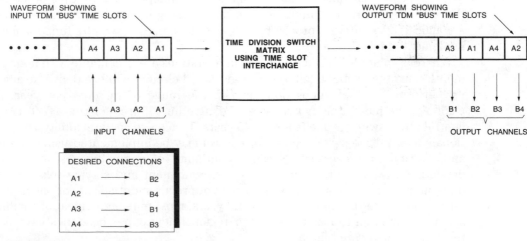

Fig. 4.2 Time-slot interchange.

switches generally use one TSI module. Large switches may use several TMS modules and a quantity of TSI modules sufficient to accommodate the number of required lines and trunks. Modular designs are an economical way to accommodate a wide range of initial requirements and to add capacity for growth without the need to replace common switch components.

Central control computer

In today's digital circuit switches, connections through matrices and other switching system operations are controlled by a central computer. Virtually all switches now use *stored program control* (SPC). With SPC, new services, special features and configuration changes can be implemented as changes in a software program versus more costly changes in hardware. The basic components of an SPC system are illustrated in Fig. 4.1. They are:

- One or more high-speed processors (central control computer) that sense input and output circuit conditions and execute stored program instructions
- Semipermanent program store memory, containing operating system and applications software program
- Erasable, temporary call store memory to record and accumulate data during call processing
- Scanners through which central control acquires input signal information such as on-hook, off-hook, etc.
- Distributors through which central control drives switch operations

Such designs are made possible by digital computer technologies and permit the operation of a switching system to be altered by software program changes. To provide the reliability essential for telephone operations, SPC processor arrangements typically use full centralized control, with independent multiple processors that can load share, and functional multiprocessing in which different functions are allocated to different processors. Over 80% of an SPC switch's functionality is implemented in software.

Whereas computers used in some data operations are acceptable with hour-per-month downtimes due to failures, well-designed circuit switches deliver availability performance that keeps downtime in the range of minutes per year. For example, the mean time between service-affecting failures for the AT&T 1AESS central office switch is one failure in 40 years. To achieve such reliability, the functionally centralized switch computer control may be implemented using multiple and distributed processors. Because continuity of some business operations depends on available telecommunications, equipment and service reliability requirements constitute an important aspect of procurement specifications. It is interesting to note that in the past several years, most telecommunications failures significant enough to make national or regional newspaper headlines have been the result of software "glitches."

Switch interfaces

The diagram shown in Fig. 4.1 applies to all types of circuit switches, i.e., central office switches, tandem switches, and business premises-located private branch exchanges (PBXs) and key telephone systems (KTSs). PBXs and KTS systems will be discussed in the next sections. Not all of the interface types illustrated in Fig. 4.1 and discussed below would necessarily be present in all switches.

The most important interfaces support lines and trunks. Reflecting today's network status of mixed analog and digital facilities, switches may be configured with both analog and digital line/trunk cards. As shown in Fig. 4.1, trunk connections between switches can be via VF analog carrier, or any of several digital carrier systems. The D-type channel banks can be used with older analog switches and provide input signals suitable for use with digital carrier facilities. Such arrangements permit the transition from analog to digital to occur at different rates and at different points for both switching and transmission systems.

Interfaces between digital switch trunk outputs and digital carrier facilities can occur directly—without analog-to-digital conversion in codecs—via multichannel digital facility interfaces, or by means of digital channel bank equipment as shown in Fig. 4.1. Direct interconnection with digital facility interfaces greatly simplifies and reduces the cost of interconnecting switching and transmission systems.

On the line side of the switch, either analog or digital interfaces can be used. Central office switches will support analog lines for the foreseeable future because of the large number of existing analog telephones. Businesses that have purchased digital PBXs may elect to use analog telephones (and therefore analog line interfaces) because they are cheaper, in spite of the fact that they forfeit many features only available with digital telephones. All digital PBX manufacturers provide, and prefer to sell, digital subscriber line interface cards and the proprietary station equipment that goes with them.

Some switches provide line side interfaces for multiplexed signals carried in digital loop carrier (DLC) systems. The types of services associated with various line/trunk facilities will be described in more detail in Chapters 8 and 9.

Subscribers who use analog telephones can selectively use lines between their offices and the switch for either voice or data modem traffic. One side effect of using digital technology is that once the decision is made to use digital telephones, only digital signals can be carried on station equipment-to-switch lines. That is, those lines can no longer support analog telephone or modem traffic.

Modem pool interfaces are provided in most digital switches as an alternative way of providing PCs and other data terminal equipment (DTEs) located in a user's office, with PBX connectivity to other DTEs, via PSTN or private voice networks. A modem pool is a group of modems connected to a digital switch. Each modem is provided with a digital input connection from PC to the switch and an analog connection from the switch to the PSTN as shown in Fig. 4.1. Each PC or DTE connection is associated with a switch extension number. In an office, the PC or DTE is connected to a data terminal connection (usually an RS-232-C interface) provided on the digital telephone.

To use the service, an office user dials an extension corresponding to the modem pool's digital connection side (manually or automatically via the DTE), the switch selects a non-busy modem and connects the modem output through the switch, usually to the PSTN or private switched or non-switched voice networks.

One advantage of modem pools is that not every user needs to have a dedicated modem in his office. However, since modems can be procured in the $100–150 cost range (less if PC plug-in modem cards are used), the upgrade to digital station equipment might not be justified unless there are reasons other than reducing the number of modems needed at a particular site.

As Fig. 4.1 indicates, some switch designs are capable of supporting direct digital connections to collocated computers and local area networks (LANs). (LANs will be discussed in Chapter 13.) Two switch-to-computer specifications advanced by the industry are known as the *computer-to-PBX interface* (CPI) and the *digital multiplexed interface* (DMI) specifications. Both specifications operate at DS1 rates of 1.544 Mbps. A single CPI/DMI connection can multiplex what would otherwise involve 24 separate switch-to-computer channels. Consequently, these PBX interfaces are applicable to installations where large numbers of data terminals are connected to host computers.

Other switch interfaces support administration and maintenance terminals and processing systems. Most larger switches include built-in test and diagnostic capabilities to automatically perform basic switching equipment and loop transmission impairment tests. Maintenance interfaces also permit connection to stand-alone test instrumentation for more extensive testing.

Digital interfaces permit data communications directly between the switch central control computer and other processors used for support functions. For example, in campus installations, some manufacturers place a central switch in one building and use *remote switch modules* (RSMs) in other locations to concentrate traffic and simplify inter-building wiring. In such cases, data communications between the central switch control computer and RSM controllers provides identical service to all stations whether directly attached to the central switch or to the RSM(s).

The final major switch interface is with primary power. For certain premises systems, loss of telephone service during public utility power outages is not acceptable. Consequently, battery systems with back-up motor-generators are installed.

Among a variety of minor line and trunk interfaces supported by most switches, are provisions for audio signals for music-on-hold, dial dictation, and loudspeaker paging systems.

Private Branch Exchanges (PBXs)

A *private branch exchange* (PBX) is a voice switching system serving a commercial or government organization and is usually located on that organization's premises. PBXs provide telecommunications services on the premises or campus, e.g., internal calling and other services, access to public switched telephone networks for outside calling, and access to private networks connecting various organizational sites.

Evolution

PBX development spans four generations. First-generation PBXs were operator-controlled plug-and-jack types. Second-generation PBXs performed switching mechanically or electronically but used analog designs. Third-generation PBXs are characterized by digital switching and control, and digital and/or analog peripheral, loop, and trunk transmission interfaces. These systems support digital telephones, integrated voice and data terminals (IVDTs), and digital interface units for directly connecting PCs and other data terminals (DTEs) through the PBX. Telephones generally use embedded codecs and exchange digital signals.

Unfortunately, the terminal-to-switch digital signal formats used by various manufacturers are proprietary. To take advantage of the features offered, a user must either buy the manufacturer's proprietary digital station equipment, (thus forfeiting competitive bids), or install analog station equipment, which might not be capable of taking advantage of all of the features supported by the PBX. Most currently available PBXs qualify as third-generation.

Fourth-generation PBXs possess all third-generation characteristics, and also directly connect with local area networks (i.e., data networks other than those supported by the PBXs themselves) and mainframe computers. Figure 4.3 illustrates a fourth-generation PBX.

Although labeling details are omitted, the figure shows the components of the PBX to be the same as those of the generic digital circuit switch described in the previous section. Interfaces for analog and digital station equipment are shown. Data terminals are connected through RS-232-C interfaces in the telephones or via standalone digital interface units. Modem pools support data communications through public or private voice networks. Although not shown on the figure, the PBX also supports IVDTs, if the manufacturer offers such products.

Digital telephones may be either single line or multiline. Multiline telephones generally incorporate visual displays to indicate busy or ringing status of multiple extension numbers. Until the mid-1970s, multiline telephones were available only with key telephone systems.

Third- and fourth-generation PBXs support a variety of analog and digital carrier connections to local and long-distance network facilities. In the example in Fig. 4.3, PBXs with DS1-type digital carrier service interfaces can support 24 voice or data channels using a single four-wire circuit, versus 24 separate two- or four-wire circuits if voice frequency (VF) analog carrier interfaces are used. Digital PBX trunk interfaces thereby offer the potential for significant savings in network connection and transmission costs. Figures 4.4, 4.5, and 4.6 illustrate representative modern digital PBX products.

Services & applications

In this book, PBX services comprise the category of voice services defined as *premises services,* in contrast with the other major category, *network services.* A comprehensive listing of voice services is provided in Chapter 8, which describes voice services at the level of detail needed to prepare specifications for procurement packages. Chapter 11 includes a section on how to go about selecting PBX and other voice systems.

In the data processing world, an *application* is a software program that performs some useful task. PBX applications are defined in similar fashion. *Functional applications* are individual tasks or telecommunications services provided by the PBX, such as station-to-station calling. At the highest level, *business applications* are

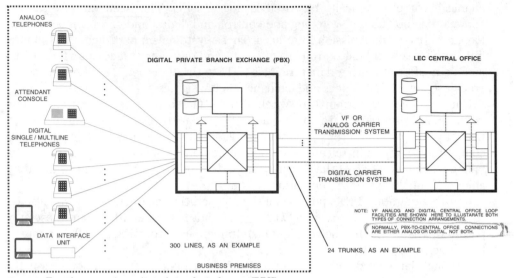

ANALOG
TELEPHONES

ATTENDANT
CONSOLE

DIGITAL
SINGLE / MULTILINE
TELEPHONES

DATA INTERFACE
UNIT

300 LINES, AS AN EXAMPLE

BUSINESS PREMISES

DIGITAL PRIVATE BRANCH EXCHANGE (PBX)

LEC CENTRAL OFFICE

VF OR
ANALOG CARRIER
TRANSMISSION SYSTEM

DIGITAL CARRIER
TRANSMISSION SYSTEM

NOTE: VF ANALOG AND DIGITAL CENTRAL OFFICE LOOP
FACILITIES ARE SHOWN HERE TO ILLUSTARATE BOTH
TYPES OF CONNECTION ARRANGEMENTS.

NORMALLY, PBX-TO-CENTRAL OFFICE CONNECTIONS
ARE EITHER ANALOG OR DIGITAL, NOT BOTH.

24 TRUNKS, AS AN EXAMPLE

Fig. 4.3 Fourth generation private branch exchange (PBX).

Fig. 4.4 Northern Telecom Inc. Meridian 1, Option II PBX (*Northern Telecommunications Inc.*).

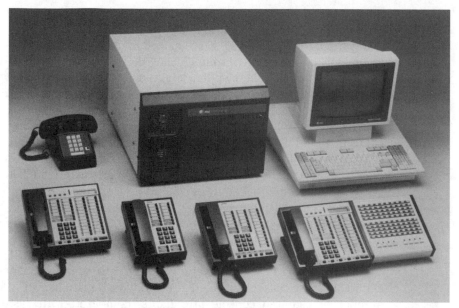

Fig. 4.5 AT&T System 25 PBX and telephone (*AT&T*).

Fig. 4.6 NEC 2400 IMS PBX and peripherals (*NEC America*).

unique aggregations of telecommunications services that satisfy particular enterprise needs, e.g., hospitality, retailing, financial, etc.

Basic dial PBX functions have remained unchanged for decades. The most fundamental is station-to-station calling, using between three and five digits. Next is the ability to connect inside telephones (stations) with public and private external networks. To automatically access these networks, a station user dials an access code (such as "9"). This characteristic trunk-side connection has historically distinguished PBXs from key systems, which were directly connected to central office (CO) lines, and therefore did not require dialing an access code to place an external call. PBXs are also characterized by attendant service, whereby an on-premises operator connects two-way trunks carrying incoming external ("listed directory number") calls with internal stations. The attendant can also use a two-way trunk to place an external call.

In addition to connecting with two-way CO trunks for attendant console operation, modern PBXs also support both *direct inward dialing* (DID) and *direct outward dialing* (DOD), using analog or digital facilities. DID allows incoming calls to a PBX to ring specific stations without attendant assistance. Similarly, DOD allows outgoing calls to be placed directly by PBX stations. DID greatly reduces the number of required console attendants, compared with systems in which all incoming external calls must be extended to PBX stations. Details on the operation of DID, DOD, and other network services furnished by LECs are provided in Chapters 8 and 9.

Other important PBX functional applications are administration, including the ability to produce reports such as *station message detail recording* (SMDR), which show calling and called numbers together with time, date, and duration of calls; station/feature software moves, adds, and changes; maintenance support; voice messaging (also called *voice mail*); automatic call distribution (ACD), which directs large numbers of incoming calls to specific departments within an organization); message center/directory services; and cable/inventory/energy/property management.

For large building, multi-building, or campus installations, premises telecommunications services can be extended from a central PBX switch in the main building to remote switch modules (RSMs) in other buildings. Figure 4.7 shows a central PBX with a switching matrix that includes a high-speed time multiplexed switch (TMS) module. The central switch connects to a group of local TSI switch modules, each sized to support station and trunk requirements for the building in which it is located, e.g., Building A.

RSMs installed in other buildings are connected to the central PBX using high-speed transmission over fiber optic cable. For example, AT&T Definity switches use a 32 MHz channel to support central TMS module-to-remote TSI module traffic, and a 4 MHz channel for control signals between the central PBX computer and remote TSI module controllers. Designs such as this provide identical services to all users regardless of their building location. RSMs can accommodate station and trunk capacity requirements on a building-by-building basis. Distances between the central PBX location and the RSMs can extend for miles, depending upon site equipment configuration.

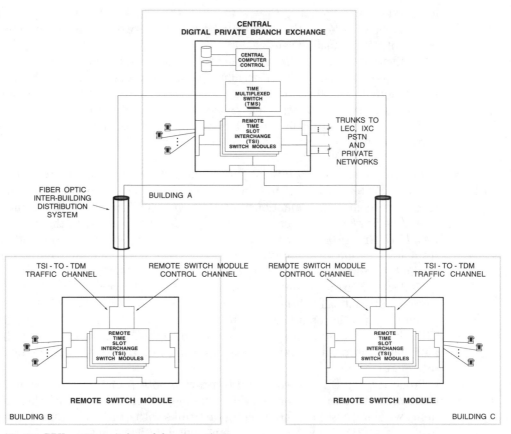

Fig. 4.7 PBX remote switch module connections.

Electronic switched network services

When an organization has PBXs in multiple locations with significant traffic be-tween sites, it might be economical to establish a *private network* (sometimes called a *tandem network*). A private network is made up of circuits and switching arrangements for the exclusive use of one organization. LECs and IXCs can pro-vide switching and transmission facilities. Organizations with PBXs at various lo-cations, need only *tie trunks* (special service circuits connecting PBXs) leased from the LEC and/or IXC(s).

In the simplest case, referred to as *point-to-point tie-line service,* users access a trunk between two locations by dialing an access code (unique to the tie trunk between those two locations), followed by the desired station extension number. For example a user on a PBX in Chicago calling a PBX in New York City might dial 8-368-xxxx. The digit 8 accesses a tie trunk group, 368 selects a dedicated tie trunk between the Chicago and New York City PBXs, and xxxx represents the called party's extension. A user on a PBX in Washington, DC, might dial 8-479-xxxx to reach that same New York City party over dedicated tie trunks between Washington, DC, and New York City. A unique exchange number would thus exist for each called location in this rudimentary private network, depending upon the location of the originating call.

A private network call from Chicago to Washington, DC, could also be completed via the New York City PBX if the calling party first accesses the New York PBX and then manually dials the same access code for Washington, DC, that a New York caller would use. This type of service is known as *manual dial tandem tie-line service* and it cannot automatically route calls through multiple PBXs to take maximum advantage of network transmission facilities, or seek alternate routes if first choice trunks are busy.

Electronic tandem switching, tariffed by AT&T as a private network service that provides customers with a uniform numbering plan and numerous call-routing features, overcomes the limitations noted above. AT&T refers to electronic tandem switching as *electronic switched network (ESN) service*. With this capability, the New York City PBX (exchange number 368 in the above example) could be called from any location on the private network, without need for dedicated tie trunks between New York and every other network location.

Under this arrangement, a network call from Denver to New York might be routed automatically through Detroit or Chicago tandem switch locations, depending on traffic conditions and the least-cost path at the time the call is placed. This alternate routing function would minimize the likelihood of network busy signals, while permitting maximum use of network tie trunks, thus reducing network usage costs.

Some larger PBXs are equipped to provide both premises services and private electronic switched network services. In effect, these PBXs create their own tandem switching system by connecting tie trunks with other tie trunks. Figure 4.8 illustrates a PBX (City A) with tandem switching capabilities, and how it relays calls between two other PBXs. Note that City A's PBX is multifunctional, i.e., it supports local station traffic, access to LEC/IXC PSTNs, and tandem switched network services.

When more than one tandem switch exists in a network, all switches must contain compatible stored program control software, which tends to limit the range of system choices to a single manufacturer. AT&T and Northern Telecom have agreed on basic ESN design attributes and are therefore a partial exception. Future national network standards may remedy the lack of full ESN interoperability.

Components

As we have noted, a modern PBX is nothing more than a specialized computer. As such, it incorporates a control portion (*central processing unit, or CPU*), a data-manipulation portion (arithmetic unit), and a peripheral connection portion (input/output unit). The operating system software transforms this general purpose computer into a PBX, and the applications software tells the PBX what functions it can perform. Peripherals such as telephones, modems, and other items of voice terminal equipment are the "eyes and ears" of the PBX, furnishing information needed to set up, connect, supervise, and tear down calls.

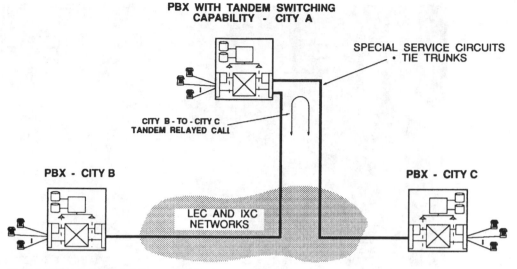

Fig. 4.8 PBX tandem switching capabilities.

Switching equipment and architecture.

Port interface capabilities. Lines and trunks are the most elementary PBX input/output ports. Over time, the meaning of the term *line,* originally denoting an extension number appearing on a telephone, evolved to also mean *station,* the telephone itself. A population of lines, each carrying a certain amount of call traffic, is supported by a smaller number of *trunks,* which connect the PBX to network resources. The percentage of trunks to lines is generally on the order of ten to thirty percent.

With the advent of digital PBXs, the distinction between lines and trunks began to blur. Most of the major PBX products are now port oriented, combining lines, trunks, attendant consoles, and utility circuits together as common sources of traffic demand. The PBX cabinets themselves are moving toward universal card slot design, where any printed circuit board (except common control) can be inserted into any card position.

Centralized vs. distributed switching. Early PBXs employed a centralized switching design—all telephones were connected to one switching entity that performed all PBX functions. As described above, a number of manufacturers have developed distributed PBX switching architectures. A user occupying a large building, campus, or office park can now elect to use multiple modular nodes (remote switch modules) which can be installed and linked by digital facilities, which, when properly designed, seamlessly emulate a centralized switch. Distributed PBXs with low port densities per module have been systems of choice in the *shared tenant service* market because initial installation costs are low and modular expansion is inherent in the system design.

Peripherals. The term *peripherals*, derived from the data processing industry, refers to items of equipment that send inputs to the system processor (PBX) and re-

Fig. 4.9. Northern Telecom Inc. Meridian Digital Telephones (*Northern Telecom Inc.*).

ceive outputs from it. Telephones, data devices, attendant consoles, and other items of terminal equipment fall into this category.

The principal PBX peripheral is the telephone. Several milestone improvements have marked its evolution—the manual dial, the integrated handset, and dual tone multifrequency (DTMF) signaling, known in the Bell System as "touch tone." Up until the early 1970s, the vast majority of telephones—both business and residential—were rotary-dial. It was not until the close of the decade that DTMF telephones, made economical by large-scale integration manufacturing techniques and the FCC's equipment registration program, became pervasive in the U.S. marketplace. Figure 4.9 illustrates some modern telephones manufactured by Northern Telecom Inc.

Adjunct systems

Adjunct systems are made up of multiple components and require systems engineering prior to installation. Some of the more popular adjunct systems are described below.

Voice message processing. The voice message processing industry was founded on two separate business needs: comprehensive call coverage with more accurate message-taking, and information exchange on a non-real-time basis.

In its most basic form, voice message processing, often referred to as voice mail, allows callers to leave recorded messages. Instead of having to hang up after dialing to a ring-no-answer or busy signal, the caller hears the telephone being answered, then a prerecorded greeting (generally in the called party's voice), followed by instructions on how to record, review, change, or erase a voice message to be left for the called party. Later, the called party can retrieve the message and perform additional annotation, distribution, and delivery verification functions.

PBX voice message processing systems can be configured in numerous ways, although the most widespread usage is as a final call coverage point, where an incoming call is forwarded to a voice mailbox, as described above.

Ease of use and number of features of a voice message processing system are related to its degree of design integration with the PBX. Integrated systems allow personalized greetings (in the called party's own voice), automatically reply with a voice message to the sender, redirect the message to other voice mailboxes, and broadcast a message within the voice mail community.

Integrated systems do not require the caller to know the called party's extension number to leave a message or to perform other functions. Also, the called party is generally notified of the message via a lamp indicator or text on a telephone visual display, rather than via muted ringing or "stutter" dial tone. Once messages have been retrieved, the system signals the PBX to deactivate the message waiting indicator.

Non-integrated voice message processing systems are unable to provide a personalized greeting for each station. All calling parties hear the same standard message and must then identify the called party, either by speaking his name or by keying in his extension number (only from a DTMF telephone). Where the caller has spoken the called party's name, an operator must later transcribe and distribute the message.

In configuring the voice message processing system, it is important to ensure that callers can easily reach a live person when additional assistance is required. This is achieved either by direct selection using a dialed code or by a timeout feature, required for callers using rotary-dial telephones.

System management. Even in the smallest configuration, a PBX nevertheless includes stations, trunks, and features, as well as telephones and associated wire and cable. As PBX size increases, a user organization is forced to devote resources to the task of monitoring and controlling PBX services, repairs, rearrangements, and configuration, including adjunct systems. This task is often referred to as *system management.*

At one time, all available system management functions were performed at the attendant console. Although the number of management operations was limited, the process was awkward because while in use for administration, the console could not be used to process calls. Later, separate *maintenance and administration panels* (MAAPS) were introduced, which were proprietary terminals designed specifically to interface with the particular PBX. Next, commercial off-the-shelf computer terminals such as DEC's VT-100 were used in the interests of cost savings and ease of operation.

Today, the trend is toward PCs for maintenance, administration, and system management. Functions performed include:

- Long-distance call costing
- Billing reconciliation
- User cost allocation

- Software feature moves, adds, and changes
- Traffic analysis
- Network optimization
- Inventory control
- Directory services
- Cable system management
- Service orders
- Trouble handling

In the past, PBX system management encompassed more functions than those associated with LEC-provided Centrex service, described in the following section. Because a user-owned PBX switch is located on-premises, whereas a LEC-owned Centrex switch is located at the CO, PBX users traditionally exercised greater control over system management. Recently, however, LECs have begun to offer tariffed and special contract system management capabilities that virtually mirror those of the PBX.

Operational characteristics

Significant PBX product improvements occurred in the 1980s, in terms of increased cabinet capacities, enhanced performance specifications, more and better features, and reduced logistical and environmental demands. In short, today's PBX products deliver more "bang for the buck" than ever. Yet, they have also become more commodity-like, with little to differentiate competitors, apart from price.

System capacities

Stations (lines) are the basic measure of PBX capacity. They correspond to electrical circuits on a printed circuit board (PCB), representing the ability to connect one telephone or data terminal. Stations may be analog or digital. Analog station PCB's tend to contain a higher circuit density than digital PCBs—often on the order of 2 to 1 (e.g., 16 analog stations vs. 8 digital stations). Digital telephones need a digital station PCB irrespective of whether voice is digitized at the telephone, or at the PCB.

PBX station traffic is "concentrated" before being connected by a trunk to an outside network. Concentration is achieved through contention of a particular quantity of stations for a lesser quantity of trunks. The concentration ratio may range anywhere from 5% trunking to 30%. Governing factors are system size, the traffic-handling requirement and specified call blocking probability (i.e., the chance for encountering a busy-signal when all trunks are in use, a performance characteristic of networks described below).

The presence of analog or digital data stations may have significant impact upon the total station quantity, the PBX's ability to handle traffic, total system size, and cost.

System performance

How well a PBX performs is largely determined by its ability to process calls, based upon the division of traffic, i.e., internal versus external incoming and outgoing calls, and provisioning of trunk *quanitities to adequately handle external traffic. To ensure satisfactory installed performance, purchase requisitions and requests for proposals (RFPs) must include trunking, signaling, and transmission quality performance specifications. This section discusses these important parameters and their cost and performance implications.

Grade of service. A PBX significantly reduces the number of transmission channels required between the business premises and the LEC central office. The quantity of PBX-to-CO trunks is normally around 10% of the number of stations, as indicated in Fig. 4.3, where 300 stations are served by 24 trunks. This reduces both the quantity of LEC transmission channels that a user must pay for and the number of PBX trunk interfaces that must be installed and maintained.

However, PBX users compete for a limited number of trunks; if all trunks are in use, the next outgoing call is *blocked* and the next incoming call receives a busy signal. So an adequate quantity of trunks must be specified to ensure that station users receive an acceptable grade of service.

Blocking probability, often referred to as *grade of service* (GOS), is an important measure of the adequacy of telecommunications networks. Other grade-of-service indicators include an estimate of customer satisfaction with a particular aspect of service, such as noise or echo. For example, the noise grade of service is said to be 95% if, for a specified distribution of noise, 95% of the people judge the service at or better than "good." GOS measurements apply to all aspects of telecommunications networks, not just PBXs. In many cases the literature equates GOS only with the probability of a blocked call. So, when used without further explanation, GOS generally refers to blocking probability.

In terms of call-blocking performance, GOS represents that portion of calls, usually during a busy hour, that cannot be completed due to limits in call-handling capabilities. For example a GOS of P=0.001 means that only one call in 1000 would be blocked. GOS is a performance factor that merits understanding and careful interpretation. For instance, in the Fig. 4.3 example, 300 trunks serving 300 stations would guarantee non-blocking access to the LEC network. Thus, a serving trunk would always be available for each station, so no contention or blocking could occur. However, if only three trunks were ordered for the Fig. 4.3 example, the GOS would be such that most calls would be blocked.

In the non-blocking example, since most people don't use phones continuously, an organization would be paying for capacity that is seldom used. In the second case, all trunks would be used almost 100% of the time, but the GOS would be unacceptable. An optimal design is one that yields an acceptable GOS and yet maintains a reasonably high level of facilities utilization.

For design purposes, GOS can be estimated from total traffic intensity. Based on assumptions regarding the randomness of call arrivals, holding times, and other factors, tables are available that relate traffic intensity and the number of servers (trunks) to the probability that a call will be blocked. In planning a new system, the best source for estimates of traffic demand is call detail records from the existing system. If no historical data is available, average industry estimates can be used.

Centi-call seconds, (CCS), is the term used to quantify traffic intensity or demand. A CCS is 100 call seconds of traffic during 1 hour. Therefore, a single traffic source that generates traffic 100% of the time produces 36 CCS of traffic per hour, or 3600 seconds of traffic every 3600 seconds. An equivalent amount of traffic could also be generated by 10 sources that only generate, traffic 10% of the time. That is, 10 sources of traffic generating 3.6 CCS, contribute the same total traffic as a single 36 CCS traffic source.

If the total traffic intensity generated by 300 subscriber stations is equal to 360 CCS, (each of the 300 stations is used two minutes out of each hour), then the trunk utilization rates and the number of trunks needed to achieve GOS levels of 0.1, 0.01, and 0.001 are shown below.

Grade of Service (GOS)	P = 0.1	P = 0.01	P = 0.001
Number of trunks (required to achieve GOS)	13	18	21
Percentage utilization of trunks	77.0%	64.0%	48.0%

If the traffic generated by the 300 subscribers totalled 2700 CCS, the results would change as follows:

Grade of Service (GOS)	P = 0.1	P = 0.01	P = 0.001
Number of trunks (required to achieve GOS)	76	89	100
Percentage utilization of trunks	98.7%	84%	75%

Two conclusions can be drawn from these calculations. First, for a given level of traffic, designing for better GOS results in poorer trunk utilization. The second is that there are economies of scale. That is, as one aggregates more traffic, greater resource efficiency is achieved while providing acceptable GOS performance. These phenomena form the basis for switching system and network design. Optimized designs can produce significant cost savings with essentially the same user quality of service.

Some telephone companies use a GOS objective of P=0.005. Smaller private networks are designed for GOS levels on the order of P=0.01. A typical large tandem switch, such as the AT&T 4ESS, can terminate over 100,000 trunks with a blocking probability of P=0.005 and channel occupancy of 70%.

Chapter 7 revisits GOS as it applies to utilization efficiency of any telecommunications switch or transmission resource, in terms of rudimentary traffic engineering principals. It also describes tariffs illustrating the relationship between performance and LEC and IXC service costs.

Signaling & transmission quality. Signaling and transmission quality are important subjective yardsticks by which users measure the performance of PBXs and other types of switching systems. The term *subjective* is used because even though standards exist governing these parameters, quality is perceived "in the ear of the listener." Often, users distinguish between "good" and "bad" service, without being able to isolate individual evaluation factors.

Key variables are station line levels, noise, distortion, and crosstalk. Line levels encompass several technical characteristics. The relative loudness of sound on the line is one such variable. Dial tone, busy tones, and DTMF tones might be too loud or too soft. The same goes for being able to hear the distant party through the handset, and to hear oneself (*sidetone*) in the handset.

Noise can be background static or hum, as well as intermittent loud interruptions, caused by electrical impulses on the line. Sometimes, feeding the PBX with poor-quality power or running wire and cable near high-voltage sources will produce static and "60 cycle hum." Even a poor plug-to-jack connection can cause annoying crackling sounds in the handset.

Distortion is a signaling and transmission problem caused either by the PBX components themselves, or by poor-quality network facilities and connections. Distortion causes DTMF tones to fluctuate, producing a warbling effect. It also creates shifts in the quality of voice reproduction due to frequency variations, sometimes making a voice sound high or low pitched, or simply unrecognizable.

Crosstalk is caused by electrical "coupling" or transformer effect, which superimposes one set of signals upon another, generally signals in adjoining cable pairs. This problem stems from faulty extension or trunk printed circuit boards (PCBs), or from improperly installed wire and cable.

A common source of degraded signal and voice quality is a poorly grounded PBX system, either because an improper ground was selected, or because the grounding design itself was faulty. Two major sources of grounding problems are electrodes tied to plastic pipe, and attempts to use electrical conduit or building steel as the system grounding electrode.

Features

Sets of features transform the PBX from a simple port-to-port connection device into a powerful information transfer system. As described previously, features are software-driven routines that enable the PBX to perform certain repetitive functions.

PBX features are categorized as system, attendant, station, and management-related. *System features* are centered around processor-oriented functions, applicable to all categories of PBX users. *Attendant features* enable a console ("switchboard") operator to answer external calls, extend them to PBX stations, serve as a call-coverage point, and assist users in placing external calls. *Station features* help the individual telephone user to communicate with people and to access other information resources more efficiently. *Management features* help the PBX administrators to review traffic information, change feature assignments, associate costs with premises and network services, and keep track of the system configuration.

Appendix C contains a complete listing of the more important PBX features. Industry standard descriptions of these features are contained in the *Master Glossary of Terminology*, published by the Aries Group/MP&SG, Gaithersburg, MD.

A 1990 study conducted by TFS CommSurv, Westford, MA, indicates that users already have equipped in their PBXs, all of the features—with the exception of ISDN—required to meet their business needs. The top features were, in order of popularity:

- Toll restriction

- Automatic route selection

- T1 network interface

- Station message detail recording

- Voice mail capability

- Automatic call distribution

- System management

- Modem pooling

The PBX market

Currently in the U.S., there are some 30 million business telephones served by PBXs. The leading U.S. PBX manufacturers are AT&T, Northern Telecom, ROLM, Mitel, NEC, Fujitsu, and Siemens, producing systems that range from as few as 25 lines to nearly 60,000 lines. These manufacturers compete not only with each other, but with advanced digital Centrex, and hybrid/KTS offerings (discussed later). The following are the leading manufacturers' products in the small and medium/large PBX categories.

Leading U.S. PBX Manufacturers (Small PBXs)

Manufacturer:	AT&T	Manufacturer:	Northern Telecom
Product:	System 25	Product:	Meridian 1 Opt 11
Max Lines:	200	Max Ports:	150
Max Trunks:	104		
Technology:	Digital	Technology:	Digital
Manufacturer:	Rolm	Manufacturer:	Mitel
Product:	9751 Mod 10	Product:	SX-200 Digital
Max Lines:	50	Max Lines:	500
Max Trunks:	5	Max Trunks:	200
Technology:	Digital	Technology:	Digital
Manufacturer:	Fujitsu		
Product:	Starlog		
Max Lines:	512		
Max Trunks:	240		
Technology:	Digital		

Leading U.S. PBX Manufacturers (Medium/Large PBXs)

Manufacturer:	AT&T	Manufacturer:	Northern Telecom
Product:	Definity	Product:	Meridian 1
Max Lines:	32000	Max Ports:	60000
Max Trunks:	6000		
Technology:	Digital	Technology:	Digital
Manufacturer:	Rolm	Manufacturer:	Mitel
Product:	9751 Mod 70	Product:	SX-2000
Max Lines:	20000	Max Lines:	10000
Max Trunks:	10000	Max Trunks:	4000
Technology:	Digital	Technology:	Digital
Manufacturer:	Fujitsu	Manufacturer:	NEC
Product:	F9600	Product:	NEAX 2400 IMS
Max Lines:	10000	Max Ports:	23000
Max Trunks:	3584		
Technology:	Digital	Technology:	Digital

Market shares

At the close of the 1980s, PBX industry statistics reflected pronounced market contraction. Of the 16 major manufacturers, three controlled over 60% of the market, according to figures published in *Telecommunications* magazine. Eight firms accounted for over 90% of the market, with the remaining eight competing for less than 8% of the market.

Continued slowing in the rate of new installations and replacements is expected to produce a nearly flat growth curve well into the 1990s. This forecast does not take into account possible increases resulting from the value-added reseller (VAR) and value-added distributor (VAD) activity.

Market share figures as of January 1990 were:

Vendor	Share
AT&T	27.7
Northern Telecom	23.2
Rolm	11.5
Mitel	9.2
Fujitsu	7.4
NEC	6.8
Siemens	4.7
Intecom	1.9
Others	7.6
Total	100.0

Fig. 4.10   AT&T 5ESS digital central office switch (*AT&T*).

Centrex (CENTRal EXchange) systems

Centrex is a LEC service offering which delivers advanced PBX-type features without the need to purchase or lease switching equipment, which greatly reduces the need for premises floor space, commercial power, and heating, ventilation and air conditioning (HVAC) required by PBXs. Centrex service can be provided from the same central office switch used for residential telephone service. Originally intended for customers with many stations, Centrex is now a candidate for users with just a few business lines. Figures 4.10 and 4.11 illustrate AT&Ts 5ESS (R), and Northern Telecom's digital Supernode digital central office switches, both of which can provide Centrex services described below.

Evolution

Centrex was introduced by the Bell System in the 1960s as a CO-based premises service offering for medium- to large-sized users. Within the Bell System, Centrex

Fig. 4.11 Northern Telecom Inc. DMS Supernode central office switch (*Northern Telecom Inc.*).

was first offered using analog, electromechanical switches (e.g., the No. 5 Cross-bar, a relay-implemented class of CO switches that at one time served more than 28 million lines). Following that, Centrex debuted in electronic switching vehicles such as the 1ESS (1965), and its successor, the 1AESS (1976). Digital Centrex was introduced in 1982 using AT&T's 5ESS, and in 1983 using the Northern Telecom DMS-100.

Early Centrex offerings provided only basic business telephone services. In the mid-1970s, PBXs outpaced Centrex in the application of microprocessor technology, producing numerous advanced features, and capturing a majority share of the premises service market. Following the breakup of AT&T in 1984, industry observers predicted the demise of Centrex. Instead, the LECs have succeeded in employing digital switching to re-establish Centrex as a viable PBX alternative, not only retaining existing customers, but attracting new ones. Centrex currently serves over 6 million (about 10%) of all business telephones in the U.S.

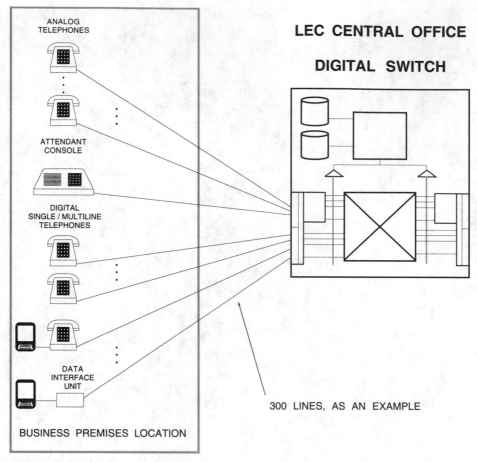

Fig. 4.12 Digital Centrex.

Figure 4.12 shows a CO implementation of digital Centrex and a typical interconnection with a business premises. Centrex station lines are similar to ordinary telephone loops, except that the line loss is usually limited to values well below the maximum for an ordinary loop to maintain a good grade of service between users on the same premises who make heavy use of Centrex services and features.

Services & applications

With digital CO switches, virtually all premises services provided by PBXs can be provided by Centrex. This includes station-to-station calling, access to PSTN and private networks, direct inward and outward dialing, and attendant and message center services. Optional adjunct services include system management comparable to that of a PBX, automatic call distribution (ACD), and voice messaging.

Centrex is a regulated service, with rates and tariffs approved by state regulatory authorities. Tariffs cover one-time and recurring charges. One-time charges cover initial installation, and any subsequent rearrangements to the service configuration. Recurring charges are those paid by the customer each month for use of the service.

Centrex is also a *discretionary service*, meaning that a LEC is under no obligation to offer it in all locations, or at all customer sizes. Centrex is attractive to a LEC where excess CO switching capacity exists sufficient to make the projected marginal sales revenue greater than the projected marginal cost of providing the service.

A large organization—spread across multiple buildings or located within a campus environment, but where all locations can be served by the same central office switch—can arrange for *uniform Centrex service* to each building. A single main listed directory number (LDN) is assigned to the organization, and a single centralized attendant location serves all buildings. For very large installations, Centrex is provided via remote switching units (RSUs) and/or digital loop carrier (DLC) facilities, installed either in one of the buildings, or on nearby LEC property.

With digital *city-wide Centrex*, it is possible to serve multiple, widely separated business locations with a single NXX (CO exchange code), using multiple, compatible LEC central office switches. In both uniform and city-wide Centrex, LDN callers are unaware that geographically separated business locations are involved.

Intra-switch, *intercom group service* is sometimes offered as a Centrex option. It may not be cost-effective, however, particularly where the same feature is also available in CPE, such as key/hybrid telephone equipment. Worse yet, activation of this feature in both Centrex and CPE can cause operational problems. Intercom groups are therefore best implemented within CPE.

Electronic switched network services

Tie trunks may be provided between Centrex switches or between Centrex switches and PBXs to support private electronic switched network (ESN) services. Like PBXs, Centrex offers *least-cost routing* (LCR), a feature that reduces long-distance charges by analyzing dialed numbers and automatically selecting the transmission service that results in least cost, giving preprogrammed advancement of a call from a low-cost path, such as a tie line, to progressively more costly routes, such as foreign exchange (FX) or direct distance dial (DDD) trunks provided by LEC and/or IXC PSTNs. (FX is a service that provides a dedicated circuit between a user telephone, PBX, or Centrex service, and a corresponding telephone, PBX, or Centrex service other than the one that serves the caller.)

It is possible to establish private networks using Centrex switches and PBXs. However, for other than simple tandem tie-line services, the challenge to achieve compatibility between a number of different Centrex and PBX switches, required for full-featured ESN operation, is presently insurmountable. Private ESN networks are best implemented by a single LEC, a single IXC, or within a private network group of PBXs with identical ESN capabilities. However, recent design agreements between switch manufacturers, such as that between AT&T and Northern Telecom, are slowly creating multisource interoperability for basic ESN services.

In the same fashion as a PBX, long distance services from several IXCs may be integrated within a Centrex alternate routing scheme. IXC networks may be accessed via LEC intra-LATA transport services as described in Chapter 8.

Components

Switching equipment & architecture. The CO switches that deliver today's Centrex services incorporate modular designs, both in upper-level architecture (e.g., CPU/memory, switch matrix, administration, etc.) and in lower-level peripheral connections. For example, individual microprocessor/memory units control groups of line/trunk ports, with each group either slaved to a supervisory microprocessor in the distributed hierarchy (digital architecture), or managed directly by a master CPU (analog architecture).

The ability to add common control intelligence and peripheral ports in modular increments translates to lower rates for Centrex customers across the entire size spectrum. By avoiding large common control step functions, Centrex switching charges exhibit nearly linear cost characteristics as the configuration grows.

System management requirements, however, particularly those involving CPE, such as cable management, call costing, voice processing, conferencing, and CPE repair/MAC work, do create cost step functions as the work performed by the LEC, or its *fully separated subsidiary* (FSS), increases.

Port interface capabilities. Each Centrex mainstation number (unique telephone number) is normally supported by a twisted pair of copper wires in the LEC outside plant. These pairs are analog current-loop facilities powered by –48 VDC directly from the CO switch. For this reason, CO service is rarely interrupted by a commercial power failure, unless that failure is caused by physical damage that also affects the CO and/or local loop.

Trunking facilities for Centrex are not the same as those in the PBX environment. Centrex trunking connections occur on the "back-side" of the CO switch, to access either tandem switches at IXC network gateways or other COs. Private line facilities are also used with Centrex, which may require terminations at the premises POT and tieback to the CO.

Centralized vs. distributed switching. The ability of Centrex to provide uniform, cost-effective service to multiple locations is contingent upon switching equipment compatibility and availability of remote modules. Remote modules are smaller versions of the main Centrex switch, tied back to the host CO by digital transmission facilities, but designed to operate in stand-alone mode if the umbilical link is cut.

Most CO switch suppliers offer several levels of remote module capability. An important issue is service degradation resulting from loss of the remote-to-host link.

If a mixture of Centrex and PBX service is used to implement a metropolitan area network, different exchange codes for each location could be encountered. This creates a situation where station-to-station dialing within the network must be seven-digit, in the absence of costly "foreign wire center" tariff arrangements. Also, the LEC may apply tariffed message unit charges to these calls.

Circumstances involving a CO "conversion" (upgrade) occurring at the time of large-scale Centrex project planning may afford the opportunity to negotiate with the LEC for a dedicated exchange code. This would allow exclusive use of up to 10,000 numbers, and network-wide four- or five-digit dialing, while avoiding message unit charges.

Where distributed Centrex is provided from multiple COs, LEC billing may not be consolidated, or even uniform, since invoices might be prepared at different data centers. This can create problems in maintaining inventory control, financial accounting, and user chargeback.

Peripherals. Figure 4.12 shows that digital Centrex can accommodate analog and digital, single and multiline station equipment. Until recently, multiline Centrex service necessitated the use of key telephone system (KTS) CPE. Centrex electronic key systems and digital single line telephones use Centrex lines over digital local loops. The CPE is proprietary to the manufacturer of the digital Centrex switch.

PCs and other data terminals can be connected via digital Centrex. Centrex lines used for digital single line and mulitline station sets are terminated at the Centrex switch on special line cards. These lines require special administration and cannot be used with analog station equipment. Centrex, like PBXs, provide modem pool service for data communications via public and private voice networks. Centrex modem pooling, however, is more complex than that for a PBX, generally because of prohibitions against locating CPE (modem pooling equipment) with the LEC CO.

Since divestiture, the RBHCs have been prohibited from manufacturing CPE, requiring their LEC sales forces to inform a Centrex customer prospect of this constraint. In October 1985, however, Bell Atlantic received a "Prime Contractor" waiver to the provisions of the MFJ. This waiver allows Bell Atlantic, as the parent RBHC company, to furnish an integrated systems proposal, which includes both Centrex switching and associated CPE. In such cases, the RBHC must ensure that the LEC maintains a neutral referral program, including other CPE suppliers, even to the extent of showing competitive equipment catalogs.

Adjunct systems

Voice message processing. Voice message processing provided with Centrex service is integrated either by connection with a voice mail system collocated at the CO or at a remote location. The voice mail system communicates with the Centrex switch over a high-speed digital data link. Where the system is located on the customer's premises or at a centralized voice mail provider's location, this data link runs back to the CO.

As in the case of PBX voice message processing, today's Centrex voice message processing is an integrated offering, providing the same ease of use and feature-richness previously associated only with a PBX-based system. The large scale of the system enables the LEC to tariff its service at a rate competitive with PBX-based voice mail.

An example of the degree to which voice message processing has become integrated with Centrex service is provided by the fact that Bell Atlantic has tariffed voice mail service using AT&T's 5ESS CO switch and AT&T's Audix system as CPE. The Audix system had previously been furnished only with AT&T PBX products.

Such trends indicate that nonintegrated voice message processing systems are fading from the scene, and that more powerful, easier-to-use systems will continue to proliferate, furnishing value-added capabilities in the voice services marketplace.

System management. System management capability enables Centrex users to control and manage features, calling privileges, and restrictions. The system management function relies on computer equipment installed on the customer's premises, connected by data links to the CO switch providing Centrex service.

System management is valuable where the following circumstances exist:

- Frequent changes in feature assignments

- Frequent internal personnel moves

- Growth mode, with frequent assignment of new stations

- Requirement for internal control over Centrex

- Need for rapid accomplishment of MAC work

- Requirement for cost control

Over the past decade, the two most widely used forms of system management have been Centrex Station Rearrangement (CSR) and Multiple Access Customer Station Rearrangement (MACSTAR). They are compatible with 1AESS, 5ESS, and DMS-100 switches.

More recently, however, a more powerful Centrex system management capability has been developed, rivaling that of the PBX. Often called *operations support system* (OSS), or *operating system control* (OSC), this system consists of a stand-alone computer, generally a UNIX-based workstation, that not only connects with the CO switch for feature management but also controls numerous other premises-related functions such as:

- Directory service

- Maintenance and repair scheduling

- Inventory control

- Cable management

- Moves, adds, and changes (MAC)

- Billing services

It is important to maintain *system management security* through frequent changing of computer passwords and limiting their distribution. Changes made to the system are used to update the LEC's billing and service records, so accuracy and control are critical to system management operations.

Costs associated with Centrex system management can be equivalent to those of a PBX. This results from the addition of management functions that relate to CPE, rather than just the CO switch. Centrex system management hardware is therefore moving into the customer premises as part of a RBHC strategic approach to capture new tariff or contract revenues from CPE-based operations, maintenance, and management services.

In the past, certain changes to the switching configuration could only be performed at the CO, while others could be made on-premises using a customer administration terminal. The terminal-based changes were not charged individually, but the CO-based changes generally were, which made frequent MAC work expensive.

Today, the trend is to enable the customer to make all necessary changes to the Centrex configuration using an RBHC-provided workstation. This lowers the cost of individual changes, which tends to reduce system operating expense. The initial capital cost of the management system, however, reflecting its new functionality, can approach that of a PBX.

Operational characteristics

The task of defining Centrex operational characteristics is complicated by issues relating to physical location and ownership of the switching equipment and the CPE. Users need to stay informed about these issues in order to properly plan and implement Centrex systems.

Decisions involving selection of terminal equipment and cable systems compatible with Centrex features and functions can be complex. Thus, the LEC needs to be brought into the planning process, and its inputs included where appropriate in specification and RFP structure.

Responsibilities created by location and ownership of equipment making up the Centrex system must be carefully identified. This will simplify not only the planning and acquisition process, but also ongoing system management. In general, the concepts described under the PBX section apply to Centrex as well. There are, however, some distinctions, which are pointed out below.

System capacities

Up to 100,000 lines of CO switch capacity means that an individual Centrex customer generally need not be concerned with accurately forecasting growth requirements. Service packages begin at several lines and can theoretically expand to tens of thousands. Type, size, and remaining capacity of the CO serving a particular location will, of course, determine the actual growth capability.

Unlike the PBX, there is no concentration of station lines prior to connection with the Centrex switch. Each Centrex extension number is served by a dedicated pair running between the CO and the customer premises. Once terminated at the premises main distribution frame, pairs are cross-connected either to single line or multiline telephones or to separate key equipment.

Centrex is dependent upon the traditional single twisted-pair copper loop. Until recently, attempts to increase digital voice and data throughput have been hampered by the bandwidth limitations of twisted pair, and constraints on flexibility imposed by analog outside plant design. Difficulties have been encountered where high speed digital services are run over the local loop, which may contain analog load coils, bridge taps, and multiple splice points—all of which can disrupt a digital bitstream.

The trend toward local loop fiber has been limited to high-density urban areas, although projects like Bell South's Heathrow, Florida, fiber-to-the-home test bed are on the rise. Fiber is also being installed between host switches and remote modules, but such enhancements are proceeding slowly.

System performance

Both analog and digital Centrex switches are engineered by the LEC to carry anticipated amounts of offered traffic at various levels of blockage. The sheer size of most LEC switches usually means that initial or growth requirements of any single user is not likely to affect the blocking grade of service. However, although it might be highly unlikely in normal conditions, during a crisis or some popular event (such as the surge in calls for reservations to a rock concert, which, some years ago, resulted in the loss of dial-tone service in the Washington, DC, area), contention could begin on the line side of the switch, caused by a physical concentration ratio of incoming lines to paths through the switch's internal network. It is therefore important that a Centrex prospect with crucial communications needs query the LEC as to what grade of service levels are being offered.

Note that calls placed within an exchange code provided by the switch do not require a trunk connection as they are station-to-station (intra-switch). This reduces blockage potential. Even where interexchange trunking is required, the large number of circuits connecting that switch to the PSN normally ensures that a trunk will be available when needed.

Where distributed Centrex arrangements are employed, involving use of remote switching modules and other peripherals, additional factors come into play. The distributed design must include sufficient node-to-host links to carry the offered traffic at the specified GOS. The remote module may be engineered to only carry intramodule traffic if the host link is lost. If network trunks are contained only at the host location, a link failure could cause loss of long distance service, call accounting, and other centralized functions.

CO switches are often engineered to a lower overall traffic intensity number than are PBXs, e.g., 4.5 CCS/line (Centrex) versus 7.0 CCS/line (PBX), simply because residential subscribers tend to present a lower level of traffic than business customers. A modern CO is engineered for between four and five CCS per line standard. Above that standard, it might be possible to obtain a guaranteed GOS level under either tariff or special contract, but at extra cost.

The use of CPE data stations usually results in higher per-station traffic intensities than telephones since holding times for data calls are longer than those for voice calls. CPE data stations tend to have less GOS impact in a Centrex versus a PBX environment because of the increased number of lines available in the CO switch. Information on all such connections should be furnished to the LEC as part of the service planning and selection process, however, if a minimum guaranteed GOS is required for either voice or data.

Features

Centrex features are provided by both analog and digital switches and are dependent upon switching and software release versions. As previously noted, digital

Centrex, where available, provides features comparable with those of fourth generation PBXs.

The analog 1AESS switch, however, nearly 20 years after its introduction, is still the principal vehicle for Centrex service, currently serving some 3 million lines. Having undergone numerous software enhancements over the years, its feature complement is very close to that of digital switching counterparts. The 1AESS is not expected to undergo widespread replacement in the near future. Instead, RBOCs are examining other alternatives, such as adding digital "front ends" to make them compatible with the emerging end-to-end digital PSN.

In general, Centrex features are roughly equivalent to PBX features in terms of type and quantity. For marketing purposes, Centrex features are often categorized by the LECs as "basic" or "enhanced." A comprehensive feature list appears in Appendix C, and full descriptions are contained in the *Master Glossary of Terminology*, published by Aries/MP&SG, Gaithersburg, MD.

The Centrex market

In the U.S., the Centrex installed base has grown by approximately 4% annually for the past two years, now representing some 6.5 million lines. Although Centrex service has been available to smaller customers for nearly a decade, current market statistics indicate that large Centrex systems are still the norm. In large business and federal government sectors alone there are approximately 145,000 switching systems of all kinds installed nationwide (including PBX and stand-alone KTS/hybrid systems), representing over 15 million lines, with an average size of 115 lines per system. Although Centrex serves only around 12,000 of these systems, it nevertheless accounts for over one-third of the total (5.5 million lines), for an average size of nearly 500 lines.

According to a recent Bellcore study, over 70% of the customers with systems of 1000 lines or more are served by Centrex. Within the Bell Atlantic region, one of the larger RBHCs, there exist some 2500 Centrex systems, representing over 1.4 million lines. Over 74% of the total number of Bell Atlantic Centrex lines serve systems larger than 1000 lines.

The following is a breakdown of Centrex users by market segment as of year end 1990.

Sector	Percent
Government	25.0
Manufacturing	20.0
Health and Education	15.0
Engineering and Legal	10.0
Wholesale	7.0
Retail	6.0
Banking and Finance	5.0
Insurance	3.0
Transportation	3.0
Other	6.0
Total	100.0

Market shares

The major Centrex switch manufacturers are AT&T, Northern Telecom Incorporated, Siemens, L. M. Ericsson, NEC, GTE, and Alcatel. In the mid-1970s, manufacturers such as GTE and TRW Vidar made the initial inroads on the Western Electric-held Centrex switch market by selling to the non-AT&T-controlled independent telephone companies (ITCs). Once the "buy-AT&T" restriction was lifted from the RBHCs by the MFJ, however, NTI's newly introduced DMS-100 switch eclipsed all comers, including AT&T's 5ESS switch.

By late 1987, Northern Telecom Inc. (NTI) was entrenched as the number-one digital CO switch supplier to the RBHCs, with some 3 million lines of DMS-100 installed or on order in the U.S. (approximately 70% of the digital CO market).

Other firms, such as L.M. Ericson, Siemens, Alcatel, ITT, Fujitsu, and NEC have achieved market penetration as Bellcore approved their products.

Key/Hybrid Telephone Systems

A *key telephone system* (KTS) is an arrangement of multiline telephones and associated equipment that permits station users to depress buttons (keys) to access CO or PBX lines and KTS features. Typical feature operations include answering or placing a call on a selected line, putting a call on hold, using an intercom path between phones at the same location, or activating an audible signal.

A KTS permits interconnection among on-premises stations without the need for central office or PBX switching systems. A multiline telephone incorporates visual indicators to show idle, busy, or ringing status of two or more lines (telephone numbers). KTSs are used by small businesses, or individual *community of interest* groups within larger businesses. Figure 4.13 illustrates a KTS and its relationship to other switching systems. Although not shown in the figure, KTSs behind Centrex service are arranged similarly to the PBX example.

KTSs are directly connected to CO, Centrex, or business lines, so it is not necessary to dial an access digit to make an outside call. Station-to-station dialing in a KTS is accomplished by one- or two-digit intercom codes, with the number of codes determined by system size. Attendant services are generally less sophisticated than those of a PBX, although hybrid and digital systems now support virtually all console functions.

Evolution

In the mid-1920s, the Bell System first provided key telephone service using custom-engineered assemblies of lamps, keys, and wiring plans. These arrangements were complex and field labor intensive and, in the early 1930s, evolved into a packaged electromechanical hardware arrangement, the 1A KTS, which standardized wiring and greatly reduced on-site labor.

The 1A2 KTS

The 1A2 KTS, first offered in the early 1960s, was even easier to install and used prepackaged components, consisting of printed circuit boards (PCBs) and miniatur-

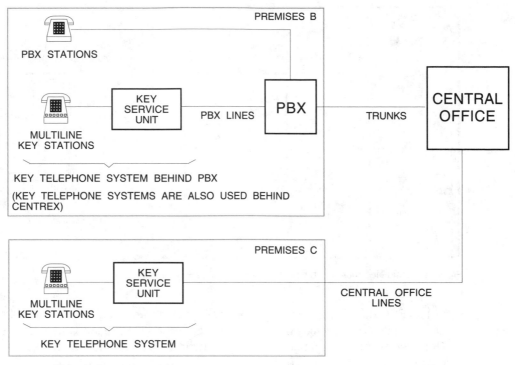

Fig. 4.13 Key telephone systems.

ized relays. This resulted in a more compact unit, while adding popular features such as line exclusion, dial intercom, music-on-hold, and paging access. Typical button arrangements provided for push-button access to 5, 9, 19, and 29 lines per telephone. 1A2 installations are characterized by one or more 25-pair "key cables" from each telephone to the central *key service unit* (KSU).

The EKTS

In the 1970s, the widespread availability of low-cost microprocessors led to the development of *electronic key telephone systems* (EKTS). EKTSs offered PBX-like features, unavailable in electromechanical designs. They also eliminated the need for key cables. This was accomplished by using PBX-like switching and control capabilities, which permitted a multiline telephone to be connected to the KSU via a "skinny-wire" one-to-six-pair cable. Thus, the distinction between KTSs and PBXs began to blur.

Hybrid systems

A *hybrid telephone system,* which may be of analog or digital design, incorporates both traditional KTS and PBX functions and features. A strict definition of a hybrid is difficult, since the mix of functions and features in a particular product is based solely on what the manufacturer believes will produce a competitive edge.

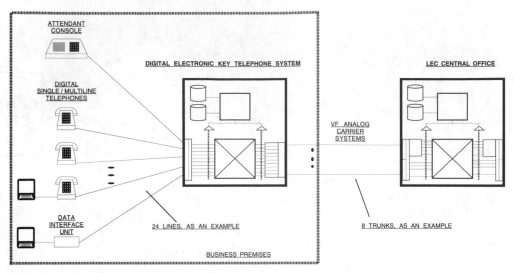

Fig. 4.14 Digital electronic key telephone system.

The FCC determines whether a system is registered as a KTS or a hybrid system based on whether a single line station can access only a single CO loop line or trunk (KTS registration), or a pool of CO loops (hybrid registration). Some systems can be configured for either type of service, and are allowed dual registration.

Digital systems

Digital KTS systems, all of which are classified as hybrids, incorporate digital PBX-type architectures. For example, the NTI Meridian Norstar incorporates a stored program control, pulse code modulation, time division multiplexed switch design, proprietary 2B+D intra-system signaling, and codecs embedded in telephones.

Figure 4.14 illustrates a digital KTS and its connection to a CO. Although the labeling details are omitted, the figure shows that the major components of the digital KTS are the same as those of the generic circuit switch previously described. The figure also indicates that, like a PBX, the range of station equipment includes single and multiline digital telephones, attendant consoles, and digital interface units.

Hybrids and digital KTSs are capable of connecting to all types of analog trunks and tie lines. Most systems, however, are not designed for direct T1 connection or tandem switching functions, which limits them to analog network end point applications. Figure 4.15 is a photograph of the NTI Meridian Norstar telephone. Figures 4.16 and 4.17 show two of AT&Ts modern key telephone systems.

Applications

KTSs have traditionally been installed either as stand-alone systems, or behind PBX and Centrex systems. With the development of stored program control and switch-driven electronic multiline telephones, functions performed by the KTS have been

Fig. 4.15 Northern Telecom Inc. Norstar modular, key system digital telephone (*Northern Telecom Inc.*).

taken over by the PBX itself. KTS behind Centrex, however, is currently a growing application. Centrex feature access is simplified by button-per-feature operation of KTS telephones, and additional functionality is contributed.

Today's KTSs are subjected to extensive product engineering to ensure high function-feature/cost performance. Competition is fierce, not unlike that in the personal computer industry. All models make heavy use of embedded microprocessors, although historically systems with 24 stations or less tend to be analog. Those in the size range of 24–100 stations have a digital core but generally rely on analog transmission between the sets and the KSUs.

KTS capacity is indicated by the maximum number of stations and PBX or CO lines that can be supported. Capacities range from a few stations and lines to about 160 total ports (a port being either a telephone or a telephone number interface).

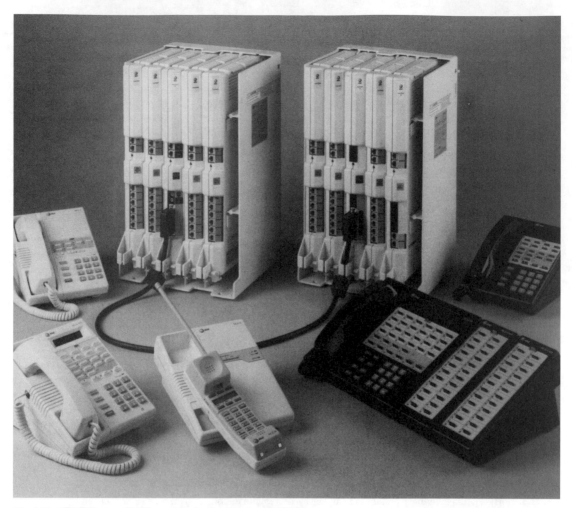

Fig. 4.16 AT&T Partner II (R) communications system (*AT&T*).

Behind-Centrex KTSs. The Centrex resurgence has administered a shot in the arm to the KTS industry. Despite the functional and financial attractions of Centrex, many customers require more in the way of features and system control than may be available under Centrex switching tariffs or special contracts.

A *non-square* small business Centrex interface enables these specialized KTSs to flexibly assign extension numbers to telephones, rather than be bound by *square* KTS arrangements, with all numbers appearing on each telephone. The heart of the behind-Centrex market currently lies in the under-35 line size.

Single-button feature access is an important behind-Centrex capability. Feature access transparency, the ability to directly invoke Centrex features with a CPE telephone not made by the CO switch supplier, will become more important as additional CO-driven features are made available to business and residential subscribers, e.g., CLASS, data-over-voice, etc.

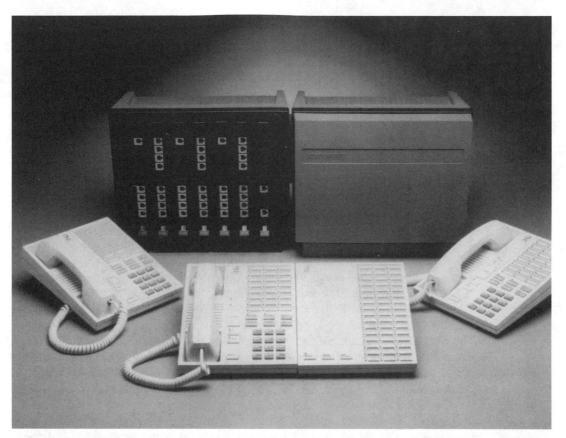

Fig. 4.17 AT&T SPIRIT communications system (*AT&T*).

1A2 Emulation. There are still many in-place 1A2 systems, particularly in large commercial and government Centrex installations. In some areas, obtaining spares and repair services is becoming difficult. Behind-Centrex KTSs that emulate 1A2 operation target this market niche. These systems normally do not offer telephones that are directly compatible with 1A2 KSUs. Instead, they either add to or replace the existing 1A2 arrangement. 1A2 emulation systems find a niche in organizations with low feature requirements, that are also apprehensive about adopting new telephone technologies.

Key/hybrid electronic switched network services

Access to public and private networks was provided by traditional KTSs but was limited to loop-start, analog facilities. Note that CO lines terminated directly on user telephones in a button-per-line fashion.

Traditional KTSs did not function in a network environment because they were not capable of supporting tie line connections. They could, however, terminate off-premises extensions (OPXs) from a main PBX location, distributing these PBX extension numbers as key telephone appearances at the distant OPX location. Ironically, most hybrid and digital systems cannot terminate OPXs in electronic telephones, because the line card data link is not passed by the LEC's OPX network interface, e.g., the OL13A/B/C interface.

Some of today's hybrid and digital systems can directly connect with analog tie lines and tie trunks, and with digital trunk facilities. These systems can only be end points on the network, however, since they do not possess tandem switching capability.

Components

Switching equipment & architecture.

Port interface capabilities.   In the traditional KTS, KSU PCBs supported individual CO directory (telephone) numbers, which were then wired to multiline telephones. This differed from the use of PBX PCBs, which supported individual stations (telephones). A significant difference between the two arrangements was that there was no upper limit to the quantity of telephone numbers that a KTS could support, as there was with PBX stations. A traditional KTS line is equivalent to a PBX trunk, since both are used to connect the systems to a host switch.

Today's KTSs are more nearly like the PBX in that they impose limits on the quantity of stations and lines that can be accommodated. If a system is outgrown, the telephones and internal PCBs are sometimes reusable, but the KSU cabinet and power supply generally have to be abandoned.

Traditional KTSs cannot connect to computers or LANs other than through a modem. Analog hybrids share this limitation. Some digital hybrids and KTSs support modem-less computer connections. None currently connect to LAN bridges or gateways. Developmental work in this area is presently underway at AT&T, NTI, NEC, and other digital system producers.

Centralized vs. distributed switching.   Traditional KTS and most hybrid/digital systems employ centralized switching designs. In this arrangement, each telephone is directly connected to a central KSU, which performs all required switching operations.

A recent, popular addition to the KTS industry is the so-called *KSU-less KTS*. These systems perform all common control functions using PCBs contained within each multiline telephone, and so do not require a KSU. They are designed for installation by the user, generally where premises wiring is already in place. Since there is no KSU, no communications closet space is required.

There is a limit to the quantity of CO lines and stations in a KSU-less system. Multiline telephones generally handle up to three telephone numbers, using between three and six pair horizontal wiring. Some telephones incorporate modular jacks for connecting ancillary equipment. At present, 12 telephones is the upper limit for KSU-less systems.

Peripherals.   In traditional KTS applications, *direct station selection/busy lamp fields* (DSS/BLFs) are used in conjunction with a multiline telephone having line appearances for all telephone numbers in the system. Upon answering an incoming call, the attendant checks the BLF. If the called party is not busy (indicator not lit), the attendant dials his intercom number simply by pressing a button on the DSS, and announces the call.

The call announcement may be received through the called party's handset, speakerphone, or over a call announcer, with either tone or voice alerting. To accept the call, the called party depresses the line button flashing on hold. If the call is not accepted, the attendant can take a message. Message waiting and electronic messaging features may also be used, and are particularly effective in situations where the called party does not answer.

Most KTS/hybrid telephones may be wall- or desk-mounted either in stock configuration or using a conversion kit. Some telephones incorporate additional features, such as long handset cord, cord swivel with strain relief, ringer volume control, and handset transmitter muting. Those with hearing aid compatibility include a control on the handset which allows the line volume to be adjusted for each call. Some systems allow storing the volume setting for future calls.

Speakerphones provide *full duplex, handsfree operation*, allowing a user to place and receive calls without having to lift the hand set. A speakerphone also supports intra-office "roaming." The speaker is normally equipped with an on/off switch. Speakerphones also support *on-hook dialing and call progress monitoring,* which allow the user to hear dial tone, dial pulses or DTMF signaling, busy tone, ringing, intercept tones and messages. The speakerphone can be used on both intercom and network calls.

Key telephones such as AT&T's Merlin assign individual basic features (transfer, add-on, and conference) to a dedicated button for ease of use. Merlin telephones also include a dedicated "drop" button for removing parties from conference mode. Fixed functions may account for over half of the total telephone button population.

A dedicated hookswitch flash button is standard on behind-Centrex telephones to ensure that in-progress calls are not cut off by users attempting to invoke features.

Soft keys on proprietary instruments allow buttons to be programmed for either CO lines or features. Some systems support *macro keys,* which store up to six different keystrokes. This allows combining features such as speed dialing and specialized common carrier service. Soft keys are generally assignable and programmable by the station user.

Adjunct systems

Voice message processing. Traditional KTSs can be connected only to nonintegrated voice messaging systems, which do not deliver the capabilities of systems tightly coupled with call processing functions. Several hybrids and digital KTSs do incorporate integral voice messaging subsystems, which also perform automatic answering and allow direct internal dialing. Functions that would be stand-alone in a PBX (such as auto attendant and voice mail) are often combined in KTS applications. These adjunct systems are designed as PCB components, physically housed within the KSU.

Interactive voice response to callers can be provided by multiple voice response unit (VRU) PCBs, housed within the KSU. The attendant can activate these VRUs as required, such as during busy periods, to invoke automated attendant service, or after hours to deliver recorded announcements. A tape unit interface can also record calls and store them as messages.

System management. Traditional and other non-stored program control (SPC) KTSs do not offer maintenance and administrative capabilities, since all features and functions are hardware oriented. Any additions or changes to these systems, as well as fault isolation and repair work, must be performed on-site. No maintenance access panel (MAP)-type terminal is available for these functions, apart from normal electronic test equipment. MAC work is accomplished largely through plug-in hardware module additions, while repairs involve module removal and replacement.

With SPC hybrids and digital KTSs, users can make certain feature changes at their own telephones, as with AT&T's Merlin system. For more complex changes, the system administrator must activate a programming switch in the KSU and use a specific telephone in the system. Telephones serve the KTS system management function more economically than would a dedicated MAP-type terminal.

Operational characteristics

Like their larger PBX counterparts, KTS and hybrid systems have shrunk in size while growing in capability. From a feature and function standpoint, there is little to distinguish among these systems. Differentiation exists more in terms of size than performance. It is therefore not surprising that KTS and hybrid systems suffer from being perceived in the same commodity-oriented perspective as PBX products.

System capacities

The KTS or hybrid system may either be *packaged,* with a certain capacity of lines and stations, or *port-oriented,* in which case lines and stations can be intermixed to a maximum total capacity. Hybrids and digital KTSs are following the PBX path to port orientation, as it is more flexible and simplifies system expansion. System sizes vary widely, based upon the market segment targeted by the product.

In the traditional KTS, the number of stations (telephones) was unimportant, so long as it did not overtax the line lamp-illumination power supply. Modern systems impose limits on the number of telephones that may be supported. These limits vary based upon the type of system. Low-end space division technology KTSs may "max out" at six or eight stations. Several hybrid and digital KTSs can accommodate up to 120 stations.

Some of today's packaged hybrids and digital KTSs impose a line/trunk-to-station percentage on the system, generally on the order of 40% to 60%. This does not offer the flexibility of a port-oriented system, which allows lines and stations to be "mixed and matched." Note that in either case, however, once the upper limit of the system is exceeded, a KSU upgrade is required.

System performance

Grade of service. In the small-systems world, GOS is largely measured—apart from voice quality—by the users' ability to access intercom paths and network connections. Early KTSs were constrained by a limited number of intercom talk paths or "links," analogous to station-to-station dialing paths in a PBX. Access to network resources, however, was non-blocking in line-per-station configurations, given the no-limit line capacity of K1A2 systems, and the ability to terminate lines and trunks directly on multiline telephones.

Ironically, the later, more fully featured systems do not share this important capability. Because of system packaging, there are fewer lines than stations, which immediately introduces blocking probability. Even where lines are pooled, line-to-station contention can prevent placement of a network call.

Probability statistics indicate that for a given line-to-station ratio, blocking potential increases as the system size decreases. Thus, a four line, six station system can be expected to experience blocking more often than a 16 line, 24 station system. Many of today's hybrids and digital KTSs provide non-blocking service on intercom, but not on network connections.

Features

KTS and hybrid features are categorized as system, attendant, station, and management related. *System features* center around processor-oriented functions performed by the KTS or hybrid, applicable to all categories of users. *Attendant features* enable the attendant or receptionist to answer calls, announce them to intercom stations, and serve as a call-coverage point. *Station features* help the individual telephone user communicate with people and access other information resources more efficiently. Management features help the system administrator to access traffic information, change feature assignments, associate costs with premises and network services, and track the system configuration.

KTS and hybrid systems offer certain features not normally available in PBXs, e.g., intercom groups, call announcing, and music-on-hold. The average system delivers around 30 features, with most equipped as standard.

The Bibliography contains a complete listing of normal KTS/hybrid features. Industry standard descriptions of these features are contained in the *Master Glossary of Terminology* published by the Aries Group/MPSG, Gaithersburg, MD.

Market

AT&T, Tie/Communications, Executone, Northern Telecom, Toshiba, and Comdial are the leading manufacturers competing for a projected 1994 $1.3 billion market for KTS/hybrid systems in the 2–100 station size.

KTSs are sold by independent telephone companies, interconnect vendors, RBHC subsidiaries, private distributors, and retailers. Because these products offer similar capabilities, warranties, and technology, often the only significant discriminator, aside from vendor installation and maintenance qualities, is price. The leading manufacturers' KTS/hybrid products are indicated below:

Leading U.S. KTS/Hybrid Manufacturers

Manufacturer:	AT&T	Manufacturer:	AT&T
Product:	Merlin II	Product:	Merlin Plus
Max Lines:	56	Max Lines:	8
Max Stations:	120	Max Stations:	20
Technology:	Digital, PCM	Technology:	Electronic Analog

Manufacturer:	TIE/Comm.	Manufacturer:	Executone
Product:	Businesscom 01	Product:	IDS
Max Lines:	16	Max Lines:	432
Max Stations:	48	Max Stations:	432
Technology:	Digital, PCM	Technology:	Digital, PCM
Manufacturer:	Northern Telecom	Manufacturer:	Toshiba
Product:	Norstar	Product:	Strata DK24/56/96
Max Lines:	80	Max Lines:	36
Max Stations:	120	Max Stations:	96
Technology:	Digital, 2B+D	Technology:	Digital, PCM
Manufacturer:	Comdial	Manufacturer:	Inter-Tel
Product:	Digitech	Product:	GMX 48
Max Lines:	24	Max Lines:	24
Max Stations:	48	Max Stations:	48
Technology:	Digital, PCM	Technology:	Electronic Analog
Manufacturer:	Panasonic	Manufacturer:	NEC
Product:	KX-T 123211	Product:	Electra 8/24
Max Lines:	12	Max Lines:	8
Max Stations:	32	Max Stations:	24
Technology:	Electronic Analog	Technology:	Electronic Analog
Manufacturer:	Iwatsu		
Product:	Omega-Phone ZT-S		
Max Lines:	6		
Max Stations:	16		
Technology:	Electronic Analog		

Market shares

Approximately 4,299,000 KTS stations were shipped in 1989. An annual growth rate of 2% (compounded yearly) is forecast through 1994. This forecast does not take into account possible increases resulting from growing value-added reseller (VAR), and value-added distributor (VAD) activity.

KTS/hybrid market share figures as of January 1991 are:

Vendor	Share %
AT&T	28.0
TIE	15.0
Executone	11.0
Northern Telecom	10 0
Toshiba	8.0
Comdial	7.0
Alcatel/ITT/Cortelco	4.0
Inter-Tel	4.0
NEC	3.0
Iwatsu	2.0
Panasonic	2.0
Others	6.0
Total	100.0

As late as 1989, NTI was not ranked among the top 10 KTS suppliers. They had not focused on this segment since the days of their Vantage product line. But the overwhelming popularity of the Norstar digital KTS/hybrid has thrust NTI into fourth place. NTI's 50 U.S. regional distributors and the RBOCs have been instrumental in this effort.

Given the low unit margins on KTS products, vendors are now concentrating on a value-added market approach. Adjunct systems and equipment such as voice mail, ACD, auto attendant, and call accounting, can support relatively high retail markups. Manufacturers are attempting to integrate these capabilities into KTSs and hybrids to enhance market appeal and to improve margins.

5

Signaling in Circuit Switched Voice Telecommunications Systems

Telephone exchange service was originally accomplished with manual switchboards and human operators. Prior to the rotary dial telephones, to make a call a user turned a crank that caused a lamp to flash on a panel at the central exchange. Responding to this *alerting* signal, an operator would then manually plug into the user's line (associated with the flashing lamp), and the user could verbally request connection with the called party. If the called party was served by a different exchange office, the operator used a similar process to alert a distant operator via trunks connecting the exchanges, and—again verbally—requested connection with the called party number.

The distant operator would check the called party's line for a "busy" condition. If not busy the distant operator would ring the called party's telephone. Both the calling and called party lamps remained lit during a call, providing the operators with indicators needed to *supervise* the use of customer lines and trunks. If the called party line was busy, the operator informed the caller.

Today, station equipment, switches and transmission systems incorporate *call handling* designs which generate and exchange signals to take the place of the above manual actions and verbal requests. The overall process is referred to as *signaling.*

Knowledge of signaling is important for several reasons. First, CPE acquired must be LEC and IXC compatible. Second, in private networks, CPE at different locations must be compatible. Third, telecommunications functionality increasingly depends on signaling capabilities, hence there is motivation to acquire business systems that ensure the ability to use new developments.

CPE – customer premises equipment

Signaling is the process of generating and exchanging information among components of a telecommunications system to establish, monitor, or release connections (call handling functions) and to control related network and system operations (other functions).

As in prior years, telephone system users remain a part of the signaling process. They participate when they elect to use the service by going "off-hook" (by lifting a handset), by dialing digits to access a service, by dialing telephone numbers, and by responding to various alerting signals such as audible dial tone, ringing signals and recorded messages.

Signaling generates and transfers the functionally categorized signals described below.

Address signals. Convey destination information such as a dialed 4-digit extension number, central office code, and when required, area code and serving IXC carrier code. These signals may be generated by telephone or other station equipment, or by the switching equipment itself.

Supervisory signals. Convey to a switching system or an operator the status of lines and trunks as follows:

- *Idle circuit*. indicated by an "on-hook" signal and the absence of existing switching system connections on that line
- *Busy circuit*. indicated by an "off-hook" signal
- *Seizure*. a request for service indicated by an "off-hook" signal in the absence of an existing switching system connection
- *Disconnect*. indicated by an "on-hook" signal subsequent to an established connection
- *Wink-start*. indicated by an off-hook/on-hook signal sequence on a trunk from a called office after a connect signal is sent from the calling office.

On-hook and *off-hook* are terms derived from the placement or removal of a handset at the telephone cradle, which closes or releases the *switchhook*, a plunger-activated switch built into the cradle. Although modern station equipment may use different switching arrangements, the on-hook/off-hook functions are still supported and the descriptive terminology is still in use. In LEC and IXC systems, supervisory signals are extended to billing and administration equipment for message accounting.

Alerting signals. Notify users, operators, or equipment of some occurrence, such as an incoming call. Included are ringing, flashing, recall, rering, and receiver-off-hook signals.

Call progress signals. Include dial tone, audible ringing tones, system-generated recorded announcements, and special-identification tones.

Signaling Interfaces & Techniques

Signaling is described in terms of interfaces among network components, as well as techniques used to transmit signaling information between interfaces. *Signaling interfaces* exist between station equipment and transmission systems, between transmission systems and switching systems, and between transmission systems themselves. *Interfaces* are common boundary points between two systems or pieces of equipment.

Loop signaling interfaces

An example of a *loop signaling interface,* one of three types of signaling interfaces, is shown in Fig. 5.1. Although other signaling techniques can be associated with loop signaling interfaces, the technique illustrated in the figure is called *direct current (dc) signaling.*

The left side of the figure shows the major parts of a telephone and its connection to a two-wire metallic VF transmission facility, which is connected on the other end to a line card in a CO switch. Loop signaling interfaces, indicated on the figure, derive their name from the metallic electrical loop formed by the line or trunk conductors (wires) and the circuits in terminating components.

Figure 5.2 illustrates signaling sequences for a typical call from one telephone to another through originating and terminating central offices. Event sequence is depicted by circled numbers. Figures 5.1 and 5.2 (as well as the text associated with them that is used to describe the principal signaling functions) are consistent with AT&T definitions and descriptions provided in Bibliography Reference 6.

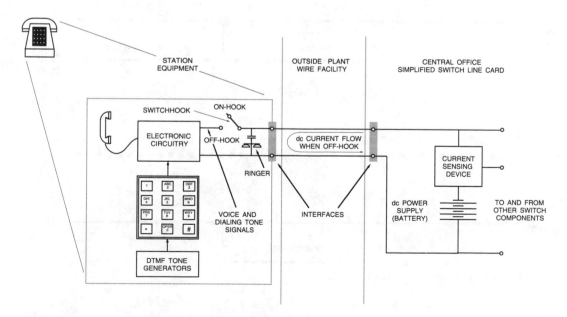

Fig. 5.1 Example of loop signaling interfaces.

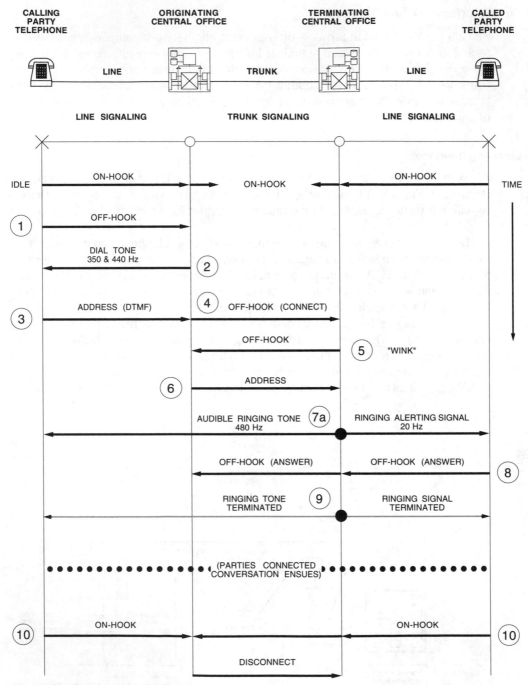

Fig. 5.2 Signaling for a typical call connection.

In Fig. 5.2, the sequence begins with both calling and called-party telephones on-hook, or in an *idle* condition. A *seizure signal* is generated by the calling telephone when the caller removes the handset from the cradle (step 1 on Fig. 5.2).

As Fig. 5.1 shows, when idle, the telephone switchhook is open, and no dc electrical current can flow through the loop. Going off-hook closes the switchhook switch and permits current to flow through the line from the battery to a current sensing device, both at the central office. The current sensing device detects the off-hook status of the telephone and provides that information to other parts of the central office switch. This type of seizure signal is called *loop-start,* a supervisory signal generated by a telephone or a PBX in response to completing the loop current path.

An alternative seizure signal associated with loop signaling interfaces is *ground-start,* a supervisory signal generated by certain coin-operated telephones and PBXs by connecting one side of the line to ground (i.e., a point in an electrical circuit connected to earth).

Wink-start is yet another supervisory signal that consists of an off-hook followed by an on-hook signal, exchanged between two switching systems. The wink-start signal is generated by the called switch to indicate to the calling switch that it is ready to receive address signal digits.

Once the CO switch recognizes the telephone off-hook status, it generates and returns a *dial tone* over the loop (step 2 on Fig. 5.2). Dial tone is a continuous tone formed by combining 350 Hz and 440 Hz tones.

Following receipt of dial tone, the user is free to dial the number of the party he or she wishes to call. In Fig. 5.1, the caller does this by pressing appropriate dialing keypads, which generate a *dual tone multiple frequency* (DTMF) address signal (step 3 on Fig. 5.2). DTMF signaling uses a simultaneous combination of one of a group of lower frequencies and one of a group of higher frequencies to represent a digit or character. Figure 5.3 illustrates the frequencies used to represent digits and characters.

Referring back to Fig. 5.1 (page 97), the telephone incorporates a DTMF tone generator, which under control of the dialing keys sends the DTMF address signals over the same metallic loop conductors used for the supervisory signals. The originating CO receives the DTMF tones and stores them in a register for extension through the network via trunks to a terminating CO. The originating office seizes an idle interoffice trunk and sends an off-hook indication and a digit register request to the terminating office (step 4 on Fig. 5.2).

The terminating CO sends a "wink" to indicate a register-ready status (step 5 on Fig. 5.2) and the originating CO sends the address digits to the terminating CO (step 6 on Fig. 5.2).

If a connection to the called party telephone can be established (working line not busy), the terminating CO generates a *ringing signal,* a 20-Hz alerting signal that causes the called telephone to ring. The terminating CO also generates an audible ringing tone and returns it to the calling party's telephone (step 7a on Fig. 5.2). This call progress signal is formed by combining 440-Hz and 480-Hz tones. When the called party is connected to the same central office as the calling party, the originating CO completes these actions.

When the called party answers (goes off-hook—step 8 on Fig. 5.2), *ring tripping*

HIGH FREQUENCY GROUP

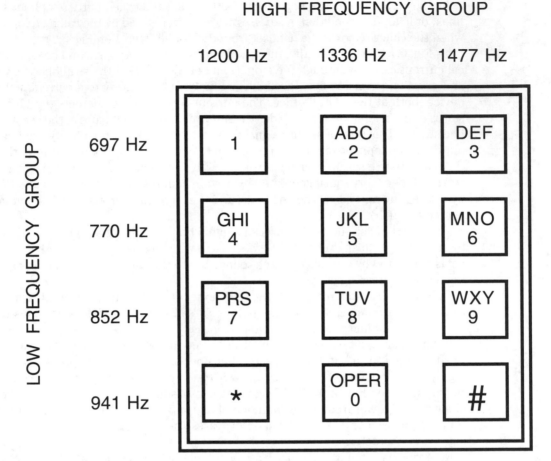

Fig. 5.3 DTMF frequency groups.

occurs immediately so that he does not hear the 20-Hz alerting signal. The terminating central office removes the audible call progress ringing tone when it detects the called party's off-hook status (step 9 on Fig. 5.2).

If the called party's line is busy, the central office sends a busy signal (another call progress signal) to the caller (step 7b, not shown on Fig. 5.2). The *busy signal* is a combination of 480-Hz and 620-Hz tones, switched on for 0.5 second and off for 0.5 second. A *trunk-busy* signal is formed by the same tones but is repeated at a faster rate, such as 0.25 second on and 0.25 second off.

Once the call connection is completed, the parties can begin their conversation. In most cases, when either party hangs up (returns to on-hook status), the connection is released. This feature is called *first-party disconnect,* and without it, CO lines can get stuck in an in-use ("high and dry") condition, disabling both calling and called party telephones.

PBX-to-central office (PBX-CO) trunks used to access public switched networks can use two-wire loops and loop signaling interfaces. PBX trunks and associated signaling may be arranged for rotary or DTMF service, as well as direct outward dialing (DOD), direct inward dialing (DID) or two-way operation. (Trunks arranged for two calling directions are called *two-way trunks.*)

Note that DOD trunks support only calls originating from PBX station equipment, and DID trunks support only calls terminating on PBX station equipment. (Trunks arranged to handle only one calling direction are called one-way trunks.) Once a connection is made, however, voice-message traffic is carried in both directions (note that one-way trunks do not imply 1-way circuit paths).

Ground-start signaling is used on two-way PBX trunks to prevent simultaneous seizure from both ends, or *glare.* Glare is a condition that can occur if loop-start signaling is used on two-way trunks and can lead to conditions that take trunks out of service. Either ground-start or loop-start signaling can be used with DOD trunks, a choice driven by tariffs and overall premises requirements. Wink-start signaling is used with DID trunks.

E&M leads signaling interfaces

E&M leads signaling interfaces, used for connections between switches and transmission systems and between transmission systems themselves, support two-way operation. *E&M leads* is an interface in which the signaling information is transferred across the interface via 2-state voltage conditions on two leads, each with a ground return, separate from the leads used for message information. The message and signaling information are combined and separated by means appropriate for the associated transmission facility.

E&M leads signaling is used in business telecommunications primarily for PBX tie trunks in private networks. Other types of tie trunks, such as one-way dial, were once common; however, FCC registration rules emphasize two-way E&M interfaces, which is consistent with most of today's analog equipment.

Circuit-associated signaling

In Fig. 5.1, voice signals to and from the handset are electronically coupled to the same loop used for supervisory, call progress, and alerting signals. Thus, *circuit-associated signaling* is a technique that uses the same facility path for both voice and signaling traffic. Historically, this approach was selected to avoid the costs of separate channels for signaling and because the amount of traffic generated by signaling is small compared to voice, minimizing the chance for mutual interference. Circuit-associated signaling is contrasted with common-channel signaling that will be discussed later.

Circuit-associated signaling uses either in-band or out-of-band signaling techniques. *In-band signaling* uses not only the same channel path as the voice traffic, but also the same frequency range (band) used for the voice message. For example, in Fig. 5.1 on page 97, DTMF addressing signals use the same frequency band as the voice signals.

Out-of-band signaling uses the same channel path as the message but signaling is in a frequency band outside that used for voice traffic. Digital time division multiplexed and carrier system signaling is considered out-of-band. As previously shown in Fig. 3.11 on page 45, for one frame out of every six, the eighth bit in each channel of a DSO signal is "robbed" for the purpose of transferring signaling information.

In D-type channel banks, the channel units provide the means to convert switch loop closure, DTMF, and other analog signaling formats to equivalent digital formats. Available channel units for digital carrier support a full set of signaling capabilities including loop-start, ground-start, and E&M signaling for PBX tie trunk applications.

Common-channel signaling interfaces

Signaling between switching offices may be provided on a per-trunk or a common-channel basis. In *per-trunk signaling,* signals pertaining to a particular call are transmitted over the same trunk that carries the call as shown in Fig. 5.4A. Interoffice signaling other than common-channel signaling falls into this category.

Common-channel signaling (CCS) is a signaling system developed for use between stored program control switching systems, in which all of the signaling information for one or more trunk groups is transmitted over a dedicated signaling channel, separate from the user traffic-bearing channel. Figure 5.4B shows an associated CCS approach in which one signaling link per trunk group is routed between the switching offices terminating the trunk group, using the same transmission facility.

Nonassociated CCS, a more economical approach that greatly reduces call setup time and enhances total network routing flexibility, is shown in Fig. 5.4C. Here signaling is routed through signaling transfer points over completely separate facilities so that the ability to complete a call on an end-to-end basis can be determined prior to the commitment of trunk and switch resources.

This is in contrast to circuit- and CSS-associated signaling, where traffic-bearing resources are already committed as the call progresses through a network. Should the final result be that a called party's telephone is busy and the call cannot be completed, with associated signaling, traffic resources have in effect been wasted during the call setup time. For greatly overloaded networks, as could occur during disasters or other call-causing events, circuits become blocked primarily from unsuccessful call attempts. This phenomenon is illustrated on the right side of Fig. 5.4B, wherein at low levels of call-attempt incidence, all calls are completed by the network (except, of course, for those cases where the called-party line is busy or does not answer). During crises, such as the Three Mile Island nuclear accident, the number of call attempts will saturate available circuits with call-attempt signaling traffic and the capacity for conversational traffic will actually diminish.

While nonassociated common-channel signaling prevents this failure (illustrated in the right portion of Fig. 5.4C), it poses new reliability and availability problems. For example, if the CCS network fails, no calls can be completed even if all of the user-traffic channels are operative. To protect against such occurrences, CCS networks incorporate redundancy so that individual-link or equipment failures are not catastrophic. In recent years, however, we have witnessed several PSTN "crashes" caused by failures in CCS software.

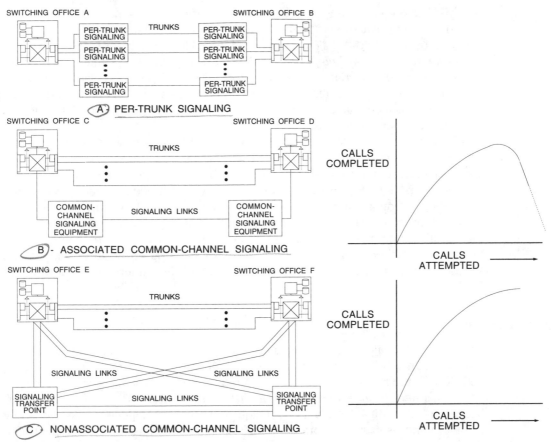

Fig. 5.4 Per-trunk and CCS configurations.

In terms of the ability of U.S. telecommunications to withstand terrorist or military attack, it is not clear whether the 15 signaling transfer points in the AT&T system would be adequate. Whereas intentional enemy destruction of 15 signaling transfer points is plausible, totally destroying associated signaling would require destruction of virtually all domestic telecommunications capabilities. National Security and Emergency Preparedness (NSEP) authorities are currently developing circuit-associated back-up mechanisms that address survivability vulnerabilities associated with CCS signaling.

CCS networks are packet switched data networks. CCS architecture and data network protocol standards are specified by the CCITT. The initial CCITT CCS standard, called *CCITT Signaling System No. 6* (or simply *SS #6*), has been supplanted by *Signaling System No. 7* (*SS #7*). AT&T introduced and implemented a version of SS #6 called *common-channel interoffice signaling* (CCIS) in 1976.

The AT&T and Bellcore versions of SS #7 are commonly referred to as CCS 7 and SS 7 respectively. Although these versions are functionally similar, gateways will be required between LECs and IXCs to account for differences—one of the penalties of divestiture and private telecommunications network ownership.

User interfaces with SS #7 are addressed in Q.930/931, part of CCITT's *General Recommendations on Telephone Switching and Signaling,* a specification for *digital subscriber signaling* systems (DSS). DSS will permit telecommunication users to capitalize on economies, service enhancements, and flexibilities offered by integrated (voice and data) digital networks. It will also define and set the standards for future business system signaling requirements.

CCS networks, capable of supporting far more than supervisory and address signals, provide generalized communications between computers. For example, they are used for accessing credit-card-calling data bases that permit special call screening and restriction, call cost accounting associated with 800 and 900 services, and support operation, administration, and maintenance (OA&M) activities.

CCS significantly reduces post-dial delay (call setup time), enables more efficient call routing methods, permits optimization of network traffic handling, and facilitates network management. It is not an exaggeration to state that CCS is the cornerstone of modern digital networks, with all the vulnerability that the term implies.

Final Remarks

There are five major signaling techniques and three types of signaling interfaces. The signaling techniques are: direct current, in-band tone, out-of-band tone, digital, and CCS. These signaling techniques, also referred to as signaling systems, are classified as either *facility-dependent* or *facility-independent.* Direct current, out-of-band tone, and digital signaling techniques are facility-dependent because they cannot operate separately from the transmission facilities used for the voice traffic that they control.

Inband tone (e.g., *DTMF, SF*, and *MF*) and CCS are *facility-independent* techniques in that their operation is independent of the facilities used to carry the voice traffic. *Single frequency (SF) signaling* is a method of conveying addressing and supervisory signals from one end of a trunk to the other, using the presence or absence of a single specified frequency. A 2600 Hz tone is commonly used. *Multiple frequency* (MF) is an interoffice address signaling method in which 10 decimal digits and five auxiliary signals are each represented by selecting two of the following group of frequencies: 700, 900, 1100, 1300, 1500, and 1700 Hz.

For networks built from a combination of tandem facilities, facility independent, in-band signaling simplifies interfaces since signaling is automatically extended as part of the voice channel. That is, any facility that supports the voice channel is also capable of transmitting signaling information. With the proliferation of digital facilities, however, CCS out-of-band signaling is rapidly becoming the dominant inter-switch signaling approach.

The three types of signaling interfaces are: loop signaling, E&M leads, and CCS. From LEC and IXC points of view, combinations of interfaces and techniques are in three application related realms—line signaling, interoffice trunk signaling and special-services signaling.

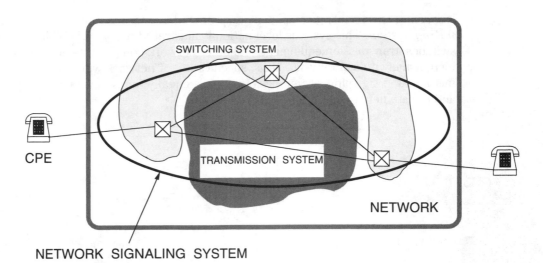

NETWORK SIGNALING SYSTEM

(components in both the switching
and transmission subsystems)

Fig. 5.5 Notional "system" level network switching, transmission and signaling representation.

Since station equipment and PBXs must be compatible with signaling arrangements of various LEC access services, line signaling is the realm of greatest importance to business telecommunications planners/users. Signaling associated with LEC access services will be discussed in Chapter 9.

Telecommunications planners and users need to understand the characteristics of interoffice signaling to evaluate new services made possible by CCS signaling systems, and to select business systems best able to take advantage of digital access signaling, such as SS #7.

Special services signaling refers to signaling used with special services, i.e., any of a variety of LEC and IXC switched, nonswitched, or special-rate services that are either separate from public telephone service or contribute to certain aspects of public telephone service. Examples include PBX tie trunks, foreign exchange (FX), and private line services. These services and their signaling arrangements are important to business telecommunication planners/users.

Finally, all signaling approaches can be classified as either *stimulus/response* or *message-based*. As stored program control switches replace electromechanical switching systems, the trend is clearly towards message-based signaling. From a total network point of view, signaling systems are emerging as data communications subnetworks linking computers embedded in station, switching, and programmable transmission/multiplexing equipment.

In earlier parts of this book, switches, transmission media and components, and signaling are discussed on an individual basis. In fact, within any telecommunications network, all of the switches taken collectively can be viewed as a *switching system*, and all of the transmission components as a *transmission system*. Similarly, all of the signaling components, distributed in these systems, comprise a *signaling system*. This systems level notion is illustrated in Fig. 5.5. While individual switches, transmission equipment, and other network components need to be specified, usually on a site-by-site basis, specification of network-wide switching, transmission, and signaling system performance is equally important in selecting public and private network services.

6

Premises Distribution Systems

A *premises distribution system* (PDS) is the transmission network inside a building or among a group of buildings, such as an office park or campus. PDS is used here as a generic term, although AT&T has used it to describe a specific product offering. The PDS connects desktop and other station equipment with common host equipment (such as switches, computers and building automation systems) and to external communications networks, as shown in Fig. 6.1.

Because a modern PDS represents between 15% and 50% of the total telecommunication system cost, it is commanding a growing share of management attention. For new buildings, costs are minimized with detailed design and planning and PDS installation during building construction. PDS upgrades in existing buildings are normally more expensive, and in some cases (such as in historical buildings), prohibitively so.

As a consequence, a prime PDS objective is to provide uniform wiring capable of supporting voice and data in multiple product/vendor environments, thus minimizing the need to modify wiring as user requirements change or when technology upgrades are justified. In the past, this has been nearly impossible. At one time, for example, IBM alone used over 50 different types of cabling for its product lines. Recent innovative developments now permit the use of a single type of unshielded twisted pair (UTP) copper wire, versions of which have been used in telephone service for years, for premises voice and data transmission. The objective is to achieve useful PDS life cycles of 10 to 20 years.

To meet these objectives, the Electronic Industries Association (EIA) and the Telecommunications Industry Association (TIA), published in July, 1991, the Commercial Building Telecommunications Wiring Standards (EIA/TIA 568 and 569) to enable planning and installation of building wiring with little knowledge of the telecommunication products that will be installed. These standards define the following PDS components (illustrated in Fig. 6.2, which includes a PBX-based system as an example):

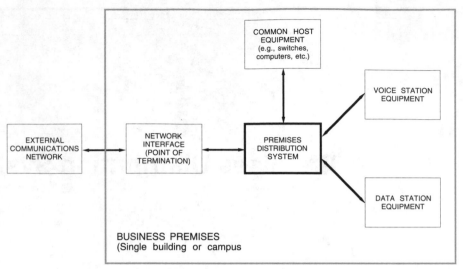

Fig. 6.1 Premises distribution system.

- *Horizontal wiring.* The connection between the telecommunications outlet in work areas and the telecommunications closet.

- *Backbone wiring.* The connection between telecommunications closets and equipment rooms within a building, as well as between buildings in a campus. Backbone wiring is also referred to as the *riser subsystem.*

- *Work area.* An area containing stations and the connections between those stations and their telecommunications (information) outlets.

- *Telecommunications closet.* An area for connecting the horizontal and backbone wiring and containing active or passive equipment.

- *Equipment room.* A special purpose room(s) with access to the backbone wiring for housing telecommunications, data processing, security, and safety alarm equipment.

- *Entrance facilities.* The point of interconnection between the building wiring system and external telecommunications facilities (LEC networks, other buildings, etc.). As stated earlier, Bellcore defines the interface with LEC networks as points of termination (POTs)/network interfaces (NIs).

- *Administration.* The requirements for specifying the labeling, identification, documentation, etc., across all of the telecommunications wiring system. Documentation includes the creation of the system and documents required to manage and maintain end-to-end connectivity. A number of PC-based PDS management systems are available.

Figure 6.2 includes main cross-connects in the equipment room, and intermediate cross-connects in the telecommunications closets. A *cross-connect* is used to connect and administer communications circuits. In a cross-connect, jumper wires or patch cords are used to make circuit connections, for example between horizontal and backbone wire and cable segments.

Older punch-down "66-type" models use only jumper wires and require special tools and training. These are still the most popular cross-connect systems. More recent, 110-type termination systems, as well as modular plug-and-jack types, use patch cords with attached plugs. These provide flexible circuit labeling arrangements, and can be easily changed and administered by nontechnical personnel.

Typically, for a building with 1000 work areas, the main cross-connect involves tens of thousands of wire connections. Many organizations require moves, adds, and changes (MACs) of station equipment at 50% of total work areas per year. Such high churn rates and the need to accommodate growth and equipment upgrades makes automated PDS management systems essential for most medium and large installations.

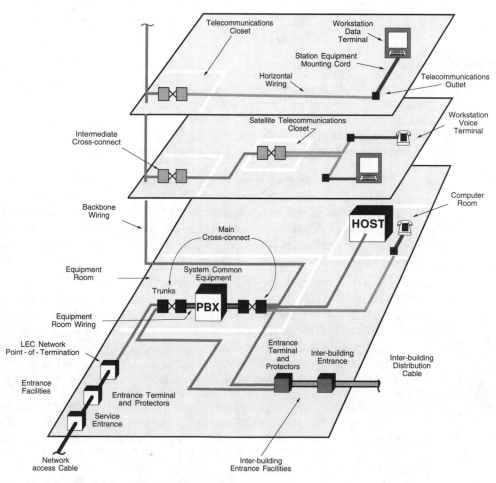

Fig. 6.2 Premises distribution system components.

To convey an appreciation for the manpower-intensive nature of wire management, consider this example. In 1984 AT&T employed over 20,000 technicians just to operate their distribution frames (cross-connects in telephone buildings).

Cross-connects are arranged *hierarchically;* that is, main cross-connects feed several intermediate cross-connects, as shown in Fig. 6.2. Special cross-connects are available for coaxial and fiber optic cable. Both of these types of cable offer greater bandwidths over longer distances than twisted pair wire and find application for high-speed data, wide-bandwidth video (e.g., cable TV) and multichannel (multiplexed) voice applications.

Fiber optic cables (lightguides) are direct replacements for coaxial and twisted pair wiring. At present they are considerably more expensive per connection but do find PDS application in multichannel backbone wiring, where the cost can be spread over large numbers of circuits. For example, fiber has significant advantages for backbone wiring, including interbuilding connections between central and remote switch modules, as well as for long runs in large, multifloor single buildings. With underground duct bank construction costs in the range of $100 per foot, a major advantage accrues when two strands of fiber displace copper pairs serving 500 or more voice channels.

Horizontal Wiring

Horizontal wiring includes telecommunications outlets in work areas, the horizontal wiring itself, termination for the wiring, and cross-connects located in telecommunications closets. The EIA/TIA 568 standard recommends two telecommunications outlets for each work area, as shown in Fig. 6.3.

The telecommunications outlets are defined by the following:

- One of the outlets shall be supported by four-pair unshielded twisted pair (UTP) cable. This medium meets Institute of Electrical and Electronic Engineers IEEE 802.3 10BaseT technical specifications. IEEE 802.3 10BaseT specifies a class of stand-alone data local area networks (LANs). Incorporating this specification into the EIA/TIA standard makes it possible to use a single telecommunications outlet for either voice or data, a significant development.

- The second outlet can be supported by either the four-pair UTP cable, by two-pair shielded twisted pair (STP) cable, or by coaxial cable. The latter cables are specified by IEEE 802.5, and 802.3 10Base2 specifications and correspond to other classes of data LANs. In late 1991, the EIA/TIA issued Technical Systems Bulletin (TSB) 36, which defined transmission performance standards for UTP cable up to 100 MHz. This major milestone has galvanized the UTP cable industry, arming them with a potent weapon in their attempts to delay the inroads of fiber to the desktop.

Neither shielded nor coaxial cables are appropriate for voice applications, and their use is not consistent with the product and vendor independence objective of uniform wiring. However, solutions in the form of adaptors called *baluns* have been developed which permit carrying IEEE 802.5 and 802.3 10Base2 LAN traffic over standard four-pair UTP wiring.

The EIA/TIA standard calls for FCC registered modular telephone jacks, flush- or

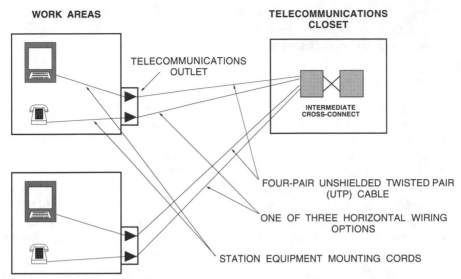

WORK AREAS

TELECOMMUNICATIONS CLOSET

TELECOMMUNICATIONS OUTLET

INTERMEDIATE CROSS-CONNECT

FOUR-PAIR UNSHIELDED TWISTED PAIR (UTP) CABLE

ONE OF THREE HORIZONTAL WIRING OPTIONS

STATION EQUIPMENT MOUNTING CORDS

Fig. 6.3 Typical horizontal and work area PDS wiring.

surface-mounted. Eventually all horizontal wiring will be identical and all station equipment moved easily using standard plugs and jacks. With user-administrable cross-connects, outlets can be reassigned to different users or reallocated between voice and data use, resulting in complete connection compatibility and rearrangement flexibility. A high-speed (100 MHz) plug and jack standard is pending at EIA/TIA at the time of publication of this book.

Optical fiber cable is an optional EIA/TIA horizontal wiring medium. This is putting fiber optic capabilities into work areas for high speed Computer Aided Design (CAD)/Computer Aided Manufacturing (CAM), high resolution graphics, video and other wideband PC terminal applications. For this reason, AT&T has introduced a *composite cable* constructed with two four-pair, 24-gauge, UTP cables, combined with two multimode fiber cables and matching work area outlets.

Fiber-distributed data interface (FDDI) is an American National Standards Institute (ANSI) standard for 100 Mbps local area data communications networks that is becoming popular within and between buildings on a campus. It is discussed in Chapter 13. While the AT&T composite and other fiber optic cables support FDDI, a number of companies have established a UTP Forum with the goal of developing ways to support FDDI rates (i.e., 100 Mbps) within premises using UTP (unshielded twisted pair) cable. Members of the forum include AT&T, Fibronics, Hewlett Packard, British Telecommunications, and others. Two schemes are being defined: MLT-3 for data grade UTP and CAP-32 for older voice grade cable.

Special electronics to achieve 100 Mbps over UTP are being developed by Advanced Micro Devices, Cabletron, Chipcom, Motorola, Synoptics, Cresendo Communications Inc., and others. The important aspect of this work is that EIA/TIA standard horizontal UTP wiring will reduce or eliminate the need for expensive fiber optic cable alternatives. Not only will this downsize investment costs for each cable run (the next section presents alternative PDS cost analyses), but it will make UTP a viable long-term mechanism for uniform wiring, reducing PDS maintenance and administration costs accordingly.

Typical PDS Costs

Figure 6.4 illustrates how PDS costs compare to per-PBX line costs. The cost figures were developed during a recent competitive bid for a feature-rich, high-technology campus system. In this program, buildings are to be added over a period of time, so vendor responses were requested to show the decrease in per-line costs as the system grew from 100 lines to more than 5000 lines.

Unlike charges for switching, voice mail, administrative, and other services and features that benefit from economies of scale, the per-line cost for PDS, being largely manpower-driven, remains relatively constant as more lines are added.

The PDS costs shown in the figure correspond to the following horizontal wiring and work area options:

A One RJ-45 modular UTP wire jack per work area and no fiber outlets
B Two RJ-45 jacks and no fiber outlets
C Two RJ-45 jacks and two unterminated ("dark") fiber outlets
D Two RJ-45 jacks and two terminated fiber outlets

The dark fiber option refers to installing the fiber cable but not initially terminating the connection. PDS cost calculations are based on an average telecommunications closet-to-work area run length of 150 feet.

The figure indicates that for large systems and EIA/TIA-recommended dual modular, eight-conductor jack configurations, PDS costs are, significantly, about 40% of other costs. For either dark or terminated fiber options, PDS costs exceed other costs. These results alone dispute the belief that wiring is the "low-tech" part of telecommunications business systems, and therefore of lesser importance. Analyses treating operations, administration, and maintenance life-cycle costs lead to similar conclusions.

Reflecting the magnitude of life-cycle PDS costs, a number of cable management system products have reached the market. These products generally incorporate computer-aided design (CAD) capabilities and support all PDS activities from design and installation through day-to-day MAC and configuration management. The more comprehensive products involve interactive graphics and relational database tools that provide automated inventory tracking for all voice, data, video, and PDS assets, initial and on-going documentation and report generation, troubleshooting, trouble ticket and work order generation, other maintenance support capabilities, and help desk functions.

Some products include high-performance graphical drawing capabilities that can electronically import building architectural engineering drawings. In these cases you can visually display floorplans and locations of equipment, user terminal outlets, equipment room layouts, conduit systems, power systems, etc. Capabilities also may include the storage and retrieval of text data in the databases that directly relate to selected graphical objects in the engineering drawings. Industry studies and surveys indicate that cable management systems can reduce costs as-

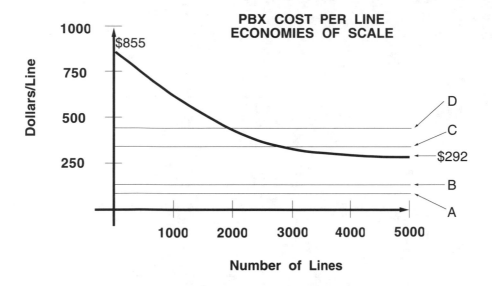

Fig. 6.4 Comparison of PDS and other costs.

sociated with moves, adds, and changes (MACs typically cost around $200 each) by as much as 30%. For locations with 1000 MACs per year, cable management system savings for MACs alone can amount to $50,000 per year. Vendors include Autotrol, ISICAD, CHI/COR Information Management Inc., Vycor Corporation, AT&T, CAD COM, and others.

Appendix B presents practical considerations for PDS planning and implementation.

Chapter

7

Tariffs & Network Design

The network design challenge is to connect combinations of switches, transmission media, and terminal equipment to allow users to exchange information. As seen in the previous six chapters, economics motivates the development and use of switches and multiplexers to share expensive transmission resources. In telecommunications, the nature of cost functions and statistically efficient use of resources governing performance favors large aggregations of traffic. That is, at equivalent performance levels, a single network resource handling a large quantity of traffic is less expensive than several resources with smaller amounts of traffic. Properly designed, networks permit traffic aggregation while efficiently routing traffic to destinations in response to user demand.

To be successful, network design must determine the type, quantity, and location of resources required to meet system-performance requirements at minimum cost. Costs include initial equipment and installation investment and recurring costs for use of telecommunications resources. Often, trade-offs are involved between these cost elements. Telecommunications carriers provide resources through tariffed services. Consequently, knowledge of both tariffs and network design is crucial in implementing networks that meet all business requirements while minimizing costs. An introduction to principles underlying tariffs is presented in the next section. Following this, principles of *traffic engineering*, a branch of network design science which predicts and measures performance and can be used to evaluate how well tariffed offerings satisfy business requirements, are introduced. Chapters 10 and 15 deal with voice and data network design in more detail.

Tariffs

A *tariff* is the published rate for a specific telecommunications service, equipment, or facility that constitutes a public contract between the user and the

115

telecommunications supplier or carrier. Tariffed rates are established by and for telecommunications common carriers in a formal process in which carriers submit filings for government regulatory review, possible amendment, and approval. Historically, tariff filings provide the most complete and precise descriptions of carrier offerings. Further, tariffed offerings cannot be dropped or changed without regulatory approval.

All IXCs were originally regulated by the FCC. In the Competitive Carrier proceeding (Docket 70-252), the FCC decided to forbear from regulating OCCs. Although this freedom from regulation no longer requires OCCs to file tariffs, most of them do file rate structures as tariffs and the ensuing discussions of tariff rates in this chapter apply to OCCs. As the dominant IXC, AT&T is still regulated by the FCC and required to file tariffs for its IXC services. LECs are also regulated and required to file tariffs for most of their LEC services. The regulating entity is normally the state Public Utility Commission, with one exception. LEC services providing access to inter-LATA services and which carry inter-LATA traffic are regulated by the FCC under a separate set of tariffs. These services provide access from a user's premises to the IXC POP on a shared or dedicated basis. Dedicated access is provided under *special access* tariffs, and shared access is provided under *switched access* tariffs.

Figure 7.1 summarizes the different types of tariffs and the regulatory body responsible for each. The jurisdictional partitioning shown in the figure assumes that inter-LATA user networks carry interstate traffic. For networks entirely within the boundary of a single state, intrastate versions of the inter-LATA tariffs apply and are under PUC jurisdiction. In the remainder of this book, inter-LATA implies interstate traffic.

Intra-LATA		Inter-LATA (Inter-state)	
LEC		LEC	IXC
Access	Transport	Access	Transport
Dedicated Services — Private Line (PUC)		Special Access (FCC)	Private Line (FCC)
Shared Services — Switched Services (PUC)		Switched Access (FCC)	Switched Services (FCC)

Fig. 7.1 Types of tariffs.

Cost recovery structures

Whether determined by regulation or by competitive pressures, the rates charged for telecommunications services are designed to recover the costs of providing the services. Historically, the correlation between the rates for a particular service and the cost of providing that service has not been high. In a monopoly environment, one service (like long distance calling) could subsidize another (like local phone service) with little overall impact on the monopoly service provider. In today's highly competitive, post-divestiture environment, subsidization is economically unwise and legally prohibited for regulated services. Hence, current rate structures reflect underlying costs and attempt to fairly distribute costs among service users.

From a pricing viewpoint, the primary distinction is between dedicated services and shared services. A *dedicated service* assigns resources to users for exclusive use on a full time basis. The rates for these services are on a monthly basis and independent of actual quantities of user traffic carried by the service. A *shared service* provides resources shared by many customers. Customers may request resources at any time, and pay in accordance with actual service usage.

Pricing for IXC-provided services is more complex. There are three distinct pricing environments for each IXC service: originating LEC access, IXC transport, and terminating LEC access. Each environment can be priced on dedicated or shared bases, and each combination defines a specific IXC service offering. Generally, high-volume users can justify dedicated resources while low-volume users are more economically served by shared resources.

When dedicated LEC services are used to access IXC services, LEC charges are normally paid directly by the user. When shared LEC services are used to access IXC services, LEC charges are usually paid by the IXC and recovered in IXC usage rates. Figure 7.2 illustrates three examples of IXC price structures and the underlying resources used to support them. These examples are discussed further in Chapters 8 through 10.

Elements of tariffs

In the remaining parts of this chapter, some mathematical formulae are introduced. Although not necessary to an understanding of tariff concepts and applications, these formulae are included to furnish additional clarification for use by telecommunications professionals.

Tariffs are based on common elements used to recover costs of specific telecommunications resources. Understanding the terms and concepts behind these elements facilitates tariff analysis.

Dedicated services require carrier interface resources and assignment of transmission facility bandwidth between user premises. Accordingly, the pricing structure for a dedicated service contains a *fixed-charge* element and a *distance-sensitive* element. The unit cost per mile may be constant or may decrease with increasing mileage (reflecting use of more efficient, high-volume long distance transmission media). Moreover, IXCs may give volume discounts in exchange for long term monthly minimum purchase commitments.

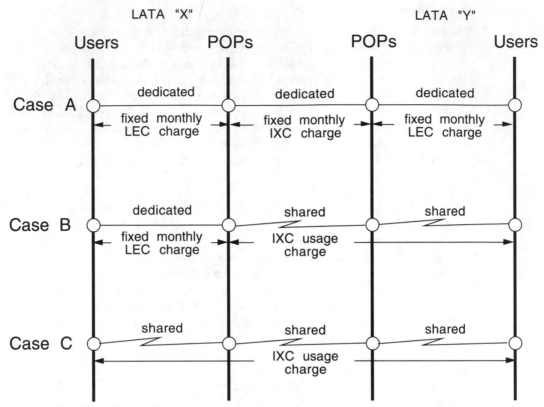

Fig. 7.2 LEC/IXC tariff examples.

Shared services also require interface resources and thus a fixed charge element, but the bulk of shared service charges reflect usage. Usage charges are derived from a *unit price* (or a set of unit prices based on distance), multiplied by the number of units actually used. These charges may be discounted by time of day or the day of the week. Carriers charge less during low-use time periods since additional traffic can be carried during those periods without adding shared network facilities. Again, certain services may include volume discounts.

Since most tariffs are distance-sensitive, there must be an established, simple mechanism for determining distances as a basis for price determination. Fortunately, it's not necessary to know the latitude and longitude of every building in the country to price telecommunication services. Rather, in modern tariffs, distances are computed only between key physical network elements employed to provide services. These elements, introduced earlier, include LEC central offices serving users, and IXC POPs serving the user's LATA. Figure 1-4 (on page 11) illustrates the physical structure, but it should be noted that several COs may reside in the same LEC building.

For tariff purposes, *wire centers,* locations of one or more local switching systems and a point at which customer loops converge, are assigned specific coordinates (*wire-center coordinates*), on a grid system. Each IXC POP is also assigned a coordinate, known as a *rate center coordinate,* on the same grid. Known as the V&H Coordinate System, the grid was developed to permit accurate call mileage determination in North America. Its origin is in Greenland and vertical (V) and horizontal (H) numerical coordinates increase toward the west and south in a somewhat skewed manner as shown in Fig. 7.3. Distance is determined using a modified version of the Pythagorean formula:

$$D = \sqrt{\frac{((V1 - V2)^2 + (H1 - H2)^2)}{10}} \tag{7.1}$$

Note: The division by 10 is performed before the square root is taken.

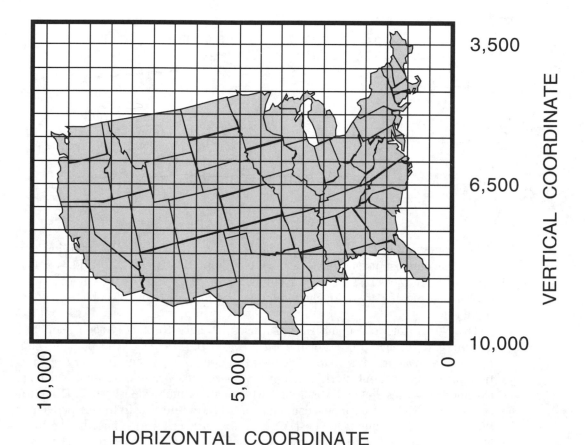

Fig. 7.3 V&H (vertical and horizontal) tariff coordinate map.

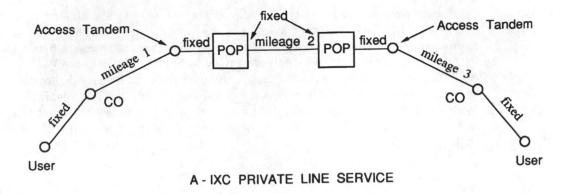

A - IXC PRIVATE LINE SERVICE

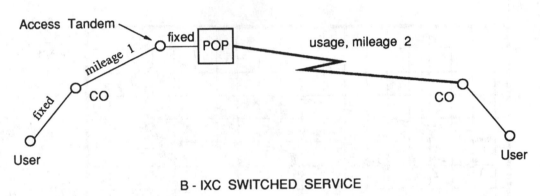

B - IXC SWITCHED SERVICE

Fig. 7.4 Examples of tariff elements.

Distances for LEC tariffs are computed between the end point serving wire centers, while distances for IXC tariffs are computed between IXC POPs. V&H co-ordinates for the AT&T POPs are published in AT&T Tariff #10, whereas wire-center coordinates are uniquely defined by the first six digits of public telephone numbers.

The North American numbering plan uses 10 digits to identify each telephone. The first three digits represent the *numbering plan area* (NPA) portion (commonly known as the *area code*). The next three digits represent the *exchange* portion of the number. In historical Bell System parlance, *N* refers to a digit which can assume the values 2–9 and *X* is a digit which can assume values of 0–9. Using this convention, NXX has become a commonly used term referring to the exchange portion of a telephone number. NPA-NXX is sufficient user location information to determine distance for the purposes of computing tariffs. Electronic databases associating NPA-NXX with V&H coordinates, LATAs, serving telephone companies, cities, and states are available from Bellcore and several commercial companies. Such databases are essential for tariffed service analyses.

Examples of tariff elements

Figure 7.4 illustrates two examples of services and the tariff elements used to price those services. The first example corresponds to end-to-end, dedicated inter-LATA private-line service, while the second example portrays dedicated private line access to shared, switched inter-LATA service. The multiple fixed and mileage-dependent elements indicate the complexity involved with large network analysis. Ordinarily, large networks must support a plurality of different end-to-end services. They involve many LECs, each with different PUCs and tariffs, and potentially a mix of alternative carriers, again with different tariffs.

While carriers render computer-based network modeling and design optimization assistance, normally they treat only their own service offerings. Organizations such as Systems Research and Applications Corporation (SRA) of Arlington, Virginia, offer comprehensive voice and data network modeling and design optimization services, and since they offer no telecommunications carriage services, they provide unbiased assessments of all carrier offerings. Chapter 10 further details tariff elements for a wide range of LEC and IXC services.

Traffic Engineering Basics

The concept of sharing telecommunications resources through switching immediately raises the engineering question of how many of shared resources must be provided to ensure adequate user "performance" (however one chooses to define *adequate*). If too many resources are provided, many are never used, extra costs are expended by carriers, and rates charged for service are higher than necessary. On the other hand, if too few resources are provided, service might not be available when requested. In such cases, users are forced to try later or join queues, waiting for a resource. Traffic engineering experience teaches that in network performance determination and analysis, the following factors are key:

- User demand for resources
- User behavior upon finding all resources busy

It is interesting to note, and helpful to an understanding of underlying principles, that these factors and the mathematics used to quantify performance apply to other than telecommunications resource sharing. Although our focus is on telephone callers and shared trunks, interactive behavior between grocery shoppers and checkout clerks or customers and bank tellers is analogous. Visualizations based on everyday events help both managers and engineers to see beyond the apparent complexity imposed by mathematical rigor and to gain a practical insight into the business consequences of traffic engineering factors and characteristics.

Characterizing demand for resources is the first important step in traffic engineering. Demand has two components; first, how often users "arrive" or request resources (termed *arrival rate*); and second, how long they use resources before releasing them (termed *holding time*). (Note: It is assumed that resources in use by one user are unavailable to others until released.) The product of these quantities is a measure of traffic demand on resources. As shown below, both components can be statistically modeled, and experience shows that model predictions correlate well with actual measured telecommunications system performance.

As users randomly arrive to request service, two outcomes are possible. If a resource is free, it is seized by the user for a period of time. If no resource is free (all resources are busy), the user may or may not leave without receiving service. This "busy" phenomenon is known as *blocking* in telephone networks and users that leave without receiving service following a busy indication are said to employ a *lost-calls-cleared* discipline.

Most voice networks operate in this manner. The measure of performance in such a system is the probability of blocking that users experience when entering the system. As discussed in Chapter 4, this probability is commonly called grade of service (GOS)—although, as previously noted, GOS can also relate to indicators of customer satisfaction with other aspects of service, such as noise or echo. In the traffic context, a grade of service of 1%, or 0.01 means that for every 100 users entering the system, on the average, one will find all resources busy.

While voice networks operate in the above manner, data networks (and grocery stores) use a different discipline when users encounter busy resources. Here users join *queues* (lines) and wait for service. In this discipline, called lost-calls-delayed, the measure of performance is average delay, that is the elapsed time between when a user enters a system until a service is completed.

Beginning theory

The discipline of computing performance for a given traffic demand, number of resources, and probabilistic distribution of arrivals and service times is called *queuing theory*. A comprehensive discussion of this topic is beyond the scope of this book, but basic assumptions and procedures yield useful answers to most traffic engineering problems, and a rudimentary understanding of the theory provides valuable insight into the relationship between network design and business operations which the networks support.

In the previous section, we characterized arrival of users and the time spent in the system as statistical quantities. Figure 7.5 helps illustrate the importance of this characterization. In the figure, the first example shows three users presenting a combined load of one hour of traffic during an hour of time. Each user makes one call with a holding time of 20 minutes. The calls are scheduled so that they occur on the hour, at 20 minutes past the hour, and at 40 minutes past the hour. With such arrival scheduling and controlled call duration, a single circuit is sufficient to carry the traffic load with no blocking.

The second example shows the same average load over the hour, but all arrivals occur during the first twenty minutes. One circuit results in two of three calls being blocked over the hour even though the average offered load is the same as the first case. The situation would be even more complicated with varying call duration. The statistical distribution of arrival times and call lengths play a major role in calculating expected performance. Fortunately, a large body of analysis has been accom-

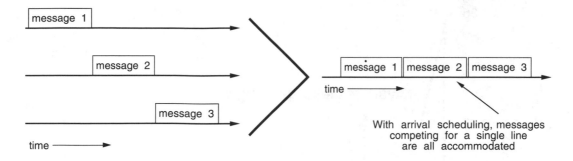

A - SCHEDULED MESSAGE ARRIVALS

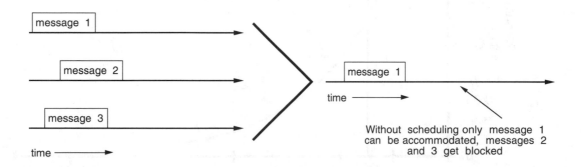

B - RANDOM MESSAGE ARRIVALS

Fig. 7.5 Message traffic examples.

plished and a workable characterization developed in terms of both call interarrival and call duration distribution.

As it turns out, both quantities have been shown to follow a negative exponential distribution shown in Fig. 7.6. For example, in the figure, the average call duration is 3 minutes. As shown, call durations with values much larger than average values can occur, but only with ever decreasing probability. Calling patterns with negative exponential distribution interarrival times are said to follow a *Poisson (random) distribution.* All the performance formulas presented herein are based on assumptions that traffic obeys these statistical distributions. As a consequence, users need only calculate average arrival rate and average holding time to characterize traffic demand. Traffic demand or intensity is commonly called *offered load.*

The international dimensionless unit of traffic demand, obtained by multiplying arrival rate by holding time, is called an *erlang,* named after A. K. Erlang, a Danish mathematician. Acknowledged to be the father of queuing theory, Erlang's for-

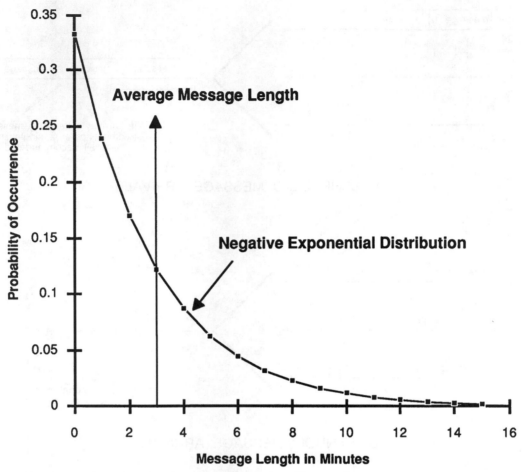

Fig. 7.6 Statistical variation of random message lengths in traffic networks.

mulae, derived in 1917, remain the backbone of queuing performance evaluation today. Quantitatively, one erlang of traffic is equivalent to a single user who uses a single resource 100% of the time. Alternatively, one erlang of traffic is generated by 10 users who each occupy a resource 10% of the time. An example traffic intensity (or offered load) calculation for a group of 100 users each making an average of three voice calls per hour, with an average length of three minutes per call is as follows:

erlangs = (Message Arrival Rate) × (Holding Time)

erlangs = 100 (users) × 3(calls per user per hour)
            × 3(minutes per call)   (7.2)
            ÷ 60 (minutes per hour)

erlangs = 15

Note that in the equation, all dimensional units cancel out. One erlang is equal to 36 CCS (*centi call seconds*), another dimensionless traffic measure defined in the example cited in the PBX section of Chapter 4. Either measure can be used in characterizing network performance, providing care is taken to ensure that consistent dimensional units of measure are used in calculations.

Besides traffic, the other quantity needed to compute performance is the number of circuits (or more generally, *resources*) available to handle traffic. Circuits act like tellers in a bank; they are "servers" satisfying shared service demand. In traffic engineering, there are two categories of circuits: those dedicated to a single user or piece of equipment (i.e., dedicated circuits or lines) and those shared among users using switches or concentrators (i.e., shared circuits or trunks). A trunk group is traffic engineered as a unit for the establishment of connections within or between telephone company switching systems in which all paths are interchangeable. A trunk group can be searched in sequence until a free trunk is found.

The basic voice traffic engineering problem is to determine the probability of blocking of a trunk group given an erlang load and the number of circuits in the group. Since each call utilizes a trunk in both directions, the erlang loads offered to both ends of a trunk group are added together to compute the total load. Erlang derived a formula to compute blocking using the above statistical assumptions. For lost-calls-cleared, the formula is called the *Erlang-B formula* and is as follows:

$$B(a,n) = \frac{\left(a^n/n!\right)}{\sum_{i=0}^{n}\left(a^i/i!\right)} \qquad \begin{array}{l} a = \text{erlangs} \\ n = \text{circuits} \\ B(a,n) = \text{Probability of blocking} \\ \qquad\text{(lost-calls cleared)} \end{array} \qquad (7.3)$$

Here $B(a,n)$ is the probability of blocking (grade of service) which n circuits provides under an offered traffic load of a erlangs.

Unfortunately, this closed-form equation is very difficult to compute for large values of a and n (even a computer will encounter overflow and underflow errors working with numbers like 100 to the 100th power divided by 100 factorial). This difficulty is overcome using a recursive form of the equation, which can be implemented using a simple loop in a computer program:

$$B(a,k) = \frac{a \times B(a,k-1)}{k + a \times B(a,k-1)} \qquad \begin{array}{l} a = \text{erlangs} \\ k = \text{circuits} \\ B(a,k) = \text{Probability of blocking} \\ \qquad\text{(lost-calls-cleared)} \end{array} \qquad (7.4)$$

Example Computer Loop

$$b = 1.0$$
$$do\ i = 1\ to\ n$$
$$b = \frac{(a \times b)}{(i + a \times b)}$$
$$enddo$$

Here $B(a,k)$ is the probability of blocking which k circuits provides under an offered traffic load of a erlangs. In statistical analysis, a recursive estimation technique is one in which successive (recursive) estimates are based on previous estimates until a sufficiently accurate value of some desired numerical quantity (in this case blocking) is determined. It should be noted that in lieu of computer evaluation of Erlang's formulae, tables relating number of servers and blocking probability to traffic load, in terms of traffic load measured in either erlangs or CCS are published in many electronic reference data books, such as the *Reference Manual for Telecommunications Engineering*, by Roger L. Freeman, published by Wiley-Interscience.

The recursive algorithm is also convenient for determining the number of circuits required to achieve a given grade of service by simply executing the recursive formula until blocking is less than a desired objective value. The formula is very nonlinear, i.e., identical changes in input parameter values do not result in proportional, or linearly scaled, changes in output parameter values. Such nonlinear performance reveals the reason why larger trunk groups are considerably more efficient than smaller ones in terms of carrying traffic at any given call-blocking performance level. At the heart of network design, this nonlinear property is discussed in more detail beginning on page 136 and illustrated in Fig. 7.15. As a final Erlang-B observation, note that performance does not depend on arrival rate or holding time individually, but only on their product. For lost-calls-cleared numerous short calls have the same effect as fewer long calls.

Lost-calls-delayed performance is characterized by the *Erlang-C equation* and associated equations for delay. The relevant quantities here are message arrival rate, average message length, and a third quantity, *circuit transmission speed* or line speed. In this case, average message length and transmission speed are treated separately since message-length distribution is a demand parameter (not controllable by the system designer), while transmission speed is a parameter within a system designer's control.

As before, a usual assumption is that interarrival time and the message length follow a negative exponential statistical distribution, and again the erlang load is the arrival rate multiplied by holding time. In this instance, however, holding time is message length divided by line speed. An example traffic-intensity (or offered load) calculation for a group of data terminals generating an average of 10.42 messages per second, with an average message length of 40 bytes and using a 2400 bps line, is as follows:

Erlangs = (Message Arrival Rate) × (Holding Time)

Erlangs = (Message Arrival Rate) × (Message Length) ÷ (Line Speed)

Erlangs = 10.42(messages per second) × 40(bytes per message) ×
  8(bits per byte) ÷ 2400(bits per second)   (7.5)

Erlangs = 1.6

Note again, that when consistent units are used all dimensional units cancel out.

For lost-calls-delayed, the probability of finding all circuits busy upon arriving and thus joining the queue is given by the Erlang-C equation:

$$C(a,n) = \frac{B(a,n)}{1 - \dfrac{a}{n} \times (1 - B(a,n))} \qquad \begin{array}{l} a=\text{erlangs} \\ n=\text{circuits} \\ B(a,n)= \text{Erlang-B equation} \\ C(a,n)= \text{Probability of blocking} \\ \qquad\quad \text{(lost-calls delayed)} \end{array} \qquad (7.6)$$

Note that Erlang-C is given in terms of Erlang-B, which can be computed recursively as before.

The *average delay time* that a message spends in a network (queuing time plus transmission time for the message) is given by:

$$T = \frac{L}{C} + \frac{C(a,n) \times L}{\left(1 - \dfrac{a}{n}\right) \times n \times C} \qquad \begin{array}{l} a=\text{erlangs} \\ L=\text{message length in bits} \\ n=\text{circuits} \\ C=\text{line speed in bits per second} \\ C(a,n)=\text{Probability of blocking} \\ T=\text{average message delay time} \end{array} \qquad (7.7)$$

It should be noted that the performance of multiple low-speed circuits is not equivalent to a single high-speed circuit. The basis for this is revealed in the two components of the delay equation presented above. The first component (L/C) takes into account the time required to transmit a single message once a circuit is available. This delay component is referred to as *transmission time* and is dependent only on line speed, not the number of circuits in a trunk group. It represents a minimum trunk group delay. The second component corresponds to the time a message spends in queues waiting for circuits to become available. This delay, *queue time,* is dependent on both line speed and the number of circuits in a trunk group. Note also that delay is not independent of message length as in the lost-calls-cleared (circuit switched) case where many short calls have the same effect as fewer long calls. In this case, for a given line speed, many short messages result in less delay than fewer long messages.

Queue lengths at opposite ends of a circuit are totally independent. Each direction acts like an independent circuit with its own traffic and performance. The only relation between two directions of transmission occurs when a circuit is purchased, when usually both directions of transmission are procured at the same bit rate.

Lastly, the average *queue length,* in messages, is given by:

$$q = \frac{C(a,n) \times \dfrac{a}{n}}{\left(1 - \dfrac{a}{n}\right)} \qquad \begin{array}{l} a=\text{erlangs} \\ n=\text{circuits} \\ C(a,n)=\text{Erlang-C equation} \\ q=\text{average number of messages in a} \\ \qquad \text{queue} \end{array} \qquad (7.8)$$

The average time a message spends in a network and queue length, in messages, are the performance measures of greatest interest to users.

Traffic Engineering in Practice

The preceding section outlined basic traffic engineering theory. While this (and much more) is required for engineers designing and administering large national carrier networks, end users typically need a more practical approach to estimating facility demands from more recognizable business need parameters. Ask telecommunications managers how many erlangs their sites generate in the busy hour and you will probably receive blank stares. However, they can usually tell you the number of call minutes on their monthly phone bills and whether their operations work nights and weekends. This section outlines a step-by-step approach to single site/trunk group traffic engineering based on typical business operation utilization characteristics.

Before analyzing examples, it is important to discuss how traffic demand varies as a function of time of the day and day of the week, and how these variations affect traffic engineering results. Since the relationship between demand and performance is nonlinear, performance under peak load conditions may degrade if the network design is based on average load. To understand the trade-offs involved, let's first examine daily and hourly voice traffic profiles for a typical corporate location.

Figure 7.7 shows the number of measured call minutes per day for a typical month. A complete month of traffic is the minimum that should be used for sizing studies. Weekends and holidays are apparent from this chart. It would also be obvious when measurement instrumentation does not capture all traffic for a day, as sometimes happens near the end of a month.

It is not necessary to explicitly examine every hour of every day. Nor, as indicated, is it advisable to base designs on average day traffic, i.e., total monthly traffic divided by the number of days. When this approach is pursued, the fact that traffic is low and performance is great on weekends when no one is calling, is of no consolation to users routinely denied service on workdays during busy periods. To circumvent such deficiencies, it is standard practice to introduce the notion of *busy days* and *busy hours*.

Ordinarily, the next step would be to determine average *business day traffic*, i.e., the total business day traffic divided by the number of business days. Generally this is acceptable if the level of traffic for each business day is nearly equal. However, for some businesses, specific days of the week are considerably busier than others (such as customer phone-in orders on Mondays). To account for this phenomenon, yet not react to isolated single-day peaks with no enduring statistical significance, it is preferable to use *busy day loads*, defined to be the average traffic of the five busiest days of the month. The effective number of busy days in a month is the total traffic in a month divided by the busy day load.

Having calculated the busy day load, the next level of analysis is to examine *hourly traffic profiles*. Again, good nighttime performance is not what users desire. An hourly

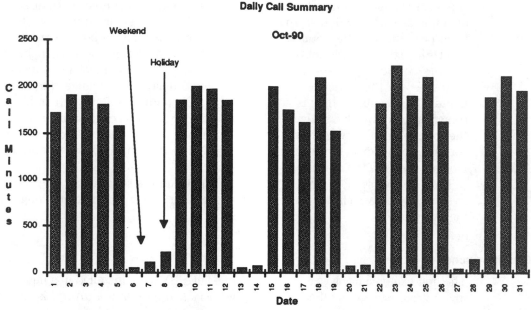

Fig. 7.7 Example distribution of call-minutes per day.

profile for a typical corporate user is shown in Fig. 7.8. Business hours and lunchtime behavior are readily determined from this chart. On the chart, calling that extends beyond the normal business day is normally due to multiple-time-zone traffic. Ordinarily, there are four busy hours, two in the morning and two in the afternoon, with approximately the same traffic level. Each busy hour normally accounts for between 12% and 16% of a day's total traffic. If no other information is available, the busy hour traffic level can be assumed to be 14% of a busy day's traffic. By convention, performance is specified over the busy hour. Hence, the erlang load of a trunk group is the number of hours of traffic offered during the busy hour of a busy day.

Different approaches may be taken to determine the traffic requirements of a system. For a completely new system with no historical traffic records, monthly minutes can be estimated and standard assumptions used to derive busy-day and busy-hour traffic levels. If a current system exists, monthly bills or even a month's call detail records in electronic form can normally be obtained. Actual daily and hourly distributions can be plotted from call detail records, available from PBXs and other switching machines, and the actual busy day and busy hour loads calculated. By far, especially in large networks with many locations, this is the preferred method for establishing quantitative requirements.

Estimating requirements for data communications networks involves additional considerations. In gathering records, the basic units to be measured are message arrival rates and message lengths. Note that line speeds of current circuits do not constitute "requirements" for future or upgraded networks, but are merely part of the current solution. Too often with data networks, managers mistakenly state requirements in terms of circuits and line speeds as opposed to actual traffic units.

Also, it is important in data networks to identify requirements for both directions of traffic at each location so that both directions can be engineered separately. A large percentage of the total volume of data traffic may be via *batch transfers* at night. Since this traffic is not time sensitive, network capacity is normally engineered for *interactive traffic*—or query-and-response traffic usually involving human operators—with the objective to achieve specified delay performance during busy hours. Once engineered, this capacity is typically sufficient for batch traffic, the transmission of which can span multiple hours after business closing times. Delay is the time interval between the instant at which a network station seeks access to a transmission channel to transmit a message, and the instant that the network completes delivery of the message.

The next two sections present voice and data examples separately, using illustrative worksheets which can serve as guides for other, more general traffic engineering exercises.

Circuit-switched example

The first example involves a 2000 telephone line PBX to be installed in a new building. A group of central office trunks, known as a *two-way trunk group,* is used for

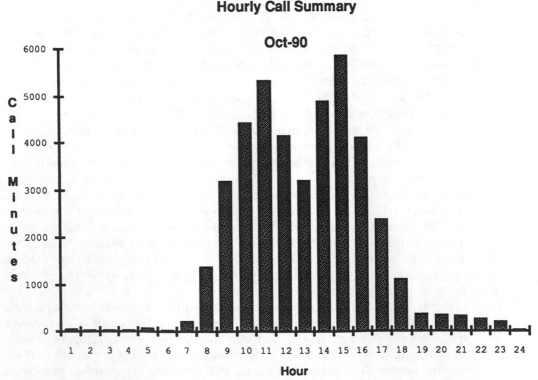

Fig. 7.8 Example distribution of call-minutes per hour.

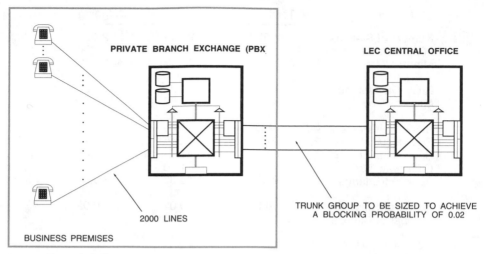

Fig. 7.9 Example PBX trunking configuration.

outgoing and incoming local and long distance calls. If incoming calls are on a separate trunk group from the CO, then each group must be engineered separately. The configuration is shown in Fig. 7.9. The worksheet shown in Fig. 7.10 documents the computation (described in the following paragraphs) of the number of trunks required in the CO trunk group to achieve a grade of service of 0.02.

For this example no measured traffic statistics are available, but experience has shown that 400 minutes per month of originating and terminating "outside" traffic per telephone line is typical for business traffic. This results in a total of 800,000 minutes per month. Note this estimate does not include station-to-station traffic supported by the PBX, since this traffic does not use trunks to the CO.

In the absence of measured or inferred business application data to the contrary, an assumption is made that half of the traffic is derived from calls originating from the site. Also in the absence of measured traffic, it is assumed that there are 22 busy days in the month. Thus, busy day load is computed by dividing 800,000 minutes per month by 22 busy days per month yielding 36,364 minutes per busy day. Similarly, using 14% as the busy hour percentage of daily traffic yields a busy hour load of 84.85 erlangs in both directions (42.425 in each direction). That is, 14% multiplied by 36,364 minutes per busy day equals 5091 minutes per busy hour; and, 5091 minutes per hour divided by 60 minutes per hour equals 84.85 erlangs (recalling that an erlang is equivalent to one circuit occupied full-time).

One final adjustment that needs to be made is the conversion from *conversation time* (also called *billable time*) to *circuit usage time*. Traffic derived from billing sources or estimated from projected usage accounts only for the time users actually engage in conversation. In telephone parlance, this is the elapsed time between *distant end answer* (or called party going off-hook at the start of a call) and either party going on-hook at the end of a call. However, circuits are actually tied up during call setup before the distant party answers. They are also occupied for uncompleted

	Total	Incoming	Outgoing
Number of Stations	2,000	2,000	2,000
Minutes/Month/Station	400	200	200
Minutes/Month	800,000	400,000	400,000
Busy Days/Month	22	22	22
Minutes/Day	36,364	18,182	18,182
Busy Hour Percent	14%	14%	14%
Busy Hour Minutes	5,091	2,545	2,545
Busy Hour Erlangs	84.85	42.42	42.42
Overhead Factor	1.10	1.10	1.10
Adjusted Erlangs	93.33	46.67	46.67
Grade-of-Service	0.02	0.02	0.02
Number of Circuits	97	53	53

Fig. 7.10 Circuit switched traffic example worksheet.

calls (busy signals or ring-no-answer calls), for which no billing records are generated. Statistics for these cases indicate an average overhead factor of 10% be added to billing time to estimate total circuit usage. This factor is applicable for circuit sizing only and should not be applied to traffic used for estimating tariff costs, since carriers charge only for connection time.

Using the Erlang-B formula, it is determined that 97 circuits are required in the two way trunk group. Had one-way trunks been required, 53 circuits would be required for each direction, indicating the efficiency of larger groups, even at groups of this size. Differences are more marked at smaller sizes, as demonstrated below. When traffic begins flowing, actual loads and blocking performance should be monitored daily to determine if design adjustments are required. Monthly load profiles should be tracked for possible seasonal variations. Using measured per station statistics, the impact of increasing the facility size (e.g., the number of served stations) can be predicted and additional trunks provisioned ahead of time, minimizing the chance for level-of-service disruptions.

Message-switched example

The second example considers an *interactive* (typically short messages) switching system. Traffic statistics noted in this example would be different if a *file transfer* (typically long messages) example were used. One-hundred terminals are connected to a concentrator that buffers terminal data until a user hits the enter key or a maximum message length is reached. The concentrator then forms a message and sends it to a host system. The average and maximum message lengths are 40 bytes and 128 bytes respectively. While these assumptions do not correlate to normal negative exponential distributions for message lengths (the fixed maximum length precludes very long messages that would otherwise occur, albeit infrequently, if the message duration statistics followed the negative exponential distribution), standard equations nevertheless yield useful upper bounds on the delay performance.

Unfortunately, no standard assumptions are available regarding terminal activity, since the traffic is strictly dependent on the particular information system applications being used. If network service already exists, again the best source for traffic information is actual characters per month (or packets per month). See Chapter 12 for data communications definitions and operational descriptions which can be derived from billing data. When historical data is not available, traffic is best derived on a per-terminal basis. This involves estimating peak bytes transferred per minute, based on specific knowledge of each major application's characteristics and organizational workloads.

The example application corresponds to data entry wherein each terminal generates 5 one-hour sessions per day. During each session, 250 bytes per minute are transferred (equivalent to a typist entering data at 50 words per minute for the hour). The host computer acknowledges each message with an "OK" message of 4 bytes' length. Under this method, we characterize the busy hour and then use assumptions to derive circuit requirements. The 100 terminals generate 25,000 bytes per minute during the busy hour. This equates to 625 terminal-to-host messages per minute, or an arrival rate of 10.42 messages per second, with an average length of 40 bytes. As in the voice example, it is necessary to account for message overhead. In this case, overhead corresponds to extra bytes needed to implement data communications protocol functions. Again, these functions are discussed in detail in Chapter 12. For now, it can be assumed that 6 bytes per message are required for overhead functions, yielding an adjusted average message length of 46 bytes.

Prior to computing data message traffic statistics, the number of circuits and line speed must be chosen. As previously described, the average time a message spends in a network and queue length, as measured in messages, are the performance parameters of interest. For short messages, line speed is not a crucial factor in providing acceptable response times or user message delays (a half-second round-trip delay is usually sufficient to avoid interactive-user annoyance). While the number of circuits in a trunk group multiplied by the line speed equals the total bandwidth available for message traffic, in multiple line arrangements the composite bandwidth is not available to any given message. Consequently, to minimize transmission time and keep message delays within bounds, long messages are better served by fewer high-speed lines than by larger numbers of low-speed lines.

For a message length of 46 bytes, Fig. 7.11 illustrates transmission times for line speeds available in North America. For our example 46-byte message lengths, line speeds at or above 2400 bits per second achieve acceptable transmission times. As usual, economics is also a factor in determining whether multiple low-speed lines or fewer high-speed lines should be used. Worksheets in Fig. 7.12A, B, and C depict the design details of this current example and correspond to the use of candidate circuit line speeds of 2400, 4800, and 9600 bps, respectively.

We next size the trunk group at candidate line speeds as a prelude to selecting one that yields acceptable queue performance. Rules of thumb under maximum expected traffic loads are to maintain:

■ Average queue times below expected transmission times

■ Average queue lengths less than a single message.

LINE SPEED   **TRANSMISSION   TIME**
(bps)   (milliseconds)

300	1227
1200	307
2400	153
4800	77
9600	38
56000	7

1 millisecond = 0.001 seconds

Fig. 7.11 Transmission times for a 46 byte message.

To continue the calculation, erlang load, which depends on line speed, is computed (see Equation 7.4 on page 125 and Fig. 7.12 worksheets for values). Since all traffic is treated as lost-calls-delayed, the number of circuits provided must be equal to or greater than offered traffic erlangs. If this is not the case, infinite queues for circuits will form (analogous to grocery store checkout lines before a snowstorm). Starting with the number of circuits determined by the erlang value (rounded up to the next integer), formulae for delay and average queue length are computed for additional circuits until both queue-time and queue-length criteria are met.

Economic factors relating to available transmission services in the United States dictate the following progression of trunk group choices:

One 2400 bps circuit
One 4800 bps circuit
One 9600 bps circuit
Two 9600 bps circuits
One 56 kbps circuit
Multiple 56 kbps circuits
One DS1 (1.544 Mbps) circuit

That is, a single 4800 bps circuit is always chosen over two 2400 bps circuits since the cost is less and the transmission and queue time performance is better.

In general, not all user CPE or carrier switching equipment can support DS1 rate circuit interfaces, so beyond performance and economics, equipment compatibility must also be ascertained during the network design phase.

The worksheets in Fig. 7.12 illustrate the required number of circuits for the three line speeds, imposing the queue length criteria. Based on progression of trunk group costs and the transmission and queue time performance, one 9600 bps circuit is the best choice for this message switched example.

Behavior of traffic systems

The previous sections focused on traffic engineering theory and the mechanics of calculating key network performance attributes. The purpose of this section is to illuminate trade-offs between performance and resource utilization efficiency, that is, *cost effectiveness*. If cost were not a factor, it is clear that for any given level of traffic, very low blocking probability and message delay can always be obtained by simply providing more—and higher speed facilities. The most valuable lessons learned with respect to practical network design and implementation reflect the nonlinearity of traffic systems, described above in the context of equations used to calculate performance, and how that nonlinearity impacts cost. To illustrate the consequences of nonlinearity and its implications, both circuit and message switched cases are treated.

A - Message Switched Worksheet

(Line Speed = 2400 bps)

	Outgoing	Incoming
Number of Terminals	100	100
Bytes/Minute/Terminal	250	25
Bytes/Minute	25,000	2,500
Mean Message Length (Bytes)	40	4
Messages/Second	10.42	10.42
Overhead Bytes	6	6
Adjusted Message Length	46	10
Line Speed (bps)	2,400	2,400
Busy Hour Erlangs	1.60	0.35
Transmission Time (sec)	0.153	0.033
Number of Circuits	3	3
Actual Queue Time	0.030	0.000
Actual Time in System	0.183	0.033
Actual Number in Queue	0.310	0.000

B - Message Switched Worksheet

(Line Speed = 4800 bps)

	Outgoing	Incoming
Number of Terminals	100	100
Bytes/Minute/Terminal	250	25
Bytes/Minute	25,000	2,500
Mean Message Length (Bytes)	40	4
Messages/Second	10.42	10.42
Overhead Bytes	6	6
Adjusted Message Length	46	10
Line Speed (bps)	4,800	4,800
Busy Hour Erlangs	0.80	0.17
Transmission Time (sec)	0.077	0.017
Number of Circuits	2	2
Actual Queue Time	0.014	0.000
Actual Time in System	0.091	0.017
Actual Number in Queue	0.150	0.000

C - Message Switched Worksheet

(Line Speed = 9600 bps)

	Outgoing	Incoming
Number of Terminals	100	100
Bytes/Minute/Terminal	250	25
Bytes/Minute	25,000	2,500
Mean Message Length (Bytes)	40	4
Messages/Second	10.42	10.42
Overhead Bytes	6	6
Adjusted Message Length	46	10
Line Speed (bps)	9,600	9,600
Busy Hour Erlangs	0.40	0.09
Transmission Time (sec)	0.038	0.008
Number of Circuits	1	1
Actual Queue Time	0.026	0.001
Actual Time in System	0.064	0.009
Actual Number in Queue	0.270	0.010

Fig. 7.12 Message switched traffic example worksheets.

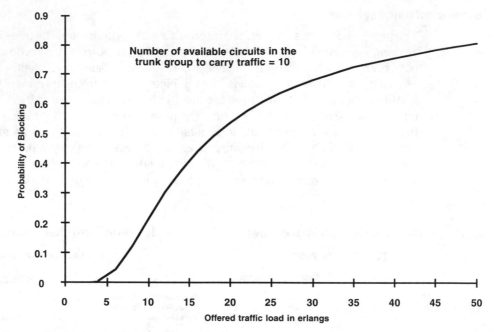

Fig. 7.13 Blocking probability versus traffic load.

Circuit-switched examples are considered first. The probability of blocking, a function of offered load and the number of circuits, is depicted in Fig. 7.13 for a 10-circuit trunk group. Below a certain traffic level (approximately 4 erlangs), essentially no blocking occurs. Blocking then rises exponentially over a range of traffic levels before approaching the limit of 100% (every call blocked). The interesting portion of this performance curve takes place between blocking probabilities of 0.01 and 0.1 (1% and 10%).

Another way of presenting this nonlinearity is to keep the load constant and calculate and plot the number of circuits required as a function of blocking. Figure 7.14 shows the results for an offered load of 10 erlangs. From this presentation some of the important nuances of nonlinear behavior begin to emerge. Surprisingly, accepting poor performance (10% blocking versus 1% blocking) results in only a 30% reduction in the number of circuits (and therefore circuit cost). The relevance of this observation is that designing for 10% blocking to reduce expenses is not a viable solution. At 10% blocking, performance degradation is so noticeable and annoying to users that typically it prompts even higher traffic load from repetitive blocked-call retries. Good practice requires a grade of service of 5% or less, and preferably a level in the 1–2% range.

The most revealing way to present traffic network performance from a cost efficiency standpoint is illustrated in Fig. 7.15. The array of numbers in the boxed area represents the amount of *carried traffic,* measured in hours per circuit per month, under varying offered load levels (the vertical variations) and varying values of network design blocking probability (the horizontal variations). Corresponding offered-load levels are indicated on the vertical axis and blocking probabilities on the horizontal axis. In the array, values of carried load correspond to traffic resulting from hourly traffic distributions with characteristics shown in Fig. 7.8.

Examination of Fig. 7.15 reveals that there are two ways to improve trunk group utilization efficiency. First, for any given offered load, accepting higher blocking probabilities yields higher per circuit utilization efficiency. Second, for any given blocking probability, aggregating more traffic yields higher per-circuit utilization efficiency.

In understanding the nature of traffic networks, the two most important lessons are:

1. For a given level of traffic, designing for a better grade of service translates directly to poorer resource utilization efficiency and higher cost per unit of delivered service.

2. There are economies of scale; at any specified level of performance, greater amounts of traffic (i.e., traffic aggregation) translates to reduced cost per unit of delivered service.

Assuming that poor GOS performance is not a viable solution, and that the offered load described for the PBX in the circuit-switched example above is determined by

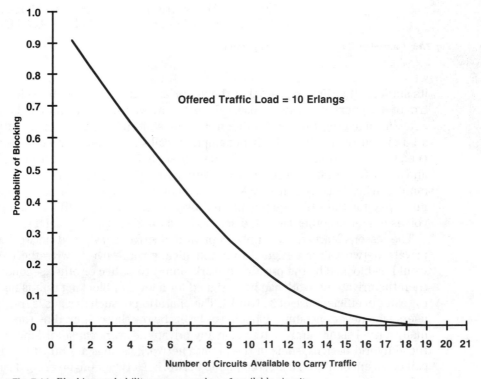

Fig. 7.14 Blocking probability versus number of available circuits.

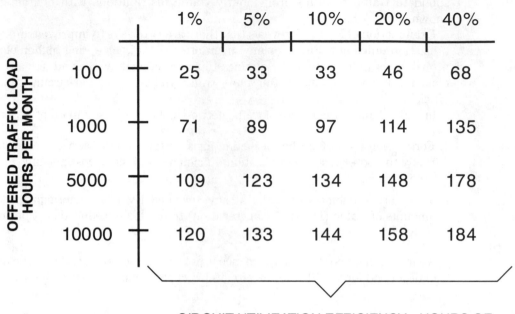

BLOCKING PROBABILITY

OFFERED TRAFFIC LOAD HOURS PER MONTH	1%	5%	10%	20%	40%
100	25	33	33	46	68
1000	71	89	97	114	135
5000	109	123	134	148	178
10000	120	133	144	158	184

CIRCUIT UTILIZATION EFFICIENCY - HOURS OF
TRAFFIC PER MONTH CARRIED BY EACH CIRCUIT

Fig. 7.15 Carried traffic, hours per circuit per month.

its stations, it at first appears that the important parameters are outside of the control of designers and network users. In fact, however, use of a PBX can be viewed as a means of aggregating local traffic from 2000 station lines and concentrating it into a 93-circuit trunk group. In this example, recall that an alternative to using a PBX (i.e., Centrex) would necessitate 2000 separate lines between a customer's premises and an LEC central office, each with a very poor utilization of only about 6 hours per month. While it is true that PBXs afford significant facility cost advantages, a user must pay for the PBX equipment, and overall the Centrex/PBX cost trade-off involves numerous other factors, discussed in more detail in Chapter 11.

The lessons learned also apply to private electronic switched network design. If private networks are designed to automatically route overflow traffic (traffic that would be blocked by the private network alone) to public or other secondary paths, then the private network can be designed for a level of blocking that is high enough to result in efficient use of its trunks. The operation of such designs is transparent to users who never encounter blocking at levels below those offered by the PSTN, normally a level significantly better than any private network can affordably support. In these hybrid designs, some of the private network savings are offset by higher per-call costs encountered using the public network. Network engineering design trade-offs are discussed further in Chapter 10.

Message-switched systems exhibit the same kind of nonlinear behavior as circuit-switched systems. Figure 7.16 depicts average message delay as a function of load for a single-circuit trunk group using the parameters of the message switched example discussed above. The nonlinear or *threshold effect* is readily observed as delay approaches infinity when the erlang load approaches 1.0. The implications of this lost-calls-delayed performance are similar to those for lost-calls-cleared voice traffic in that designing a network for acceptable performance under "peak loads" usually means that resources are poorly used during "average load" traffic conditions.

For most situations this means that networks are inefficiently used most of the time, and only become "efficiently loaded" during busy hours. In voice networks, peak-to-average traffic ratios are normally in the order of 2 to 1. Historically the required network "overbuilding" could be justified by large-volume users, particularly when alternate routes for busy-hour traffic peaks are provided.

As will be described in Chapters 14 and 15, emerging metropolitan and wide area requirements for LAN-to-LAN and certain video teleconferencing involves peak to average traffic ratios as high as 30–100 to 1. Such ratios may make the costs of responsive private data communications networks prohibitive and even require new public network technologies to provide *elastic* or *bandwidth-on-demand,* as will be discussed in the cited and other later chapters.

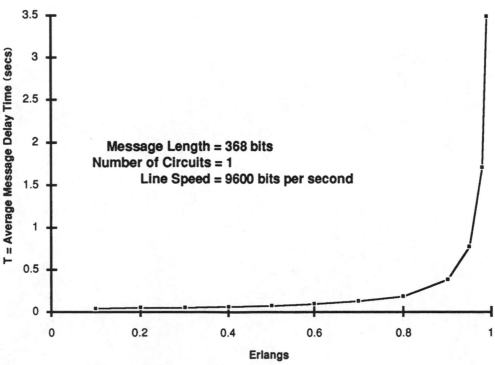

Fig. 7.16 Message delay as a function of offered traffic load.

Part

2

Voice Services

8

Voice Services & Networks

This section describes voice services and the underlying customer premises equipment (CPE) and network facilities that support them. Figure 8.1 shows the segmentation of voice services, first into premises and network categories, with network services further subdivided into access and transport services.

PBXs, Centrex, and key/hybrid telephone systems are alternative equipment used to deliver premises services. In the PBX scenario, one can talk about station equipment access to the private branch exchange services, and transport via the PBX and premises transmission to network interfaces with external networks. Similar descriptions apply for key/hybrid telephone systems.

For Centrex service, station equipment, other CPE, and the premises distribution system provide access to the LEC NI. The LEC provides access to CENTRal EXchange services rendered by a Centrex switch located in a central office. The LEC may supply Centrex service using switching equipment shared with other Centrex customers or even with equipment used for LEC public switched network services.

In any case, each customer's Centrex service is partitioned, can be customized to individual customer needs (even on a line-by-line basis), and thus provides "private exchange" services equivalent in nearly all respects to PBXs. Most business customers, for example, have little concern about information privacy, segregation of billing and administrative data, or the ability of the LEC to prevent abbreviated station-to-station dialing or intercom connections among different Centrex customers.

Returning to the PBX example in Fig. 8.1, for simplicity we illustrate only originating and terminating central office switches but imply the interconnection of all elements of LEC and IXC networks needed to complete calls. In the figure, the equipment used to support LEC *access services* includes loop transmission facilities from the NI to central office line termination equipment (e.g., distribution frame cross-connects).

For *switched access services,* equipment includes line-termination cards (or carrier and multiplexer equipment for multichannel facilities), as well as portions of

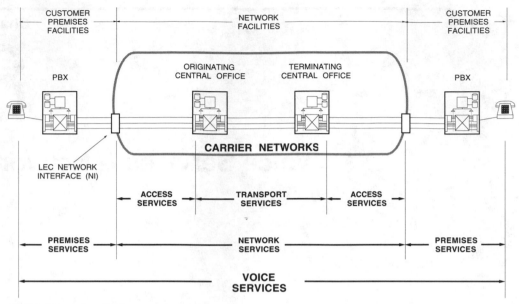

Fig. 8.1 Categories of voice services.

central office switching that support addressing, supervisory, alerting, call progress, and other signaling functions described in the Chapter 5.

Access services to *LEC special services,* or any of a variety of services separate from basic public switched services (such as private lines, PBX tie trunks, etc.), may or may not involve switching and signaling services. Examples of access services to public switched telephone network (PSTN) and private switched and nonswitched special services are provided below.

Transport services are network switching, transmission and related services that support information transfer between originating and terminating access facilities. (In Fig. 8.1 these are in the originating and terminating central offices.)

In LEC public switched telephone (intra-LATA) networks, transport services include local calling, *message telecommunications service* (MTS), directory and operator assistance, 911, intra-LATA *wide-area telecommunications services* (WATS), and access service to IXC networks.

Local service uses in whole or in part the public switched telephone network (PSTN). Examples include public telephone service, mobile radiotelephone (Cellular) service, air-to-ground service, etc. WATS permits customers to make (OUT-WATS) or receive (INWATS) long distance voice or data calls and to have them billed on a bulk rather than an individual call basis. The service is provided by means of special private access lines connected to WATS equipped central offices. A single access line permits inward or outward service but not both.

Moving beyond the LEC example of Fig. 8.1, transport services are segmented into intra-LATA and inter-LATA categories, with each category divided into switched and nonswitched transport services. Finally, switched and nonswitched transport services are offered as public switched telephone network, dedicated facilities-based private switched network, and virtual private network services.

For facilities-based private switched network services, LECs and IXCs dedicate physical switching and transmission facilities for the exclusive use of a particular customer. The carriers may elect to provide separate switches and transmission equipment or to partition facilities used for public services. Recent trends in carrier pricing are shifting users with requirements justifying private telecommunications networks to virtual private network options.

Virtual private networks provide services intended to be identical (indistinguishable by customers) to those provided by facilities-based switched private networks. Public network facilities are actually used, with additional call processing and routing computers making virtual private network services appear as if delivered from dedicated facilities. Customers can define, change and control network capabilities with the same or more flexibility as afforded with facilities-based private networks.

LEC and IXC *nonswitched transport services* can be used with PBX or other telecommunications equipment to structure private *electronic switched networks* (ESNs), where all network switching functions are accomplished in PBX customer premises equipment (CPE). This operation is described in Chapter 4. For this arrangement carriers provide only transmission services.

Figure 8.2 presents a taxonomy of voice services, each of which is described in detail in Chapter 9.

Figures 8.3 through 8.8 illustrate combinations of premises and network services. They represent a cross-section of popular business arrangements, and clarify the distinctions and capabilities across the range of available voice services.

Figure 8.3 illustrates Centrex, intra-LATA PSTN, private line, and foreign exchange (FX) services. Note that the FX and private line services are rendered via special service circuits and are terminated, but not provided with call-by-call switched services at the originating central office. FX is a service that provides a circuit between a user telephone and a central office other than the one that normally serves the caller. In the Fig. 8.3 example, private-line service provides a direct connection between two telephones at different locations. Note for this example, only a single LEC and LATA are involved.

Figure 8.4 is identical to Fig. 8.3 except that it illustrates inter-LATA PSTN, private line and FX services. That is, two LATAs are involved for provision of PSTN, FX and private line services. The FX and private lines terminate in remote LATAs.

LEC-access tandems, described in Chapter 1, provide traffic concentration and distribution for inter-LATA traffic originating and terminating within a LATA. In this example LEC intra-LATA transport service provides business users with access to IXC PSTN and special services. Conversely, access tandem connections to IXC points of presence (POPs) provide IXCs with access to LEC networks and customers.

Note again that FX and private lines are not provided with call-by-call switched service at any intervening LEC or IXC switching office.

Figure 8.5 illustrates PBX, intra-LATA PSTN and PBX tie trunk services. In this figure, combinations of direct inward dial (DID), direct outward dial (DOD) and

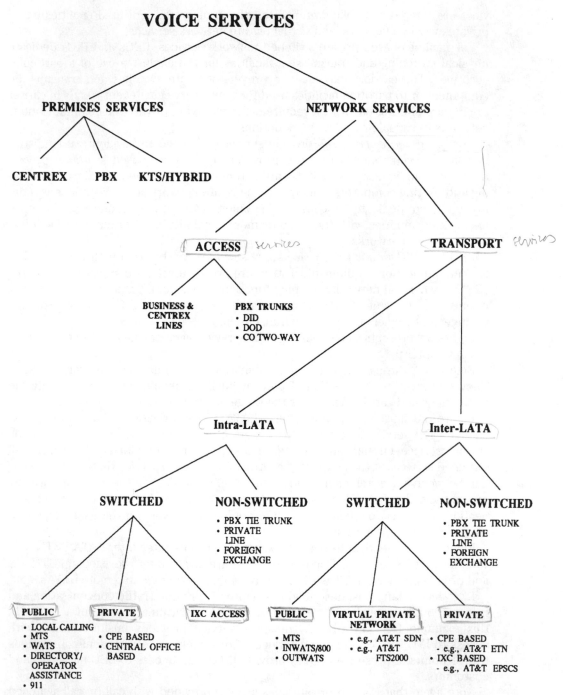

Fig. 8.2. Taxonomy of voice services.

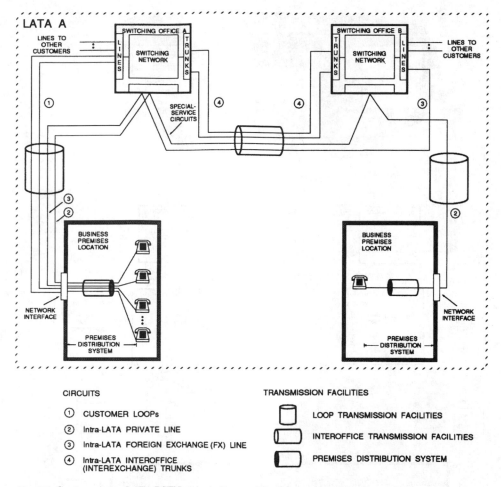

CIRCUITS

① CUSTOMER LOOPs

② Intra-LATA PRIVATE LINE

③ Intra-LATA FOREIGN EXCHANGE (FX) LINE

④ Intra-LATA INTEROFFICE
(INTEREXCHANGE) TRUNKS

TRANSMISSION FACILITIES

LOOP TRANSMISSION FACILITIES

INTEROFFICE TRANSMISSION FACILITIES

PREMISES DISTRIBUTION SYSTEM

Fig. 8.3 Centrex, intra-LATA PSTN, private line, and foreign exchange service examples.

two-way PBX trunk access services provide business users with access to LEC PSTN transport services.

Figure 8.5 is also an example of the use of special nonswitched LEC tie trunk transport services and PBX-derived transport services to establish a private intra-LATA electronic switched network (ESN). PBXs must be able to accommodate E&M leads tie trunk signaling services and be capable of supporting tandem switching functions as described in Chapter 4.

Figure 8.6 illustrates PBX premises and intra-LATA services identical to those of Fig. 8.5. In Fig. 8.6, however, the private ESN is realized using central-office-based LEC private switched network services. Here special circuits are connected to a central office switch, partitioned to segregate public and private traffic, as indicated by the dotted line on the figure.

For this case the PBX need not be capable of tandem switching or other electronic switched network functions.

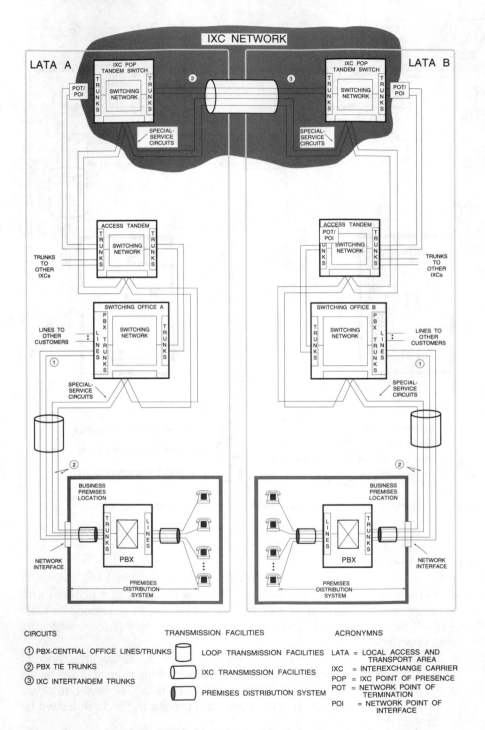

Fig. 8.4 Centrex, inter-LATA PSTN, private line, and foreign exchange service examples.

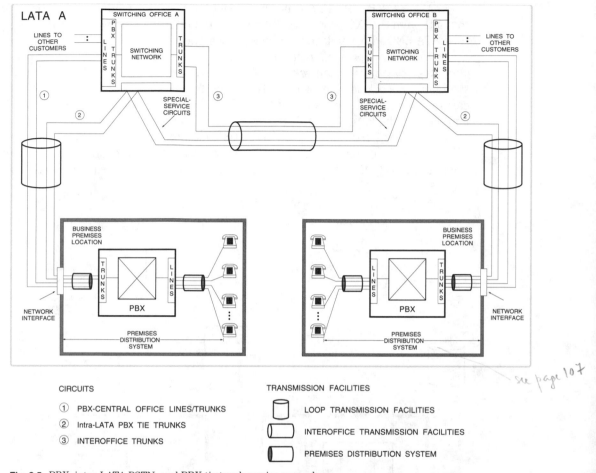

see page 107

CIRCUITS

① PBX-CENTRAL OFFICE LINES/TRUNKS

② Intra-LATA PBX TIE TRUNKS

③ INTEROFFICE TRUNKS

TRANSMISSION FACILITIES

LOOP TRANSMISSION FACILITIES

INTEROFFICE TRANSMISSION FACILITIES

PREMISES DISTRIBUTION SYSTEM

Fig. 8.5 PBX, intra-LATA PSTN, and PBX tie trunk service examples.

Figure 8.7 illustrates PBX premises, inter-LATA IXC PSTN and PBX tie trunk services. In this figure combinations of direct inward dial (DID), direct outward dial (DOD) and two-way PBX trunk access services provide business users with access to LEC PSTN transport services, which in turn provide access to IXC PSTN transport services.

Figure 8.7 is also an example of the use of special nonswitched IXC tie trunk transport services and PBXs to establish a private inter-LATA ESN. The PBXs must be able to accommodate E & M leads tie trunk signaling services and be capable of supporting tandem switching functions as described in Chapter 4.

Figure 8.8 illustrates PBX premises and inter-LATA services identical to those of Fig. 8.7. In Fig. 8.8, however, the private inter-LATA ESN is realized using IXC-based private switched network services. Here special circuits are connected to an IXC office switch, partitioned for public and private traffic, as indicated by the dotted line on the figure.

For this case the PBX need not be capable of tandem switching functions. Note that although the IXC provides call-by-call switched private network services, the intra-LATA LEC PBX trunk services are not switched.

Common-control switching arrangement (CCSA) and *enhanced private switched communication service* (EPSCS) are two examples of IXC-based private switched network services which AT&T has offered in the past. As indicated earlier carrier pricing for facilities-based private network and virtual network services has emphasized the latter.

For simplicity, in the above examples intra- and inter-LATA services are treated separately. Clearly, combinations of each are always implemented; all business systems provide access to both LEC and IXC public switched telephone network services.

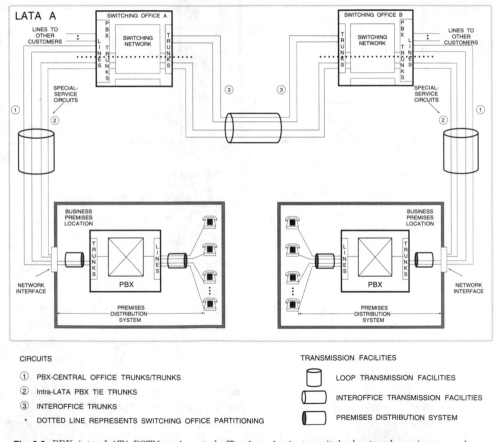

CIRCUITS

① PBX-CENTRAL OFFICE TRUNKS/TRUNKS

② Intra-LATA PBX TIE TRUNKS

③ INTEROFFICE TRUNKS

* DOTTED LINE REPRESENTS SWITCHING OFFICE PARTITIONING

TRANSMISSION FACILITIES

LOOP TRANSMISSION FACILITIES

INTEROFFICE TRANSMISSION FACILITIES

PREMISES DISTRIBUTION SYSTEM

Fig. 8.6 PBX, intra-LATA PSTN, and central office-based private switched network service examples.

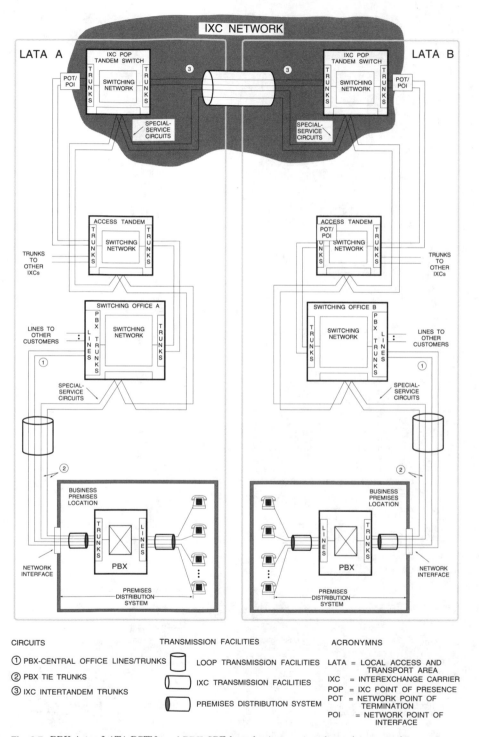

Fig. 8.7 PBX, inter-LATA PSTN, and PBX CPE-based private network service examples.

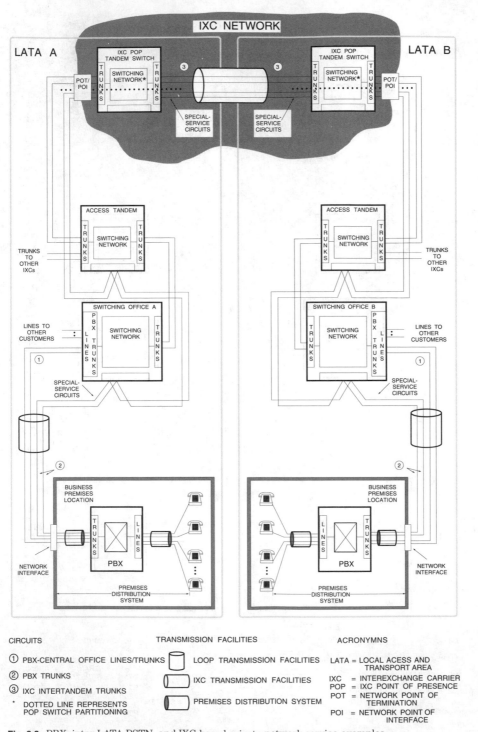

Fig. 8.8 PBX, inter-LATA PSTN, and IXC-based private network service examples.

Network Services

This chapter discusses an array of individual voice telecommunications services that can'be arranged and combined to support the end-to-end services described in the previous chapter, as well as a wide range of business applications. It relies on the introductory material of Part 1 for terms of reference and background, and on the voice services definitions and descriptions presented in Chapter 8. The sections below treat intra- and inter-LATA network services available from local exchange carriers (LECs) and interexchange carriers (IXCs). Network services categories discussed include access and transport, public and private, switched and non-switched. The use of PBX customer premises equipment and LEC and/or IXC transmission facilities to result in private switched network services is also addressed.

The discussion addresses the taxonomy of voice services illustrated by Fig. 8.2. The presentation centers around a graphic depicting each network service. The graphic identifies telecommunications facilities associated with each service and the point of termination (network interface, NI) between customer premises equipment and the service providing organization's (LECs, IXCs or bypass service providers) facilities. Principal applications supported by the service, price elements (how the service is paid for) and technical characteristics are summarized on the graphic and discussed in the narrative.

Access Services

Business line & Centrex access service

Figure 9.1 depicts LEC business line and Centrex services. The underlying facilities include LEC voice frequency (VF) loops that connect central office switches to business premises' network interfaces (NIs), customer premises equipment (CPE), and premises distribution systems that connect CPE to the network interfaces.

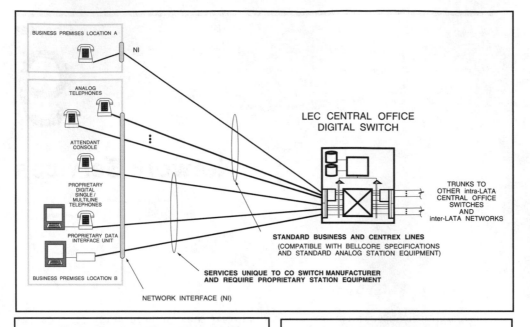

Fig. 9.1 Business line and Centrex service.

PRINCIPAL APPLICATIONS

Business Line:
- Access to the intra-LATA PSTN
- Access to the inter-LATA PSTN via the LEC intra-LATA PSTN
- Access to custom calling and CLASS* type features

Centrex:
- Access to the intra-LATA PSTN
- Access to the inter-LATA PSTN via the LEC intra-LATA PSTN
- Station-to-station calling
- Attendant services
- Access to PBX-like features
- Access to CLASS* and other enhanced features
- Access to OA&M and adjunct applications and features

* Custom local area signaling service

PRICE ELEMENTS

- Installation charges
- Monthly line charges
- Feature charges
- In most serving areas, discounts via special contracts are available to large Centrex customers willing to make long term and minimum size commitments

TECHNICAL CHARACTERISTICS

Access Facility Description:
- Single channel VF analog transmission facility
- Two-wire
- Metallic, twisted-pair interface at the NI

Call Direction Modes:
- Two-way

Signaling Interface:
- Loop signaling

Signaling Technique:
- Facility dependent, direct current (dc)
- Supervisory
  - Loop-start origination
- Addressing
  - Dial pulse
  - Dual-tone multiple frequency (DTMF)

The principal business applications supported by these services are:

- Access to intra-LATA local and toll calling and inter-LATA toll calling
- Access to LEC central office (CO) features, e.g., custom calling, CLASS
- For the Centrex case, access to station-to-station calling, feature packages, and attendant services.

Custom calling features include call forwarding, call waiting, and others, depending upon the serving office. *Custom local area signaling service* (CLASS) is a set of local calling enhancements to basic exchange telephone service. CLASS features use digital switching and signaling to provide automatic callback, automatic recall, selective call forwarding and call rejection, customer originated call trace, calling number identification and bulk calling number delivery. Chapter 4, Centrex Services, provides details.

Costs for business and Centrex line access services include one-time installation, monthly line, and feature charges, in accordance with local tariffs. (Message unit charges are allocated to LEC and IXC transport services and discussed later.) *Customer access line charges* (CALC), fees mandated by the FCC to assure that costs for local loop service are adequately provided for, are charged monthly on a per-line basis.

Through special contracts, large Centrex customers can acquire discounts in most serving areas in exchange for long-term and minimum-line-size commitments.

Technical characteristics associated with these access services are as follows. The loop facility is a VF analog carrier transmission system, consisting of bundled two-wire, copper twisted-pair wiring terminated at the NI. For large customers the LEC may elect to use either digital loop carrier or remote switch modules located on or adjacent to the customer premises. These alternatives are transparent to the customer; the appearance at the NI is identical.

The individual voice circuits are two-way. They support both call origination and termination. Loop-start supervision and either dial pulse or DTMF tone address signaling are supported by business and Centrex line access service.

Analog standard and proprietary phones can be used with these services. Although Fig. 9.1 indicates Centrex support for digital telephone and data interface units, this is accomplished using analog loop facilities (as described above), special CO line interface cards, and station equipment designed for the switch. Where available, *integrated services digital network* (ISDN) central office capabilities permit the use of digital station equipment from any vendor complying with CCITT ISDN standards. Note that whereas standard analog central office switch interface cards permit either voice or data modem traffic, proprietary digital or ISDN models do *not* support modem traffic (see Chapter 16 for an ISDN discussion).

PBX-to-central office trunk access service

Figure 9.2 depicts LEC PBX trunk service. The underlying facilities for this service are basically the same as for the business line/Centrex access services. Since the trunks are connected to a switch at the customer premises end, there might be an additional requirement for facilities to pass addressing information to the premises switch.

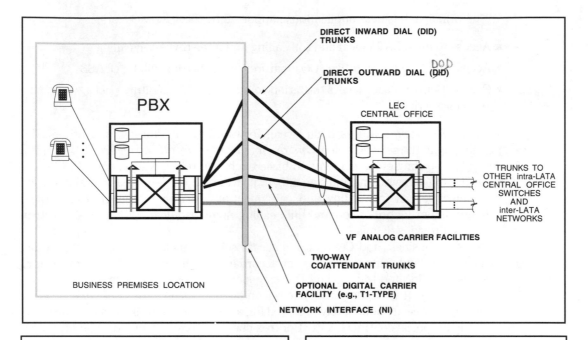

Fig. 9.2 PBX-to-central office trunk service.

PRINCIPAL APPLICATIONS

Direct Inward Dial Trunks:
- Call termination access to the intra-LATA PSTN
- Call termination access to the inter-LATA PSTN via the LEC intra-LATA PSTN

Direct Outward Dial Trunks:
- Call origination access to the intra-LATA PSTN
- Call origination access to the inter-LATA PSTN via the LEC intra-LATA PSTN

Central Office/Attendant Trunks:
- Call origination/termination access to the intra-LATA PSTN
- Call origination/termination access to the inter-LATA PSTN via the LEC intra-LATA PSTN

Note:
- Access to custom calling and CLASS features is not available with PBX-to-Central Office trunk service.

PRICE ELEMENTS

- Installation charges
- Monthly line charges
- Station number charges (DID only)

TECHNICAL CHARACTERISTICS

Access Facility Description:
- Single channel VF analog transmission facility
- Two-wire
- Metallic, twisted-pair interface at the NI
- Optional DS1 capable carrier service, e.g., four-wire twisted-pair T1 line or fiber FT3

Call Direction Modes:
- Two-way for CO/attendant trunks
- One-way for DID and DOD trunks

Signaling Interface:
- Loop signaling

Signaling Technique:
- Facility dependent, direct current (dc)
- Supervisory
  - Ground or loop-start origination for DOD trunks
  - Ground-start for CO/attendant trunks
  - Wink-start for DID trunks
- Addressing
  - Dial pulse
  - Dual-tone multiple frequency (DTMF)
- Optional robbed-bit signaling for direct digital carrier connection to PBXs

The principal business applications supported by this service are access to intra-LATA local and toll calling and inter-LATA toll calling. Three different types of trunks are normally furnished by the LEC: *direct inward dial* (DID), *direct outward dial* (DOD) and *two-way CO trunks.* DID allows incoming calls to ring specific stations without attendant assistance. DOD allows outgoing calls to be placed directly from PBX stations. Two-way trunks support attendant console operation so on-premises operators can place and receive calls.

Access to *custom calling* and *custom local area signaling service* (CLASS) features is *not* possible with PBX-to-central office trunk service.

Costs for PBX-to-central office trunk access services include one-time installation, monthly line, and DID station number charges, in accordance with local tariffs. (Message unit and other usage charges are LEC and IXC transport services charges, discussed later.) Since PBX trunks carry more traffic per circuit than business lines, they are normally more expensive than business lines. However, much fewer of them are required, as shown in Chapter 7. LECs normally don't grant discounts for large customers because they are often in competition with the vendors who offer PBXs as an alternative to LEC Centrex.

For this type of access, the CALC is charged monthly on a per trunk basis. Here, as with other charges, PBXs have an advantage over Centrex since the number of PBX trunks is normally less than the number of Centrex lines by 70% to 90%. However, PBX acquisition, operation, administration and maintenance (OA&M) must be factored into any overall life cycle cost trade-off.

Technical characteristics of PBX trunks are similar to business lines. For sites with heavy traffic, customers may elect to use LEC digital carrier transmission services.

Typically the price of 24-channel DS1 service is less than that for 24 separate analog channels. Also, with digital PBXs, a single T1-type carrier connection supports 24 voice channels, reducing interface complexity and OA&M costs. Some LECs are offering fractional T1 (that is, digital service for less than 24 channels) at lower than full T1 prices making the move to digital carrier even more attractive.

For analog PBXs, D4 channel banks can be used with digital carrier service. In cases where not all of the channels are needed for voice service, spare DS 1 capacity can be used for data or other services. Digital channel banks provide analog interfaces for the PBX identical to those associated with LEC VF carrier service described above.

Call direction, signaling interface, and signaling techniques used with PBX-to-CO trunks are depicted in Fig. 9.2.

Transport Services

Public switched intra-LATA transport services

Figure 9.3 illustrates LEC public switched services within a LATA. The underlying facilities include LEC central office and tandem switching, and interoffice transmission. For LEC serving areas that include more than one LATA, the Modified Final Judgment precludes LEC provision of transmission services between any of its central office switches if the connected switches are located in different LATAs.

The principal end user applications supported by these services are intra-LATA local and toll calling. Local services include local calling, 411 directory and operator as-

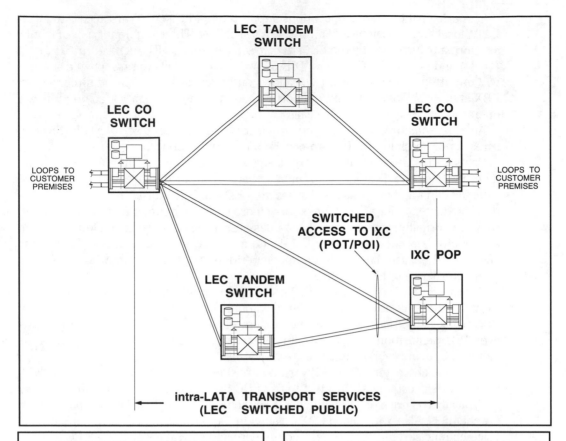

Fig. 9.3 Public switched services within a LATA.

PRINCIPAL APPLICATIONS

Local Service:
- Local calling
- Directory assistance (411), operator assistance
- Emergency service (911)
- Reliable transport between LEC COs

Toll (intra-LATA) Service:
- Message telecommunications service (MTS)
- Directory assistance (555-1212), operator assistance
- Wide-area telephone service (WATS)
- Reliable transport between LEC COs

Switched Access to IXC Services:
- Economical access for low volume users

PRICE ELEMENTS

- Local message unit charges (banded)
- Toll usage charges
- WATS (discounted toll)
- Switched access usage charges

TECHNICAL CHARACTERISTICS

Facility Description:
- Central office and tandem switches
- Trend to digital switching and transmission systems
- Increasing use of fiber optic transmission
- Redundant paths with alternate routing

Signaling Techniques:
- Older systems use primarily multiple frequency (MF) inband signaling
  - Addressing
  - Supervisory
- Trend to common-channel signaling (CCS)
  - Addressing
  - Supervisory
  - Feature messaging

sistance, and 911 emergency service. As we saw in Chapter 7, switched access to inter-LATA toll calling is normally purchased from the LEC by IXCs and the costs passed through to the end user in the IXC rates.

Intra-LATA toll service includes *message telecommunications service* (MTS), 555–1212 directory and operator assistance, and intra-LATA wide-area telephone service (WATS). In all cases the redundancy and sizing of LEC facilities results in transport service with grade of service and reliability performance significantly better than affordable private facilities.

Pricing elements include distance banded, message unit charges for local calling, toll charges and WATS charges. *Toll charges* are usage charges based on duration and distance. *WATS charges* are essentially discounted toll charges—that is, the service billing is on a bulk basis rather than for individual calls. Rates for these services are governed by tariffs established by local public utility or service commissions. Switched access charges are usage based. For interstate traffic carried on these facilities, the switched access rates are governed by tariffs established by the FCC, rather than by local commissions.

Normally, public switched intra-LATA transport services are offered by a single Bell-operating company (BOC) or by an independent telephone company (ITC). However, it is conceivable that various wireless communications systems (e.g., cellular mobile and personal communications systems) and bypass companies now focused on nonswitched transmission services (so-called *competitive access providers*, CAPs) could lead to increased competition at the local exchange level and eventually impact pricing.

While there is a trend to digital switching and transmission in intra-LATA networks, the sheer number of existing analog facilities makes the migration a slow process. Instead of hundreds of IXC switches, we are talking about tens of thousands of LEC central offices as well as the bulk of the PSTN transmission facilities. Nevertheless, cost and other savings and performance enhancements are spurring digital upgrades in most metropolitan areas. There is also increasing use of fiber optics in the outside plant.

Signaling between stored program control switches is strongly moving toward common-channel signaling for supervisory, addressing, and messaging required to support advanced CO-based features.

Private switched intra-LATA transport service

Where traffic among a company's business locations is sufficiently high, it may be possible to reduce costs by shifting the internal traffic from public switched networks to private switched networks.

Figure 9.4 depicts alternative ways of implementing private switched service within a LATA. In the CPE-based approach at the top of the figure, the underlying facilities include PBXs with tandem switching—*electronic switched network* (ESN)—capabilities, and LEC PBX tie trunk transmission.

Shown directly below is a LEC-based approach which uses LEC switches to implement tandem switching functions, and LEC tie trunk transmission. The LEC switches are also used for public service. The switches are partitioned to segregate

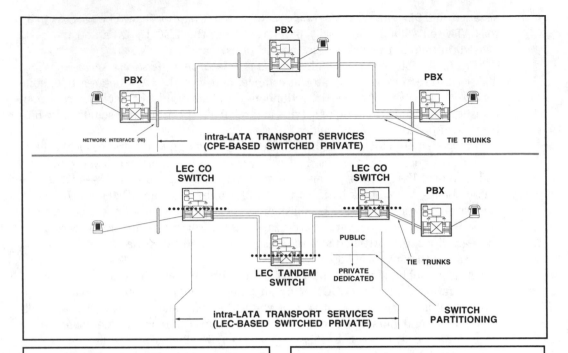

PRINCIPAL APPLICATIONS

CPE-Based Networks:
- Private alternative to the intra-LATA PSTN
- Private numbering plan
- PBX features, transparent across network
- Additional least cost routing option
- Potential for end-to-end digital connections (requires use of LEC digital special services)

LEC Central Office Based Networks:
- Same as above but without the requirement for dual function, (i.e., electronic switched service-ESN-capable) PBXs.

PRICE ELEMENTS

- Installation charges
- Monthly transmission charges (see nonswitched intra-LATA transport service)
- Switch port charges (LEC-based only)
- Feature charges (LEC-based only)

TECHNICAL CHARACTERISTICS

Access Facility Description:
- CPE-based
  - Four-wire PBX-to-CO tie trunks
  - VF analog transmission facility
  - Metallic, twisted-pair interface at the NI
  - Optional DS1 capable carrier service, e.g., four-wire twisted pair T1 line or fiber FT3
  - "Hardwired" for PBX-based, i.e., no call-by-call switched service (see LEC nonswitched service description below)
- LEC Central office based
  - Two-wire loops (without no premises located PBX)
  - Dedicated or partitioned CO switches
  - For premises with PBXs, the transmission options are the same as for CPE-based

Call Direction Modes:
- Two-way

Signaling Interface:
- E and M leads for PBX-based

Signaling Technique:
- Facility dependent, direct current (dc)
- Supervisory
- Addressing (dial-pulse or multiple frequency-MF)
- Optional robbed-bit signaling for direct digital carrier connection to PBXs

Fig. 9.4 Private switched service within a LATA.

public from private traffic and to guarantee availability and exclusive use of certain switch resources. Bellcore defines such networks as common-control switching arrangements (CCSA), one of a variety of special switched service networks (SSNs).

The LEC-based approach does not require PBX CPE with tandem switching capabilities, and the design, operation and maintenance of the transport network is handled by the LEC.

The principal private switched network business application is an alternative to public switched network intra-LATA local and toll calling. Advantages can include lower costs, private numbering plans, and PBX features transparent across the network (if all PBXs are of identical or compatible designs), with the option of digital end-to-end services if digital switching and digital LEC transmission are available.

Private networks cannot be sized economically to achieve grades of service equivalent to public networks or to include enough redundancy to equal public switched network reliability.

Consequently, alternative routing through the PSTN is normally provided to off-load private network traffic to public networks during peak loads or under failure conditions.

Pricing elements include installation charges, mileage-dependent transmission charges, and for LEC-based approaches, switch port and feature charges. Unlike public switched transport service, there are no usage charges. Therefore, economic justification for private networks, versus PSTN service, involves comparing both fixed monthly charges and estimated PSTN traffic usage charges. Rates for intra-LATA private switched network and nonswitched transmission are governed by tariffs established by local public utility or service commissions (see Chapters 7 and 10 for additional discussions contrasting network performance and cost).

Access facilities for PBXs are VF analog carrier transmission. For tie trunks with heavy traffic, customers may elect to use digital carrier transmission services.

For the PBX based approach, analog voice PBX tie trunks are in most cases two-way, fourwire circuits. The signaling interface is E&M leads. Wink start supervisory, and either dial pulse or MF addressing techniques are normally employed.

Nonswitched intra-LATA transport service

Figure 9.5 depicts LEC nonswitched services within a LATA. The underlying facilities include voice frequency (VF) or digital carrier loops, which connect central offices to business premises' NIs, and connections between originating and terminating central offices. The distinguishing characteristic of these services versus switched services is the dedication of the circuits to the customer whether there is traffic usage of the facilities or not.

Most LECs use programmable digital cross-connect systems (DCS) to "groom" connection arrangements on a per-customer basis. Grooming refers to establishing central office cross-connections that correspond to individual customer's needs. For example, a customer may require multiple channels of dedicated transmission facilities to be used for combinations of voice, data and other applications. He may also specify routing to establish connections between specified business premises, which may be different for each application, and vary on a site-by-site basis.

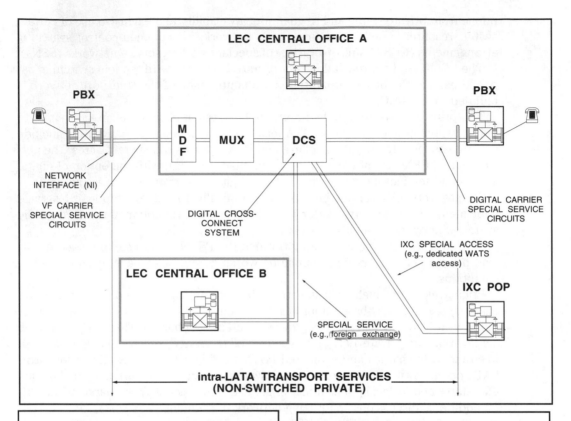

PRINCIPAL APPLICATIONS

Private Lines:
• Voice, teletypewriter, alarm, and other applications

Tie Trunks:
• Direct PBX connections
• Requires signaling service

IXC Access Trunks:
• PBX-to-IXC POP connections for access to:
  - IXC public switched services
  - IXC private special service circuits

Foreign Exchange (FX) Trunks:
• PBX access to local calling in another CO area

PRICE ELEMENTS

• Installation charges
• Monthly termination charges
• Mileage dependent monthly transmission charges
• Signaling charges when signaling is required
• Usage charge (FX only)

TECHNICAL CHARACTERISTICS

Access Facility Description (Service Dependent):
• VF analog transmission facility (as described in above examples)
• DS1 capable carrier service, e.g., four-wire twisted-pair T1 line or fiber FT3

Call Direction Modes:
• Service dependent

Signaling Interface:
• Service dependent (some services do not require or provide signaling)

Signaling Technique:
• Service dependent Central office termination.
• Most LECs use programmable digital cross-connect systems (DCS) to "groom" connection arrangements on a per-customer basis. DCS connections:
  - Are not changeable on a call-by-call basis,
  - Can change on a preset time basis throughout a day,
  - Offer rerouting protection in case of disasters

Fig. 9.5 Nonswitched within a LATA service.

The LEC can use DCS to combine traffic from multiple customers, over common, high-speed digital network facilities.

Such "static" connection arrangements could be accomplished with manual cross-connects, but the use of programmable DCS reduces cost and enhances performance and flexibility. As discussed in Chapter 3, DCSs are processor-controlled devices, which permit individual DS0 channels to be cross-connected among multiple DS1 or higher (i.e., DS2, 3 or 4) level terminating signals. The operation is illustrated in Fig. 9.5 where, for example, the DCS could establish some DS0 tie trunk voice connections between the PBXs, foreign exchange (FX) connections to a remote central office, and dedicated PBX to IXC POP connections for access to inter-LATA IXC services.

DCS connection arrangements can be programmed to change on a time-of-day or day-of-week basis. However, they do not provide call-by-call switching. DCSs offer flexible rerouting capabilities in case of disasters or failures. With them carriers can offer excellent dedicated, point-to-point network availability and reliability.

The principal applications supported by nonswitched services are:

- Private lines for voice, teletypewriter, alarms, data, etc.
- Tie trunks for connecting PBXs
- Trunks for dedicated access to IXC public and private services
- FX trunk groups
- FX lines

The basic technical characteristics associated with nonswitched services are the same as for other LEC services. In some cases, enhanced capabilities, such as sophisticated network management, and control capabilities may be provided.

Pricing elements include installation charges, central office termination charges, mileage dependent transmission charges, signaling charges where applicable, and, for FX, usage charges. Rates for LEC-provided nonswitched transmission services are governed by tariffs established by local public utility commissions, except for circuits used for IXC access. As before, these services are regulated by the FCC.

Normally public switched intra-LATA transport services are offered by a single Bell-operating company (BOC) or an independent telephone company. However, Metropolitan Fiber Systems, Teleport Communications, LOCATE and other companies are emerging to compete in the local exchange market. Most of the business of these competitors has been to provide alternative or *bypass* nonswitched service between business premises and IXC points of presence.

Public switched inter-LATA transport service

Figure 9.6 depicts IXC public switched services between LATAs. The underlying facilities include IXC tandem switching, interswitch transmission, and access facilities to LEC public switched telephone networks. IXCs buy access services from LECs to reach customers via the LEC public switched network. Businesses with heavy traffic can alternatively use dedicated access transmission facilities to IXC POPs.

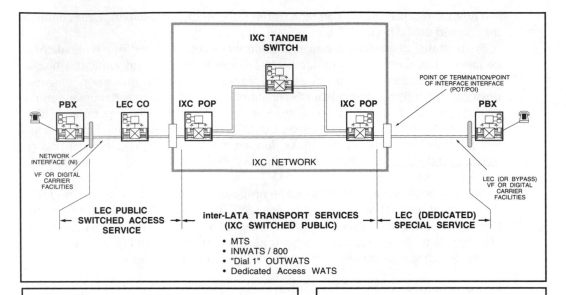

PRINCIPAL APPLICATIONS

Message Telecommunication Service (MTS):
- Reliable transport between POPs
- IXC carrier buys access service from LECs

"Dial 1" WATS Service:
- Volume discounted billing of MTS-like service, (i.e., bulk MTS)

Dedicated-Access WATS:
- IXC leases dedicated customer premises-to-IXC POP access (e.g., from a bypass carrier), or
- Customer provides dedicated customer premises-to-IXC POP access (e.g., from a bypass carrier or a LEC)

PRICE ELEMENTS

MTS and Dial 1 WATS:
- One-time connection charges
- Call duration and mileage-based usage charges
- Time-of-day and week discounts
- Volume discounted (WATS only)

Dedicated-Access WATS:
- Installation charges
- Call duration and mileage-based usage charges
- Volume, time-of-day and week discounts
- Special access dedicated line charges

TECHNICAL CHARACTERISTICS

Facility Description:
- Digital stored program control tandem switches
- Mostly digital IXC inter-switch transmission
- Increasing use of fiber transmission facilities
- Redundant, diverse paths with alternate routing

Signaling Technique:
- Mostly common channel signaling
  - Trend is to versions of ANSI SS7 (Bellcore)
- No CCS to PBXs in the USA

Fig. 9.6 Public switched services between LATAs.

The principal applications supported by IXC public switched services are inter-LATA MTS, dial-1 WATS, and dedicated-access WATS. MTS is standard (direct distance dial) or operator assisted inter-LATA long-distance service. For calls originating from business premises, the IXC carrier of choice is preselected by customers. Callers can override IXC preselection by dialing additional IXC carrier identification code digits. IXC portions of customer bills reflect LEC charges to IXCs for access service to their networks. As indicated in Chapter 1, LEC access charges to IXCs can represent half of IXC revenues.

With *dial-1 WATS,* all inter-LATA calls are completed as MTS calls but customer billing is on a bulk rather than call-by-call basis. Dial 1 WATS calls are processed through intervening LEC public switched networks as shown on the left side of Fig. 9.6. 800/900 services are variations of INWATS.

Dedicated-access WATS is illustrated on the right side of Fig. 9.6. Here dedicated special service trunks between business premises and IXC POPs bypass LEC switched access facilities. Customers pay for this nonswitched service as described in Fig. 9.5. Bypass carriers can also provide nonswitched services. IXCs can lease and administer dedicated access facilities on behalf of customers. IXCs submit separate bills to customers to cover these costs. See Chapter 10 for examples and further discussion.

These applications are described in terms of generic IXC services. Each IXC offers a menu of similar services with variations under private label names, e.g., AT&T's MEGACOM is a dedicated-access WATS-type service.

For these services, IXCs use high capacity, digital, stored program control tandem switches and digital network transmission facilities. Microwave radio and satellite carrier systems are also employed, but the trend is to fiber cable with Sprint now operating an all fiber network. The signaling trend is towards common-channel signaling versions of CCITT SS #7, for example AT&T's CCS 7 and Northern Telecom's SS7.

Pricing elements for MTS and dial-1 WATS include connection or installation charges, call duration and mileage based usage charges, time of day and day of week discounts, and for WATS, volume discounts. Dedicated-access WATS involves installation charges, special access dedicated line charges, call duration and mileage based usage charges, time of day, and day of week and volume discounts. As stated previously, rates for the LEC nonswitched intra-LATA transmission services used to access IXC services are governed by tariffs established by the FCC. Rates for AT&T IXC services are governed by tariffs approved by the Federal Communications Commission (FCC). Rates for independent local and other long distance carriers are not subject to regulation.

Private CPE-based switched inter-LATA transport service

As in the intra-LATA case, companies with high levels of internal traffic among business premises located in different LATAs may be able to reduce costs by shifting that traffic from public switched to private switched networks. Since LECs and IXCs can groom their facilities to accommodate voice, data and other traffic, all internal traffic must be considered in evaluating the private network decision.

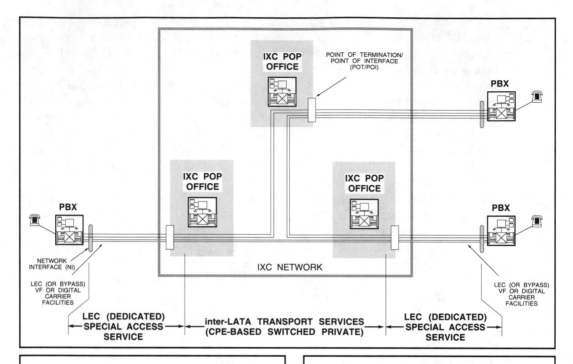

Fig. 9.7 Private CPE-based switched services between LATAs.

PRINCIPAL APPLICATIONS

CPE-Based Networks:
- Private-alternative to the inter-LATA PSTN
- Private numbering plan
- Features transparent across network
- Additional least-cost routing option
- Potential for end-to-end digital services
  (requires availability of LEC and IXC digital special services)

PRICE ELEMENTS

- Installation charges
- Fixed transmission charges
  - Dedicated trunks from customer premises to IXC POP
  - Dedicated IXC trunks
  - Includes signaling charges

TECHNICAL CHARACTERISTICS

Access Facility Description:
- Four-wire PBX-to-CO tie trunks
- VF analog transmission facility
- Metallic, twisted-pair interface at the POT
- Optional DS1 capable carrier service, e.g., four-wire twisted-pair T1 line or fiber FT3
- "Hardwired," i.e., no call-by-call LEC or IXC switched service

Call Direction Modes:
- Two-way

Signaling Interfaces:
- E&M Leads

Signaling Technique:
- Facility dependent, direct current (dc)
- Supervisory
- Addressing
  - Dial pulse or multiple frequency (MF)
- Optional robbed-bit signaling for direct digital carrier connection to the PBX

Figure 9.7 depicts a CPE-based private switched service between LATAs. In the CPE-based approach, the underlying facilities include PBXs with tandem switching—electronic switched network (ESN)—capabilities, and nonswitched PBX tie trunk transmission. By not showing connections to IXC switches, the figure illustrates that IXCs do *not* provide call-by-call switched service for this application. Like LECs, IXCs do use DCS and other crossconnect mechanisms to groom their nonswitched transmission facilities to meet individual customer requirements, as illustrated in Fig. 9.5.

The principal private switched network application is an alternative to public switched network toll calling. As with the intra-LATA example discussed above, advantages can include lower costs, private numbering plans, and PBX features transparent across the network (if all PBXs are of compatible designs), with the option of digital end-to-end service, if digital switching and digital nonswitched transmission are available.

Also, as in the intra-LATA case, with alternative routing through the PSTN is normally provided to off-load private network traffic to public networks during peak loads or under failure conditions.

Pricing elements include installation charges, termination charges, mileage-dependent transmission charges, and signaling charges. Unlike public switched services, there are no usage charges.

Private IXC-based switched inter-LATA transport service

Figure 9.8 depicts an IXC-based private switched service between LATAs. In the IXC-based approach, the underlying facilities include IXC tandem switching and nonswitched trunk transmission. Both PBX and IXC-based approaches are facilities-based private networks, in contrast to virtual private networks (VPNs) to be described in the next section.

The IXC-based approach does not require PBX CPE with tandem switching capabilities, and the design, operation and maintenance of the transport network is handled by the IXC.

The IXC switches are also used for public service. The switches are partitioned to segregate public from private traffic and to guarantee availability and exclusive use of certain switch resources. In the past, AT&T offered such network services under the *common-control switching arrangements* (CCSA) name. Later AT&T offered *enhanced private switched communications service* (EPSCS), a private network service that, like CCSA, provides a uniform dialing plan for customers with geographically dispersed locations.

Recent tariffs, however, favor *virtual* over *dedicated facilities-based* private networks. One legitimate reason for this is that, as discussed in Chapter 7, in networks, economy of scale always favors aggregating higher levels of traffic on common facilities. Consequently, service providing organizations can offer lower tariffs when they can carry private traffic on public network facilities, versus carrying private traffic on smaller dedicated segments.

Another factor that has tended to make post-divestiture facilities-based private networks less attractive is the escalating cost of dedicated inter-LATA transmis-

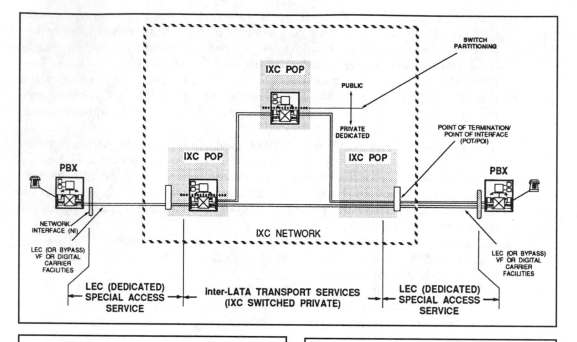

Fig. 9.8 Private IXC-based switched service between LATAs.

PRINCIPAL APPLICATIONS

IXC Based Networks:
- Private alternative to the inter-LATA PSTN
- Private numbering plan
- Additional least-cost routing option
- Potential for end-to-end digital connections (requires availability of LEC and IXC digital special services, and digital PBXs)
- No requirement for dual function, (i.e., electronic switched service - private switched network capable) PBXs.

PRICE ELEMENTS

- Installation charges
- Fixed transmission charges
- Switch port charges
- Feature charges

TECHNICAL CHARACTERISTICS

Access Facility Description:
- Four-wire PBX-to-CO tie trunks
- VF analog transmission facility
- Metallic, twisted-pair interface at the POT
- Optional DS1 capable carrier service, e.g., four-wire twisted-pair T1 lines or fiber FT3
- Dedicated or partitioned CO switches

Note: IXCs do not maintain a tandem switch at every LATA POP. Consequently, inter-LATA access lines may be required for dedicated-access to IXC private switched network service.

Call Direction Modes:
- Two-way

Signaling Interfaces:
- Access service dependent
- Mostly CCS within IXC networks

Signaling Technique:
- Facility dependent, direct current (dc)
- Supervisory
- Addressing (dial pulse or multiple frequency-MF)
- Optional robbed-bit signaling for direct digital carrier connection to the PBX

sion for access to IXC transport services. AT&T, and the other IXC carriers, do not have tandem switches at every POP, or even in every LATA. The right side of Fig. 9.8 shows a PBX connected to an IXC POP that does not have a collocated switch. (This is not atypical. Even AT&T has its digital tandem switches—4ESSs—in only half of the LATAs.)

As a consequence, customers often must lease LEC dedicated transmission from their business premises to the closest POP, and then additional dedicated IXC inter-LATA transmission from that POP to the nearest IXC tandem switch. While the same backhauling is required for public and virtual private network IXC access, charges are shared by all the service tariffs, e.g., MTS, WATS and VPN. Thus the same access facilities are priced in a way that favors virtual private networks.

Generally, private network savings are proportional to the quantity of private traffic carried. Because of initial capital investment, system design engineering and increased OA&M involved with private networks, they tend to prove out only for larger organizations. VPN was structured to make the break-even point with public network service occur at much smaller system sizes. While facilities-based approaches might still be superior to VPN for very large systems, recent analyses that take into account IXC special contracts for combinations of services (e.g., AT&T under Tariff 12) and progressively higher volume discounts, indicate that is not the case.

The principal business applications for IXC-based private switched networks are the same as for PBX-based approaches. Pricing elements include installation charges, termination charges, mileage-dependent transmission charges, signaling charges, and switch port and feature charges.

Virtual private switched inter-LATA transport service

Figure 9.9 depicts IXC virtual private switched service between LATAs. The underlying facilities include the IXC public network augmented by network control point and service management system facilities. Traffic is routed through the public network under computer control in a manner that makes virtual private network (VPN) service indistinguishable from facilities-based private networks.

A *network control point* (NCP) is a centralized data base that stores a subscriber's unique VPN information. Highly sophisticated, this data base screens every call and applies call processing control in accordance with customer-defined requirements. A *common-channel signaling* (CCS) network connects all network elements, and even after implementing customer defined call control, decreases post-dial delay by 5 to 15 seconds relative to non-CCS signaling.

The *service management system* (SMS) used to build and maintain a VPN data base, lets customers program specific functions to accommodate unique business applications. The SMS contains complete specifications of customer-defined private network specifications including location data, numbering plan, features, screening actions, authorization codes, calling privileges, etc. This information is *downloaded* (transmitted) to NCPs which implement instructions on a customer-by-customer basis.

Access to VPNs can be via dedicated special access or LEC switched access. Special access can be via dedicated VF carrier or DSN digital carrier service. IXCs will

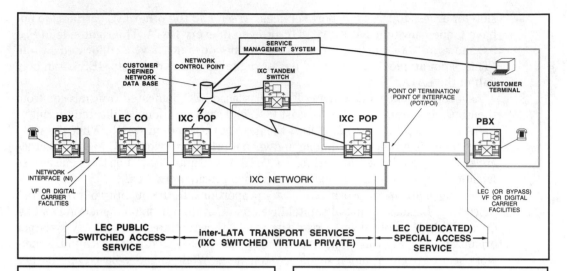

Fig. 9.9 Virtual private switched between LATAs service.

PRINCIPAL APPLICATIONS

Virtual Private Networks (VPN):
- Alternative to the inter-LATA public or facilities-based private networks
- Switched or dedicated access to VPN services
- Private numbering plan
- Features transparent across network
- Additional routing options
- Available end-to-end digital connections
- On-premises access to OA&M information systems
- No requirement for dual function, (i.e., electronic switched service - private switched network capable) PBXs.

PRICE ELEMENTS

- Installation charges
- Fixed transmission charges for dedicated access
- Call duration, mileage, time-of-day/week, type of access-based usage charges
- Feature charges
- IXC charges heavily discounted or waived for volume and long-term commitments

TECHNICAL CHARACTERISTICS

Access Facility Description:
- Four-wire PBX-to-CO tie trunks
- VF analog transmission facility
- Metallic, twisted-pair interface at the POT
- Optional DS1 capable carrier service, e.g., four-wire twisted-pair T1 line or fiber FT3

Call Direction Modes:
- Two-way

Signaling Interface:
- Access service dependent

Signaling Technique:
- Access service dependent
- Access to all VPN features requires digital PBXs which support special signaling; e.g., AT&T requires compatibility in accordance with their technical publication 41458

groom crossconnections on a per-customer basis to permit allocating DSO or analog voice channels among various IXC voice and/or data services. Dynamic access arrangements permit use of DSO channels alternatively for either voice or data on a per-call basis. Within the AT&T VPN (called *software defined network,* SDN), dynamic access requires a digital PBX which meets the parameters specified in AT&T technical publication 41458.

LEC switched access can be used to bring onto the network low volume locations that cannot justify special access. With presubscription to the VPN carrier, users can simply dial 1 to gain access to the virtual private network. Otherwise, users can dial carrier VPN codes for access.

The principal business applications supported by VPN include all those previously described for facilities-based private networks.

VPN pricing elements include one-time service establishment charges, duration and mileage dependent call charges with time of day, day of week and other discounts, and nonrecurring and monthly feature charges. Dedicated special access charges are as described at Fig. 9.5. Special access charges are only incurred between customer premises and the IXC POP, regardless of whether the POP has an IXC switch. Depending upon network specifics, SDN rates may be subject to both FCC and state regulation.

Service establishment charges can be waived or reduced if customers are willing to commit to a minimum service duration and minimum usage levels. For example, with AT&T's SDN basic service, service establishment charges are waived in return for three year duration and two million call minute per year commitments.

Nonswitched inter-LATA transport service

IXCs offer nonswitched services similar to those intra-LATA services offered by LECs and depicted in Fig. 9.5. The IXC circuits are provided between the IXC POPs in the respective LATAs. The applications, pricing elements, and technical characteristics generally correspond to LEC intra-LATA nonswitched services. LEC special access circuits are required to connect the customer premises to the IXC POP at both ends of the circuit.

Summary

Access services

Access services of greatest importance to business telecommunications users are those provided within or to customer premises. For convenience, Fig. 9.10 summarizes the popular analog and digital LEC access options discussed above. The choice of analog or digital access services is largely transparent to end-users.

For new installations involving PBXs and larger key telephone systems, the selection of digital versus analog access service can influence administration and maintenance and life cycle cost. This results from the fact that the monthly cost for 24-channel DS1 service is less than the cost for 24 separate analog channels and the existence of simpler interfaces.

Up-front costs of D-type channel banks or digitally equipped PBX/KTSs are recouped and savings accrue from that point on. When requirements for private voice, data and other user services exist in addition to the need for access to public services, the cost and technical benefits of digital technology become even more favorable.

For business lines and Centrex, except in areas where central offices offer ISDN *basic rate service* (BRI) digital service is available only with station equipment

LEC ACCESS SERVICE	CALL DIRECTION	VF FACILITY	SIGNALING TECHNIQUE	DIGITAL FACILITY
BUSINESS LINE	TWO-WAY	TWO-WIRE	LOOP START	N/A[1]
CENTREX	TWO-WAY	TWO-WIRE	LOOP START	N/A[1]
COIN OPERATION	TWO-WAY	TWO-WIRE	GROUND START	N/A[1]
PBX TRUNKS				
DIRECT INWARD DIALING (DID)	ONE-WAY	TWO-WIRE	WINK START	FOUR-WIRE[2] (24 CHANNEL DS1 SIGNALS)
DIRECT OUTWARD DIALING (DOD)	ONE-WAY	TWO-WIRE	GROUND OR LOOP START	FOUR-WIRE[2] (24 CHANNEL DS1 SIGNALS)
CENTRAL OFFICE	TWO-WAY	TWO-WIRE	GROUND START	FOUR-WIRE[2] (24 CHANNEL DS1 SIGNALS)
SPECIAL SERVICE CIRCUITS				
FOREIGN EXCHANGE (FX) LINES	TWO-WAY	TWO-WIRE	GROUND START	N/A[1]
OFF-PREMISES EXTENSION (OPX) LINES	TWO-WAY	TWO-WIRE	GROUND OR LOOP START	N/A[1]
TIE TRUNKS	TWO-WAY	TWO-WIRE OR FOUR-WIRE	E & M LEADS	FOUR-WIRE[2] (24 CHANNEL DS1 SIGNALS)

1   Digital voice access service is not normally available from LECs except with proprietary station equipment from the CO manufacturer. Digital access is also possible where ISDN basic rate interface (BRI) service is available.

2   Robbed-bit signaling is normally used for direct digital carrier connections to PBXs. Available digital channel units for digital carrier, provide a full set of signaling features, including loop-start, ground-start and E & M leads options. COs with SS #7 capabilities may offer **digital subscriber signaling** system (DSS) signaling in accordance with Q.920/921 and Q.930/931 recommendations.

Fig. 9.10 Summary of LEC access services.

proprietary to the CO switch manufacturer. The analog versus digital issue may arise, however, in those situations where a business is evaluating Centrex and PBX alternatives.

Where ISDN is available, it is, and will remain for the foreseeable future, more expensive than analog alternatives. Since life-cycle cost is not a motivation for supplanting analog services, the use of ISDN BRI must be justified in terms of new voice/data capabilities, and other enhancements.

For businesses with existing local area networks (LANs) supporting data applications, the data capabilities of ISDN BRI may not hold great appeal. At present, ISDN BRI capabilities are heavily oriented toward voice applications. Compared against today's Centrex, it may be difficult to justify ISDN's cost for an equivalent level of service.

In the United States, where telecommunications services are governed by the market, services that are not competitive are simply ignored. In Europe, state-run postal telephone and telegraph (PTT) administrations legislate telecommunications service offerings, so the migration to ISDN BRI service can be expected to occur much more rapidly. Chapter 16 and Appendix A provide insight into ISDN and its use and evolution within the United States.

Figure 9.10 depicts the array of selected LEC access services. Special services, for example, are so numerous that they are sometimes defined as everything except residence, coin and measured business telephone service. In private line services alone, Bellcore defines thirteen different voice and data categories. The figure shows those services of greatest potential for business applications.

Transport services

Differences among transport service alternatives may be less discernible to users than those of access service alternatives. In most cases, a user doesn't care whether public or private networks are used to complete calls and although digital facilities improve grade of service, the existing analog network is so well engineered that the difference is largely imperceptible to users. The primary motivation involved with selection of transport and signaling alternatives is cost reduction.

Cost analyses for large, complex, system alternatives requires technical, management and financial expertise as well as computer modeling tools. These capabilities are available from telecommunications service-providing organizations. Independent consulting firms having no product affiliations, are generally able to include a broader range of competitive offerings in their analyses. They are also less biased than firms associated with public or private carriers. An analysis of tariffs and their relation to network design is presented in the next chapter.

10

Selecting Voice Network Services

Network design has two principal objectives. First, networks must provide specified services at performance levels (grades-of-service) that satisfy business requirements. Since there is always a range of technically acceptable network design alternatives, the second objective of network design is to select the least expensive approach. As discussed in Chapter 7, from a purely technical point of view, aggregation of traffic (higher traffic volumes) always leads to more efficient use of telecommunications resources. It can be argued, therefore, that users with large traffic volumes deserve lower unit prices for public network services since they contribute disproportionately to the total traffic volume, making those facilities more efficient for all users.

Indeed, current tariffs are structured to encourage businesses to select service alternatives that minimize costs at higher levels of traffic. However, it is incumbent upon users to design their networks to make the most cost-effective use of the wide range of public (shared) and private (dedicated) carrier service offerings. This chapter examines tariffs used to price LEC and IXC network services, building on the basic concepts presented in Chapter 7. We shall now explore tariff structures, how tariffs are applied, and, most importantly, the relationship between tariffs and network designs, which minimizes costs while satisfying business needs.

As noted earlier, a primary distinction among tariffs is jurisdictional. In the post-divestiture era, there are two classes of tariffs, one pertaining to LEC intra-LATA services and the other pertaining to IXC inter-LATA services. Predivestiture intrastate and interstate service distinctions remain significant in terms of actual rate determination. Recall that under terms of the Modification of Final Judgment (MFJ), only IXCs may carry traffic between LATAs, even when those LATAs are in the same state.

Within each jurisdiction, the second major tariff breakdown is between *switched services* and *dedicated private line services*. As described previously, switched services enable users to share expensive resources, such as switches and transmission facilities, in public networks. However, users with sufficient traffic to occupy resources between two points in a network a high percentage of the time will find it economical to obtain those resources on a full-time (dedicated) basis. The two network points might be either two user premises, or a user premises and a carrier POP. In the examination of various tariffs that follow, break-even points and other trade-offs become evident.

In this chapter and in Chapter 15, the figures that illustrate tariffs are taken from CCMI's *Guide to Networking Services* (reference 29 in the bibliography).

LEC Switched Services

The following sections focus on *nonlocal services*, or services beyond the local subscriber calling area as defined by the LEC. The primary nonlocal LEC intra-LATA switched services are standard MTS and LEC versions of WATS. With MTS, calls use the same premises-to-LEC CO exchange access lines (loops) used for local calls. Tariffed rates are mileage-sensitive and discounted during evening and nighttime hours. *Evening hours* are defined as 5:00 P.M. to 11:00 P.M., and *nighttime hours* are from 11:00 P.M. to 8:00 A.M. Nighttime rates apply on Saturday and until 5:00 P.M. on Sunday. Note that tariffs in different jurisdictions may vary in both interval definition and discount rates.

There is a charge for an initial time period (usually one minute) that is higher than the charge for additional periods. Figure 10.1 illustrates typical rates for intra-LATA MTS. Note that daytime usage charges average about 25 cents per minute. Some jurisdictions permit special calling plans with lower per-minute costs, in exchange for customer volume commitments. Mileage for MTS calls is ordinarily calculated using serving wire center V&H coordinates, explained in Chapter 7.

Intra-LATA WATS rates are based on volume rather than on distance. Figure 10.2 shows a representative rate table for outward-calling WATS. The "monthly hours of usage" column refers to total call-hours per month for a group of WATS circuits at a

Mileage	First Minute ($)	Added / Minutes ($)
1-10	0.17	0.09
11-16	0.21	0.12
17-22	0.23	0.15
23-30	0.29	0.19
31-40	0.30	0.22
41-55	0.32	0.25
56-70	0.33	0.26
71-124	0.34	0.27
125-196	0.35	0.28
197-292	0.36	0.29
293-400	0.36	0.29

Evening Discount: 25%
Night Discount: 50%
Billing Basis: Rate Period Specific

Non-Discountable Surcharges
Operator-Assisted: $1.00
Person-Person: $2.50
Credit Card: $0.40

Fig. 10.1 Intra-LATA MTS rate example (SOURCE: *CCMI*).

| Monthly Hours of Usage | DAY | | EVENING | | NIGHT | |
	Hourly Rate ($)	Base Rate ($)	Hourly Rate ($)	Base Rate ($)	Hourly Rate ($)	Base Rate ($)
0 -15	9.25	---	8.10	---	4.36	---
15.2-40	8.25	138.75	6.76	121.50	3.63	65.40
40.1-80	8.00	345.00	5.31	290.50	2.86	156.15
Over 80	7.50	665.00	4.02	502.90	2.16	270.55

Fig. 10.2 Intra-LATA WATS rate example (SOURCE: *CCMI*).

particular location. Call hours are totaled separately for each rate period (day, evening, night) and then priced in accordance with intersecting usage cost bands (rows) and applicable rates for each period (columns). *Total cost per month* is the sum of usage charges plus fixed charges for "access lines"—which, incidentally, can only be used for WATS traffic. Fixed access rates are independent of customer location and priced at around $25 per month per circuit. An additional fixed cost per WATS group, typically $9, is also charged.

The following calculation illustrates the application of the rate table in Fig. 10.2:

Day cost calculation:
15 Hours	@ $9.25/hr	= $138.75
25 Hours	@ $8.25/hr	= $206.25
40 Hours	@ $8.00/hr	= $320.00
10 Hours	@ $7.50/hr	= $ 75.00

(90 day hours/month)

Evening cost calculation:
15 Hours	@ $8.10/hr	= $121.50
10 Hours	@ $6.76/hr	= $ 67.60

(25 evening hours/month)

Night cost calculation:
5 Hours	@ $4.36/hr	= $ 21.80

(5 night hours/month)

Total Usage:
    120 Hours at a total cost of $950.90

The daytime usage cost in this example is 13.7¢/minute. If three access lines are ordered to carry this traffic, then additional access costs are 3×$25+$9=$84, or 1.2¢/minute. The resulting unit cost of 14.9¢/minute is clearly preferable to MTS alternatives for customers with the above daytime traffic volume. Note that savings occur only when customers have enough traffic to offset fixed access and WATS group charges, with break-even points occurring at about 10 hours of traffic per month. Note also that a poor busy-hour grade of service over a small number of WATS access lines can be improved by switching equipment that will automatically route WATS calls away from busy access lines, back to MTS service. While this forfeits some cost savings, it minimizes user annoyance and high call retry rates, which often further degrade GOS.

The rate schedules for intra-LATA 800 service (incoming WATS) are similar to the outward WATS example. In some cases, access lines can be combined, but for rate purposes, volumes are usually tallied separately.

LEC Private Line Services

Private line services can be supported by LECs using either voice-grade (analog) or digital facilities. Technical characteristics (for example, 2-wire/4-wire, frequency response, signaling arrangements, etc.) distinguish analog facility and service offerings. Figure 10.3 relates voice-grade service designators to applications in terms of technical characteristics.

Tariffed service rate elements vary from state to state, and equivalent rate elements may bear different service designator names. The trend, however, is to common cost-based rate elements and designators. For example, in most jurisdictions, one of the private line service pricing elements is the *local channel rate* element. This element specifies charges for local loops from the customer's premises to serving wire centers, as was noted in the WATS example above. Local channel charges are normally not mileage-dependent. Where local channel elements are mileage-dependent, users rarely have access to the premises-to-wire center distance information (1 mile on the average) needed to independently calculate costs and must therefore rely on estimates.

A local channel charge is assessed at both originating and terminating COs. An interoffice channel connects two serving COs (more accurately, serving wire centers). The interoffice channel private line service rate element is mileage-dependent, and wire center V&H coordinates are used to calculate mileage charges. If interoffice channels are used to support switched network services, rate element charges for signaling equipment are assessed at each end. Other charges are assessed for special options—for example, line conditioning to improve circuit performance for data communications applications.

Unfortunately, the realm of tariffs has developed (over many decades) a unique vocabulary that neither conforms to terminology emanating from the MFJ, nor is consistent among the separate LEC and IXC tariff domains. For instance, in contrast with the definition in LEC tariffs cited above, in IXC tariffs, interoffice channels connect IXC serving points of presence (POPs), and local channels include all LEC or bypass network components used to connect customer premises to serving IXC POPs. Other unique and often contradictory tariff terminology will be defined as encountered in the information that follows.

Returning to the LEC private line service discussion, Fig. 10.4 shows representative local channel, interoffice channel, and signaling rate elements for intra-LATA

SERVICE	APPLICATION
Type 2230	2-wire interface with 2-wire facilities - voice transmission, private line, mobile radio, supervisory use
Type 2231	2-wire interface with 2- or 4-wire facilities - PBX OPX, signaling required
Type 2432	2- or 4-wire interface with 4-wire facilities, tie line: PBX-PBX signaling required
Type 2434	2- or 4-wire interface with 4-wire facilities, tie lines, centrex- centrex (with E&M signaling)
Type 2435	4-wire interface with 4-wire facilities - voice transmission, multipoint service
Type 2260	2- wire interface, 2-wire facilities - half duplex data services
Type 2261	2-wire interface, 2-wire facilities- dataphone select-a-station or telemetry alarm bridging service
Type 2462	4- wire interface with 4-wire facilities - dataphone select-a-station or telemetry alarm bridging service
Type 2463	4- wire interface with 4-wire facilities - analog data services, multipoint service
Type 2464	2-wire interface with 4-wire facilities- analog data services, multipoint provided

Fig. 10.3 Characteristics of voice-grade private lines (*SOURCE: CCMI*).

voice-grade, analog circuits. Attention should be paid to the mileage table for interoffice channels. In most states, and in all inter-LATA tariffs, total interoffice mileage is used to compute mileage dependent components, which are then added to the fixed monthly charge. As an example, an interoffice channel charge for a 50-mile circuit would be:

$32+50×$1.95/mile=$129.50

Some states still employ an older approach in which each mileage band is used, similar to the volume calculations in the WATS example. In this case, charges for a 50-mile circuit would be calculated as:

$32+8×$2.05/mile+17×$2.00/mile+25×$1.95/mile=$131.15.

Not all states use serving wire centers as the basis for pricing. Some still use *exchange*, or rate-center pricing in which connections are considered *intra-exchange* if they are in the same rate center. *Inter-exchange* circuits have mileage-based pricing components calculated on the basis of rate-center coordinates. Careful tariff reading is required to ascertain how to apply mileage-based rate tables.

It is also important to distinguish between the use of "intra-" and "inter-exchange" as used in intra-LATA and inter-LATA tariff contexts. Whereas today the term *interexchange,* as in "Inter<u>X</u>change <u>C</u>arrier," applies strictly to inter-LATA ser-

Local Channels	Nonrecurring Charge		Monthly Rate
Per point of Termination	First ($)	Additional ($)	($)
Type 2230	345.00	115.00	25.00
Type 2231	345.00	115.00	25.00
Type 2432	400.00	145.00	45.00
Type 2434	160.00	83.00	10.00
Type 2435	370.00	130.00	45.00
Type 2260	415.00	160.00	30.00
Type 2261	575.00	245.00	24.00
Type 2462	565.00	235.00	38.00
Type 2463	415.00	160.00	50.00
Type 2464	415.00	155.00	50.00

Interoffice Channels	Nonrecurring Charge	Monthly Rate	
Fixed and Mileage Charges Applicable	Per Channel ($)	Fixed ($)	Per Mile ($)
1-8 Miles	105.00	32.00	2.05
9-25 Miles	105.00	32.00	2.00
Over 25 Miles	105.00	32.00	1.95

Signaling (Per Local Channel)	Nonrecurring Charge		Monthly Rate
	Initial ($)	Subsequent ($)	($)
Manual Ringdown	215.00	40.00	11.00
Automatic Ringdown	72.00	15.00	10.00
E & M Signaling Type A:	185.00	45.00	10.00
(0-199 ohms)	135.00	43.00	6.00
Type B: (200-899 ohms)	135.00	42.00	6.00
Type C: (900 or more ohms)	135.00	12.00	3.00

Signaling (Per Local Channel)	Nonrecurring Charge		Monthly Rate ($)
	Initial ($)	Subsequent ($)	
Types 2463 and 2464			
Type C1	81.00	9.00	3.00
Type C2	90.00	23.00	3.00
Type D1	85.00	15.00	3.00

Fig. 10.4 Intra-LATA voice-grade private line rate examples (*SOURCE: CCMI*).

vices, the central office connotation of exchange, as reflected in local and state tariffs and many telephony reference books, dates back almost a century.

Digital circuits have tariff structures similar to analog circuits. Differences involve physical facilities used to provide circuits. Today, most facilities are derived from

1.544 Mbps DS1 facilities, described in Chapter 3. In intra-LATA networks, these circuits are provided directly to users at the 1.544 Mbps rate under generic names such as, "high-capacity," or trademarked names like "Megalink Service." Digital local channels and interoffice channels can often be substituted for analog counterparts. Figure 10.5 displays rate tables for digital facilities from the same jurisdiction as the voice grade example above. The interoffice DS1 channel cost for a 50-mile circuit is $85+50×$31.00/mile=$1,635—equivalent to the cost of just 13 analog voice-grade circuits. Since DS1 service provides 24 voice-grade circuits, it is more cost-effective than analog service whenever more than 13 channels are required. This *crossover point* varies from state to state, but, in general, this comparison is typical.

In addition to point-to-point use between user premises, LECs now offer CO multiplexing and dedicated 1.544 Mbps DS1 services as multichannel interoffice backbones for multiple lower-speed circuits from customer premises. This configuration is shown in Fig. 10.6. LEC charges take the form of port charges for each low-speed circuit terminated on the input side of the multiplexer and the 1.544 Mbps circuit connected to the output side of the multiplexer.

Historically, lower-speed digital data services (circuits with line speeds of 2.4, 4.8, 9.6, and 56 kbps) have been provisioned with separate, synchronous facilities using special hubs and end-to-end timing. More recently, 56 kbps service is offered on channels derived from T1 facilities. Both options may exist in the same service area under different service names.

	Nonrecurring Charge ($)	Month To Month ($)	36 Months ($)	60 Months ($)	84 Months ($)
Digital Local Channel					
Each					
First 1/2 Mile	300.00	82.00	81.00	81.00	81.00
Each Additional 1/2 Mile, or Fraction Thereof	---	35.00	34.00	32.00	30.00
Interoffice Channels					
(Furnished between COs)					
Each Channel 0-8 Miles					
Fixed Monthly Rate	100.00	65.00	65.00	65.00	65.00
Each Airline Mile, or Fraction Thereof	---	35.00	34.00	32.00	30.00
Each Channel 9-25 Miles					
Fixed Monthly Rate	100.00	70.00	70.00	70.00	70.00
Each Airline Mile, or Fraction Thereof	---	33.00	32.00	30.00	28.00
Each Channel Over 25 Miles					
Fixed Monthly Rate	100.00	85.00	85.00	85.00	85.00
Each Airline Mile, or Fraction Thereof	---	31.00	30.00	28.00	26.00

Fig. 10.5 Intra-LATA 1.544 Mbps circuit rate example (*SOURCE: CCMI*).

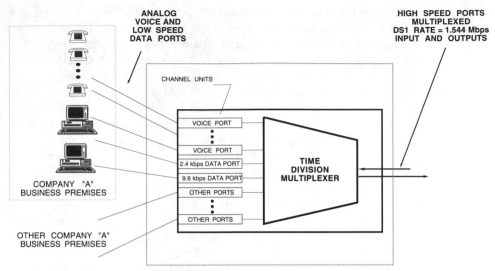

Fig. 10.6 Central office multiplexing service example.

Lower-speed synchronous digital services are engineered to superior bit-error-rate and error-free-seconds specifications when compared to analog circuits with modems or T1-derived circuits, but they are more expensive. Users generally find that performance improvements do not justify the more expensive facilities, although careful examination of technical specifications for each application is mandatory. Figure 10.7 shows a rate structure for synchronous digital data service. Here, circuits must pass through at least one LEC node office for test and maintenance purposes, increasing mileage and termination charges.

Carrier circuits used for digital transmission require user-premises *network channel terminating equipment* (NCTE) to provide for digital signal processing and to protect the network from harmful signals. Following the FCC's NCTE Order in 1983 (see Appendix A), this equipment is no longer supplied by the LEC. Users must provide the equipment, which has to be designed in accordance with LEC interface specifications. For digital circuits, channel service units (CSU) and data service units (DSUs) are required, and, as described in Chapter 3, these functions are usually combined in a single piece of equipment. For analog facilities, users must provide modems, described in Chapter 2.

The decision to include dedicated (private) facilities in business telecommunications networks is nearly always driven by the desire to reduce costs, thus, traffic volume and applicable tariffs ultimately dictate the design, apart from privacy and other reasons. To ensure high utilization of private facilities, during busy hours, traffic that would be blocked on private facilities may first be routed to WATS, and lastly to MTS. In this way, businesses can employ a hierarchy of services to minimize costs while maintaining grade of service. PBXs and other switching equipment offer *least-cost routing* features, which support the automatic selection of preferred service options in response to offered traffic.

Digital Local Channel	Nonrecurring Charge		Monthly Rate
	First ($)	Additional ($)	($)
Per Local Channel			
2.4, 4.8, or 9.6K bps	340.00	118	50.00
56 kbps	340.00	118.00	70.00

Node Channel Termination

	First ($)	Additional ($)	Monthly Rate ($)
Per Local Channel			
2.4, 4.8, or 9.6K bps	48.00	42.00	11.00
56 kbps	48.00	42.00	30.00

Digital Interoffice Channel
Furnished between serving wire center and Node CO or between Node COs.

	Nonrecurring Charge ($)	Monthly Rate	
		Fixed ($)	Per Mile ($)
2.4, 4.8, or 9.6K bps:			
0-8 Miles	79.00	20.00	2.05
9-25 Miles	79.00	20.00	2.00
Over 25 Miles	79.00	20.00	1.95
56K bps:			
0-8 Miles	79.00	40.00	4.10
9-25 Miles	79.00	40.00	4.00
Over 25 Miles	79.00	40.00	3.90

Fig. 10.7 Intra-LATA synchronous digital private line rate example (*SOURCE: CCMI*).

IXC Switched Services

Inter-LATA switched services, provided through IXC POPs, enable sharing of IXC facilities and involve rate structures based on usage. Services and rate structures differ primarily in terms of the way in which users access POPs (i.e., through private line versus switched LEC facilities) and in accordance with user monthly traffic volume.

Inter-LATA MTS is the long distance service of choice for low-volume users. Here, access to IXC facilities in originating and terminating LATAs is gained through switched LEC network facilities. Users employ the same local loops used for local and intra-LATA toll switched service to access IXCs and complete inter-LATA MTS calls. LEC charges for this switched access are paid by the IXC and are included in the IXC MTS rates. Because of this, the IXC appears to offer service from originating to terminating CO, and mileage is measured accordingly (from originating to terminating CO rate centers).

In reality, of course, this is not the case, and the IXC provides end-to-end service using LEC originating and terminating transport, with IXCs fully compensating LECs for the use of their switched intra-LATA network services. Since these LEC services provide access to interstate services, they are provided under Switched Access tariffs which fall under FCC, not state public utility commission jurisdiction.

Figure 10.8 depicts a rate structure with mileage bands that have become standardized among competing IXC carriers, facilitating inter-LATA MTS rate comparison. Note that the daytime rates average between 21¢ and 23¢/minute over longer distances than those encountered in the intra-LATA examples above (which averaged 25¢/minute). The fact that the inter-LATA marketplace is largely deregulated and competitive, while the intra-LATA service is still monopolistic and regulated, drives the lower IXC rates. Certainly, keen competition and freedom from the regulatory mandate to provide universal local service has been a factor in the steadily downward trend in post divestiture inter-LATA rates.

As user traffic volume increases, network designs with originating and terminating shared access WATS can be considered as an alternative to inter-LATA MTS service. Physically, this service uses the same facilities as MTS. However, in exchange for volume commitments, IXCs offer discounted rates. Currently, WATS pricing is based on mileage and time of day. Hourly rates are presented in Fig. 10.9 for each time period, along with the applicable time period discounts for users subscribing to multi-location calling plans. WATS access line charges of $50 per line and a service group charge of $12 also apply.

To illustrate a typical shared access WATS application, consider the 120 hours/month user with calls in the 500 mile average range:

Day cost calculation:
    5400 Minutes @ $0.238/min = 1285.20 × 0.82 = $1053.86

(90 day hours per month)

Evening cost calculation:
    1500 Minutes @ $0.172/min = 258.00 × 0.92 = $ 237.36

(25 evening hours per month)

Direct Dial Rates

Rate Mileage	Day ($)		Evening ($)		Night/Weekend ($)	
	Initial Minute	Additional Minute	Initial Minute	Additional Minute	Initial Minute	Additional Minute
10	0.1700	0.1700	0.1200	0.1100	0.1051	0.1051
22	0.1800	0.1800	0.1300	0.1300	0.1139	0.1139
55	0.1900	0.1900	0.1300	0.1300	0.1208	0.1208
124	0.2100	0.2100	0.1450	0.1450	0.1208	0.1208
292	0.2100	0.2100	0.1450	0.1450	0.1223	0.1223
430	0.2300	0.2300	0.1457	0.1457	0.1256	0.1256
925	0.2300	0.2300	0.1495	0.1495	0.1306	0.1306
1910	0.2440	0.2440	0.1496	0.1496	0.1331	0.1331
3000	0.2459	0.2459	0.1496	0.1496	0.1357	0.1357
4250	0.3000	0.2900	0.2077	0.2010	0.1650	0.1650
Over 4250	0.3300	0.3200	0.2211	0.2144	0.1750	0.1750

Fig. 10.8 Inter-LATA MTS rate example (*SOURCE: CCMI*).

Intra-Mainland, Mainland-Alaska and Mainland-Hawaii

Rate Mileage	Initial 30 Seconds Or Fraction			Each Additional 6 Seconds Or Fraction		
	Day ($)	Evening ($)	Night ($)	Day ($)	Evening ($)	Night ($)
55	0.1020	0.0735	0.0565	0.0204	0.0147	0.0113
292	0.1120	0.0810	0.0630	0.0224	0.0162	0.0126
430	0.1190	0.0860	0.0670	0.0238	0.0172	0.0134
925	0.1190	0.0860	0.0670	0.0238	0.0172	0.0134
1910	0.1240	0.0900	0.0700	0.0248	0.0180	0.0140
3000	0.1240	0.0900	0.0700	0.0248	0.0180	0.0140
4250	0.1240	0.0900	0.0700	0.0248	0.0180	0.0140
Over 4250	0.1240	0.0900	0.0700	0.0248	0.0180	0.0140

Discount		
Day	Evening	Night
18 %	8 %	4%

Fig. 10.9 Inter-LATA shared access WATS rate example (*SOURCE: CCMI*).

Night cost calculation:
    300 Minutes @ $0.134/min =   26.80 × 0.96 = $ 25.73

(5 night hours per month)
Total Usage:
    120 Hours at a total cost of $1316.95

The daytime usage cost in this example is 19.5¢/minute. If three WATS access lines are ordered to carry this traffic, then 3×$50.00+$12=$162.00 or 2.3¢/minute is spent for access. The total unit cost is 21.8 cents/minute, an improvement over IXC MTS costs for customers with 120 inter-LATA hours/month of volume, but a less dramatic difference than that for the intra-LATA MTS/WATS example above.

Still larger users, with more than 1000 hours of traffic per month, can benefit from yet lower dedicated access WATS rates. Volume and term discounts are also offered as IXCs strive to attract large customers for extended periods. Figure 10.10 shows a typical, undiscounted dedicated WATS tariff in which call rates are based on mileage. In this case, mileage is measured from the carrier POP rate center to the terminating LEC CO rate center. Note that charges for terminating LEC transport between IXC POPs and called-party premises are covered under switched access tariffs (described earlier in this chapter). These latter charges are incorporated in IXC tariff rates, shown in Fig. 10.10.

For access from customer premises to the IXC POP, DS1 service (supporting 24 voice channels) costs approximately $1000/month and is adequate for 2000 hours of traffic. On a per-call basis, this access cost amounts to 0.8¢/minute. From Fig. 10.10's undiscounted entries, the average daytime usage rate is about 15¢/minute. Under a 36-month term, multi-location calling plan, this traffic level gains a 19% discount, yielding a net unit charge of 12.2¢/minute. Adding access charges, the total daytime unit cost of 13.0¢/minute represents a considerable improvement over other alternatives at this volume level.

The use of dedicated private line access to IXC POPs sets the stage for their flagship *virtual private network* (VPN) service offerings. Instead of leasing private lines between carrier POPs, the user shares the IXC carrier's network, pays usage charges, and yet enjoys equivalent benefits of facilities-based private networks, with the exception that no physical facilities are dedicated exclusively to any user.

The VPN rate structure is similar to dedicated WATS service, except that now there are more end-to-end call connection alternatives. As in any private network, users at locations served by dedicated access to IXC POPs can call each other, but unlike dedicated WATS, the dedicated access lines carry both originating and terminating traffic. Calls between such locations are referred to as *on-net* calls. Calls using switched access to IXC VPNs for either origination, termination, or both are termed *off-net*. This leads to four possible kinds of traffic:

1. On-net to on-net

2. On-net to off-net

3. Off-net to on-net

4. Off-net to off-net

Rate schedules are published for each traffic type. Figure 10.11 shows A, B, and C rate schedules for AT&T's Software Defined Network. Schedule A pertains to traffic type 4 and includes charges for both originating and terminating LEC switched access service. Schedule B pertains to traffic types 2 or 3 and includes a single LEC switched access service charge for either call origination or termination. Schedule C is used for traffic type 1 and includes no LEC access charges. In this case LEC private line access services are separately arranged and paid for.

Intra-Mainland, Mainland-Hawaii/Alaska, Hawaii-Alaska

Miles	Initial 18 Seconds Or Fraction			Each Additional 6 Seconds Or Fraction		
	Day ($)	Evening ($)	Night ($)	Day ($)	Evening ($)	Night ($)
55	0.0354	0.0249	0.0225	0.0118	0.0083	0.0075
292	0.0417	0.0291	0.0261	0.0139	0.0097	0.0087
430	0.0453	0.0318	0.0282	0.0151	0.0106	0.0094
925	0.0495	0.0345	0.0309	0.0165	0.0115	0.0103
1910	0.0531	0.0372	0.0330	0.0177	0.0124	0.0110
3000	0.0564	0.0393	0.0345	0.0188	0.0131	0.0115
4250	0.0564	0.0393	0.0345	0.0188	0.0131	0.0115
Over 4250	0.0564	0.0393	0.0345	0.0188	0.0131	0.0115

Fig. 10.10 Inter-LATA dedicated WATS rate example (*SOURCE: CCMI*).

SCHEDULE A

	Initial 18 Seconds Or Fraction			Each Additional 6 Seconds Or Fraction		
Miles	Day ($)	Evening ($)	Night ($)	Day ($)	Evening ($)	Night ($)
55	0.0492	0.0489	0.0489	0.0164	0.0163	0.0163
292	0.0543	0.0489	0.0489	0.0181	0.0163	0.0163
430	0.0579	0.0489	0.0489	0.0193	0.0163	0.0163
925	0.0603	0.0489	0.0489	0.0201	0.0163	0.0163
1910	0.0636	0.0489	0.0489	0.0212	0.0163	0.0163
3000	0.0636	0.0489	0.0489	0.0212	0.0163	0.0163
4250	0.0636	0.0489	0.0489	0.0212	0.0163	0.0163
Over 4250	0.0636	0.0489	0.0489	0.0212	0.0163	0.0163

SCHEDULE B

	Initial 18 Seconds Or Fraction			Each Additional 6 Seconds Or Fraction		
Miles	Day ($)	Evening ($)	Night ($)	Day ($)	Evening ($)	Night ($)
55	0.0327	0.0270	0.0270	0.0109	0.0090	0.0090
292	0.0487	0.0312	0.0312	0.0129	0.0104	0.0104
430	0.0426	0.0339	0.0339	0.0142	0.0113	0.0113
925	0.0465	0.0366	0.0366	0.0155	0.0122	0.0122
1910	0.0501	0.0393	0.0392	0.0167	0.0131	0.0131
3000	0.0528	0.0411	0.0411	0.0176	0.0137	0.0137
4250	0.0579	0.0447	0.0447	0.0193	0.0149	0.0149
Over 4250	0.0600	0.0462	0.0462	0.0200	0.0154	0.0154

SCHEDULE C

	Initial 18 Seconds Or Fraction			Each Additional 6 Seconds Or Fraction		
Miles	Day ($)	Evening ($)	Night ($)	Day ($)	Evening ($)	Night ($)
55	0.0186	0.0132	0.0132	0.0062	0.0044	0.0044
292	0.0246	0.0171	0.0171	0.0082	0.0057	0.0057
430	0.0282	0.0198	0.0198	0.0094	0.0066	0.0066
925	0.0324	0.0228	0.0228	0.0108	0.0076	0.0076
1910	0.0360	0.0255	0.0255	0.0120	0.0085	0.0085
3000	0.0390	0.0273	0.0273	0.0130	0.0091	0.0091
4250	0.0438	0.0306	0.0306	0.0146	0.0102	0.0102
Over 4250	0.0456	0.0321	0.0321	0.0152	0.0107	0.0107

Figure 10.11 Virtual Private Network (VPN) rate example (SOURCE: *CCMI*).

As described in Chapters 8 and 9, VPN is a true network service with the objective to be indistinguishable, by users, from facilities-based private networks. As such there is a wide range of selectable features, each with defined charges. Since this service targets the largest customers, high-volume and long-term commitment plans are aggressive. For example, in exchange for a four-year, 200-million-minute-per-year commitment, AT&T discounts monthly schedule rates by up to 37% (dedicated access is never discounted). For normal traffic mixes, very large corporate users can attain average unit charges of 9.5 cents per minute, including access. While the break-even point between VPN and other alternatives occurs at much lower levels of traffic, and IXCs actively pursue smaller VPN customers, it is important to note that only the largest attain unit charges as low as the above example.

IXC Private Line Services

Like their intra-LATA counterparts, tariffs for IXC private lines are categorized as either analog or digital. Along with all other inter-LATA services, both types share the local channel/interoffice channel physical and tariff structures. In dedicated private line service, IXCs only provide physical facilities for interoffice channels between POPs in different LATAs. Dedicated local channel access connections between customer premises and IXC POPs must be provided by LECs or by competitive bypass carriers. Local channel service is provided by the LECs under Special Access tariffs. As with switched access services, since these services are used for accessing interstate services, the tariffs fall under FCC jurisdiction.

Users can order local channels directly from LECs; however, it is common practice for IXCs to offer rate schedules which pass along LEC charges. In these cases, for a small monthly fee, IXCs assume responsibility for ordering and coordinating local channel circuits, providing users with a single point of contact for the service. AT&T's Tariff #11 compiles local channel rates for Bell and independent LECs in all 50 states. Rates in Tariff #11 closely reflect LEC special access rates and, while Tariff #11 is readily available, individual LEC access tariffs are not as accessible.

Figure 10.12 shows physical and tariff rate elements for IXC private line services. There are three rate elements. First, mileage-dependent charges are assessed for interoffice channels (IOCs). Next are charges for central office connections (COCs), which are charges for connecting IOCs to other facilities such as LEC provided local channels. As shown in the illustration, a COC is required at each end of an IOC. Finally, charges are assessed for certain interoffice channel options such as line conditioning, signaling, diverse routing for higher reliability, etc. AT&T's Tariff #9 is used below to represent IOC, COC, and optional elements of IXC private line tariffs. AT&T's Tariff #9 basically addresses those elements of inter-LATA private line services supported by IXC owned facilities.

Figure 10.13 depicts local channel rate elements of inter-LATA private line service supported by LEC-owned facilities and is consistent with AT&T's Tariff #11. Local channel charges are mileage-dependent. *Access coordination fees* (ACFs) cover IXC costs for providing LEC coordination, as mentioned above. *Access multiplexing charges* apply to lower rate (2.4–56 kbps) digital access circuits which must be multiplexed into higher rate, such as DS1 transport services between IXC POPs.

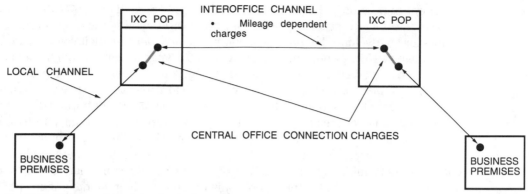

Fig. 10.12 IXC interoffice channel rate elements.

A radical change is occurring in facilities provisioning for dedicated circuits as the underlying IXC physical plant evolves from analog to digital transmission. In the past, analog circuits and low-rate digital circuits provided over a separate, synchronous network were the norm. Today DS1 service facilities dominate, and most analog voice private line service is actually delivered using 64-kbps-channel DS0 channels derived from DS1 digital facilities. Consequently, IXCs now offer 64 kbps digital private line service to users at analog prices. While performance specifications are not as high as with the older synchronous networks, 64 kbps DS1-derived circuits have proven adequate for data communications at a much lower cost. IXCs have expanded these offerings to include multiples of 64 kbps DS0 service in increments up to 768 kbps. The resulting service, commonly known as Fractional T1 service, is described in Chapter 3.

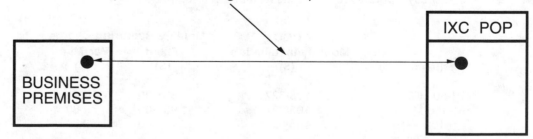

Fig. 10.13 Local channel rate elements.

Heavy competition in the inter-LATA arena has precipitated up to 57% discounts for high volume, long-term commitments for private DS1 digital services. In contrast, discounts for high-volume, long-term commitments for lower-rate digital and analog services range to 15%. At present, the situation is so dynamic that thorough readings of current tariffs, and prudence in considering long-term commitments are mandatory.

The rate tables for IXC private line services follow the same format as the newer intra-LATA rate tables; that is, the applicable mileage band is determined and fixed and mileage charges are computed. Figure 10.14, a representative table for DS1 circuits, illustrates IOC and COC charges. Note that originating and terminating COC charges as shown in Fig. 10.12 and any optional charges must be added to IOC charges to determine total inter-LATA costs. Some IXC discount plans apply to total inter-LATA charges, while others apply only to the IOC components. Figure 10.15 is a representative rate table for fractional T1 services, including equivalent analog voice-grade 64 kbps DS0 service.

Figure 10.16 shows the entries for a typical state from AT&T's FCC Tariff #11 offering voice-grade local channels. Rate Schedule 1 reflects the BOC rates. The designation *1W* is for one-way circuits, and *2W* is for two-way circuits (calls may be made both into and out of a location). In almost all applications, the 2W rates will apply. Rate Schedule 2 is for independent LECs and has rates that vary by LATA. Note that Bell-operating companies follow Rate Schedule 1, whereas independent telephone companies follow Rate Schedule 2.

As in the intra-LATA case, multiplexing is offered as an option at an IXC POP and DS1 interoffice channels can connect to 24 voice-grade local channels through multiplexing.

IXC Special Contract Services

There is an increasing trend to bundle both switched services and dedicated services for large customers into a single, integrated package, offered under a special contract at rates generally beneath discounted rates for each service individually. These contracts are usually the result of competitive procurements. In the deregulated IXC environment, only AT&T is required to file tariffs justifying its rates. The FCC has granted considerable leeway in competitive situations, allowing AT&T to file special contracts under FCC Tariff #12 and, more recently, as Contract Tariffs. The exact boundaries as to what can be included in AT&T special contracts and what cannot are constantly changing. Nevertheless, it is clear that large users are negotiating increasingly favorable terms as the result of IXC competition.

Mileage	COC Monthly/Installation ($)	Monthly Recurring Charge Fixed ($)	Per Mile ($)
1-50 Miles	106/327	1,861.00	10.60
51-100 Miles	106/327	1,901.00	9.80
101 +	106/327	2,111.00	7.70

Fig. 10.14 Inter-LATA DS1 interoffice channel rate example (*SOURCE: CCMI*).

INTEROFFICE CHANNEL CHARGES

Channel (bps)	Mileage Band	COC Monthly/ Installation ($)	Fixed ($)	Per Mile ($)
9.6K/56K/64K	1 - 50	21.25/207.00	75.72	3.00
	51 - 100	21.25/207.00	156.22	1.39
	101+	21.25/207.00	261.22	0.34
128K	1 - 50	31.50/527.00	145.13	5.69
	51 - 100	31.50/527.00	298.13	2.63
	100+	31.50/527.00	496.13	0.65
256K	1 - 50	37.00/527.00	291.32	11.35
	51 - 100	37.00/527.00	596.32	5.25
	101+	37.00/527.00	991.32	1.30
384K	1 - 50	42.25/527.00	436.44	17.04
	51 - 100	42.25/527.00	895.44	7.86
	100+	42.25/527.00	1,489.44	1.92
512K	1 - 50	47.50/527.00	552.13	21.52
	51 - 100	47.50/527.00	1,131.63	9.93
	101+	47.50/527.00	1,881.63	2.43
768K	1 - 50	52.75/527.00	828.71	32.26
	51 - 100	52.75/527.00	1,697.21	14.89
	100+	52.75/527.00	2,821.21	3.65

Fig. 10.15 Inter-LATA fractional T1 rate example (*SOURCE: CCMI*).

Network Design Optimization

Previous sections of this chapter provide a sampling of the wide range of available public and private network services. Each example describes business applications that capitalize on particular service capabilities and cost structures. In many cases, the examples interpreted performance and cost benefits relating to single-user locations. For most large users, the problem is to simultaneously consider all corporate locations, and design networks to optimally meet both site specific and inter-location telecommunications requirements. It is evident that this involves selecting the best mix of public and private, premises and network services, on a total, enterprise-wide basis.

Optimization criteria derive from the basic traffic engineering principles and lessons set forth in Chapter 7. Quickly recapitulated, cost-effectiveness in telecommunications occurs when resource utilization is high. With large amounts of traffic,

Rate Schedule 1 - BOC LEC Rates

			Monthly		
LATA	Element	Mileage	Fixed ($)	Per Mile ($)	Installation ($)
438	1W 1LNK9	0	68.69	N/A	733.11
440		1-4	73.89	3.24	821.03
442		5-8	75.68	3.24	821.03
444		9-25	77.44	3.24	821.03
446		26-50	79.21	3.24	821.03
		Over 50	80.97	3.24	821.03
	2W 1LNL9	0	108.24	N/A	733.11
		1-4	113.44	3.24	821.03
		5-8	115.22	3.24	821.03
		9-25	116.98	3.24	821.03
		26-50	118.76	3.24	821.03
		Over 50	120.52	3.24	821.03

Rate Schedule 2 - Independent LEC Rates

			Monthly		
LATA	Element	Mileage	Fixed ($)	Per Mile ($)	Installation ($)
438	1W 1LNK9	0	64.82	N/A	464.89
444		Over 0	64.82	10.53	
446					
	2W 1LNL9	0	101.31	N/A	464.89
		Over 0	101.31	10.53	
440	1W 1LNK9	0	83.29	N/A	653.42
		Over 0	136.36	3.63	
	2W 1LNL9	0	129.47	N/A	653.42
		Over 0	182.54	3.63	

Fig. 10.16 Voice-grade local channel rate examples (*SOURCE: CCMI*).

high resource utilization and good grades of service go hand in hand. Without large amounts of traffic, high resource utilization is only achieved at the expense of grade of service. As seen in several of the examples above, during busy hours, least-cost routing among combinations of private and public network services permits large users to achieve savings and maintain good GOS in situations where the use of private services alone would have resulted in unacceptable GOS. In particular, obtaining the proper mix of private line and switched services is often the key to a cost-efficient network.

Chapter 4 describes how PBXs, which concentrate user premises traffic, permit users to efficiently employ a smaller number of access lines than is required when each telephone is connected separately to a CO. Large users may employ private *hub switches* to concentrate traffic within a metropolitan area. Hub switch functions can include switching traffic among the PBXs connected to it in the same city and forwarding traffic to hub switches in other cities. Due to the concentration of traffic from satellite PBXs at hub switches, private line connections among satellite PBXs via the hub switch prove economical at traffic volumes far less than would be needed to interconnect all PBXs directly (*full mesh connectivity*). Similarly, since hubs aggregate traffic from multiple satellite PBXs, private line intercity services between hubs may produce savings where intercity private line service between individual satellite PBXs would not have sufficient traffic to prove economical. Aggregation of traffic at hubs may also justify private line connections to IXC POPs, taking advantage of lower cost intercity services such as dedicated WATS, where again no single site may have sufficient traffic alone to warrant such connections.

As noted, the trend is toward LECs, IXCs and competitive access providers offering virtual private network, and other integrated switched and private line special contract alternatives for user-owned and -operated satellite and hub switch enterprise networks. In large networks, trade-off factors are complex and interrelated, spanning technical, management, maintenance, and financial realms. Normally, the user's focus should be first to quantify requirements, as outlined in Chapter 7, and then to obtain analyses from service providers and/or third-party network design professionals with modern computer aided tools, to select the best mix of services. The purpose of this book is not to enable readers to directly accomplish complex network designs, but rather to ensure that they are aware of the trade-off areas and are able to judge the quality of service proposals.

11

Selecting Premises Services

This chapter discusses practical aspects of planning and implementing premises telecommunications products and services. Principal activities in this regard include determining user requirements, defining the project, preparing technical specifications and *requests for proposals* (RFPs), evaluating proposals, and selecting approaches that best meet business needs. Following requirements determination and project definition, the next step is RFP preparation. Three key RFP ingredients are required to ensure the success of the balance of the selection and implementation process, namely:

- Accurate, unambiguous technical, management and cost requirements articulated in industry-accepted terminology
- Explicit instructions on specifying mandatory proposal information content
- Clearly defined proposal format instructions

For the specified system to conform with user requirements, RFP statements must be consistent with those needs and phrased in a manner that minimizes possible misinterpretation. Next, it is crucial to ask the right questions of prospective bidders. If information needed to conduct thorough and systematic bid evaluations is not explicitly requested, it probably will not be included in the proposal, particularly if it might expose the less competitive aspects of a bidder's offering. Finally, lack of uniformity in proposal formats often creates insurmountable obstacles in comparing and evaluating competing bids.

Although an in-depth treatment of the telecommunications product/service selection process is beyond the scope of this book, this chapter outlines major RFP sections, with example exhibit and bidder formats for key areas. The outline and the formats have evolved over years of successful use in actual procurements. The chapter closes with important lessons learned from recent procurements.

RFP Example Outline

RFP size and composition varies with the size and complexity of the telecommunications project. An RFP for a large installation (greater than 1000 workstation locations) covering voice, data and premises distribution services and systems includes the following:

```
1.0  Introduction
     1.1  Purpose of the RFP.
     1.2  Site description, referencing any attached drawings.

2.0  General Requirements
     2.1  Project scope.
     2.2  Project definitions.
     2.3  Proposal format.
     2.4  Treatment of proprietary data.

3.0  Evaluation Criteria
     3.1  How the proposal will be evaluated, including areas of focus, prece-
          dence, weighting, and necessary bidder qualifications.

4.0  Specific Requirements and Technical Specifications
     4.1  Voice System.
     4.2  Data System.
     4.3  Premises Distribution System.
     4.4  Fire/Life Safety/Security System (as required).
     4.5  Video System (as required).

5.0  Environmental
     5.1  Requests details on support needs of proposed system, required of
          the buyer or the seller.

6.0  Installation
     6.1  Implementation schedule.
     6.2  Project plans.
     6.3  Contractor responsibilities.
     6.4  Project logistics.
     6.5  Contractor compliance issues.
     6.6  Testing and acceptance goals.
     6.7  Training.
     6.8  Project documentation.

7.0  Maintenance and Support
     7.1  Maintenance concept.
     7.2  Failure categories.
     7.3  Response times.
     7.4  Spare parts.
     7.5  Maintenance information.
     7.6  Tools and test equipment.
     7.7  Equipment installed base.
     7.8  Contractor work scheduling.
     7.9  Progress reporting.
     7.10 Contingency planning.
     7.11 Moves, adds, and changes (MAC) work philosophy.
     7.12 Service intervals.
     7.13 Post-warranty services.
     7.14 Additions to warranty and maintenance agreements.

8.0  Acquisition
     8.1  Payment schedule.
     8.2  Contractual additions.
     8.3  Assumption of risk.
     8.4  Buyer indemnification.
```

Appendices and Attachments

This portion of the RFP contains additional information necessary for bidders to submit a proposal, e.g., site survey schedule, architectural, mechanical/electrical/interior layout (furniture) drawings, together with details of any unusual or critical customer requirements.

As outlined above, the RFP calls out technical specifications to be addressed in bidder proposals covering the engineering, installation, testing, and cutover (placing into operational service) of the particular telecommunications project. Each major portion of the required deliverable is described in the RFP. RFP instructions should direct bidders to demonstrate how their proposed telecommunications solutions address the following general characteristics:

- Incorporation of industry standards
- Application of modular hardware/software designs
- Use of commercial off-the-shelf products and services
- Ease of management
- Optimized life-cycle costs/benefits

Content of the sections in the above list are implicit in the titles. It has become common practice in the last several years to break Section 4, Specific Requirements and Technical Specifications, into at least three separate parts, one each for voice, data, and premises distribution requirements. Less than a decade ago, voice and data vendor products demanded unique wiring and cabling. Consequently, premises distribution system requirements were incorporated separately into voice and data system requirements sections.

As noted in Chapter 6, the current approach favors *universal wiring*, intended to support vendor voice and data systems as well as future upgrades to new technologies. Hence most contemporary RFPs include stand-alone PDS sections, which indirectly impose compliance requirements on voice and data systems. Recall from Fig. 6.4 on page 113 that per-line PDS costs can be a major portion of per-line PBX system costs, and that in cases where fiber cable to desktops is required, PDS costs can exceed PBX costs—another factor justifying separate PDS section treatment.

Within the voice part of the Specific Requirements section of the RFP, there should be one or more pages devoted to a detailed description of the required system size at cutover, and its expansion capacity throughout the operational life cycle. A size description for a typical digital telephone system is shown in Fig. 11.1. Note that the specification can be filled by either a PBX switching system or by Centrex service. It is important not to slant the description, in order to retain maximum flexibility of choice.

Beyond sizing, where records are available, bidders should be provided with traffic data (such as busy day/busy hour data provided in Figs. 7.7 and 7.8, found on page(s) 129 and 130), and the overall division of voice traffic (and data traffic, where applicable), both for cutover and projected expansion capacity levels. The division of traffic is expressed in terms of *intrasystem* (station-to-station), *incoming* (calls originating outside the system), and *outgoing* (calls terminating outside the system).

Information on division of traffic makes it easier for the bidders to propose the most efficient, cost-effective trunking arrangements meeting individual business needs. If this information is not available, bidders will generally presume an equal division of traffic: one-third intrasystem, one-third incoming, and one-third outgoing.

System grade of service (GOS), discussed in Chapters 4 and 7, should also be specified in the RFP. In cases where records from a current system exist, GOS requirements can be expressed in terms of available traffic statistics. In the absence of traffic statistics, a rule of thumb is to specify a GOS of P=0.005 (1 call in 200 blocked) per station for intrasystem calls, and P=0.01 (1 call in 100 blocked) per station for incoming and outgoing calls.

Recall from Chapters 4 and 7 the relationship between GOS and centi-call seconds (CCS). CCS expresses the average time during the busy hour that a station line is busy (in hundreds of seconds). A line used 100% of the time represents 36 CCS. As stated in Chapter 7, typical per-line voice and data offered traffic loads are 4-6 and 18-36 CCSs, respectively. Line and trunk CCS traffic loading is the basis on which GOS is specified and measured.

Bidders should be requested to state the combined (based on division of traffic) per-voice (or data) station CCS capacity proposed in the wired-for system configuration. The wired-for configuration represents system traffic handling resources available at cutover to support future expansion without adding shelves or cabinets. For large systems, bidders should also be asked to state the maximum number of busy hour call attempts that the proposed system can handle.

The average telephone system is engineered to provide between four and seven CCS per line at P=0.01. For greater traffic-handling capacity, optional pricing may apply, even for Centrex service. Bidders should be requested to state any optional CCS levels available in the proposed system, together with associated price premiums.

Next, it is important to pay close attention to the structure of the voice system dialing plan. Cutovers have been delayed because of clashes between telephone numbers and access codes for features and network facilities. Simple dialing plans are best, with single digit access codes used wherever possible. Feature access codes generally include a * or # prefix, followed by one or two digits. So long as these prefixes are used, clashes with station numbers and network facilities will be avoided. Station numbers vary in digit count, based upon the size of the system. A three-digit dialing plan supports approximately 800 stations; a four-digit plan supports 8,000; a five-digit plan supports 80,000. Most PBXs employ three-digit or four-digit station-to-station dialing. A example PBX system dialing plan is shown in Fig. 11.2.

ITEM	AT CUTOVER		SYSTEM CAPACITY
	EQUIPPED	WIRED	
Stations:			
• Analog	1000	2000	**
• Digital	200	2000	**
Total	1200	4000	**
Off-Premises Stations			
(All analog)	100	300	**
DID Trunks	40	120	**
DOD Trunks:			
• Band 5 WATS	20		
• Local, DDD, IDDD	15		
Total	35	100	**
Two-Way Trunks:			
• LDN	20		
• DISA	10		
Total	30	90	**
Radio Paging Trunks	10	30	**
Tie Lines:			
• Incoming	20	60	**
• Outgoing	20	60	**
Total	40	120	**
PFCT Trunks (All Two-Way Trunks)	2	6	**
Direct DS1 Digital Trunk Terminations			**
Attendant Consoles	2	4	**

*   Bidder shall quote prices for the following options:
-   All analog stations
-   Analog/digital mix shown above
** Bidder shall state the maximum size to which the system can be expanded, e.g., through cabinet additions, etc.

> **NOTES:**
> 1. Equipped = Installed and working or spare.
> 2. Wired = Can be equipped by adding PCBs only.
> 3. PFCT = Power failure cutthrough trunks.
> 4. DISA = Direct inward system access
> 5. LDN = Listed directory number.
> 6. All tie lines shall be analog 4-wire E&M.

Fig. 11.1 Example RFP telephone system sizing exhibit.

DIGITS	**SERVICE**
1	Reserved
2-XXX	Stations
3-XXX	Stations
4-XXX	Stations
5-XXX	Stations
6	Reserved
7	Reserved
8-XXX-XXXX	On-Net
9-XXX-XXXX	Local or On-Net
* 9-XXX-XXX-XXXX	Long Distance or Off-Net
0	PBX Attendant

* May require dialing "1" or "0" if not automatically inserted by switching equipment.

Fig. 11.2 Example PBX dialing plan.

The station-to-station dialing feature of a Centrex switch has potentially wider scope than that of a PBX. The CO switch's normal seven-digit dialing plan (single NXX) can serve around 800,000 separate telephone numbers. Thus, if a Centrex customer has several offices, all served by the same CO switch, it is relatively easy to develop an abbreviated station-to-station dialing plan of five or fewer digits, if sufficient station (-XXXX) numbers are available within one of the exchange (NXX) codes provided by that CO switch.

Under such a uniform dialing plan, the customer would make local calls using seven digits, intracompany calls using three, four, or five digits, and intercom calls using one or two digits (generally using CPE telephone equipment).

Dialing-plan capacity can give Centrex an advantage over a multiple PBX configuration, which requires additional hardware in the form of tie lines and/or OPXs (both of which are expensive to operate and maintain) and additional tandem network software to integrate multiswitch operations.

The PBX system dialing plan shown in Fig. 11.2 also applies to the Centrex environment. As previously noted, in a distributed (city-wide) Centrex system using multiple COs, planning for a universal dialing plan could be complicated by the lack of available CO exchange codes, station numbers within exchange codes, and differences in features and operations from one CO switch to another.

The Voice part of the Specific Requirements and Technical Specifications section of contemporary RFPs usually includes subsections for requirements in the following areas, some of which were just discussed:

- System line and trunk sizing
- Cutover date
- System expansion
- Division of traffic
- Grade of service
- Dialing plan
- Survivability
- Station equipment
- Network connections
- System management
- Types and quantities of deliverables
- Computer common control architecture
- Power supplies (including any requirements for uninterruptible power supplies)
- Network interface characteristics
- Operational parameters
- Peripheral equipment
- Reliability/maintainability/availability
- Regulatory compliance
- System security
- Features

Feature specifications normally include *required features* and *bidder-suggested features*. The RFP should provide a brief description of each required feature to clarify expected general characteristics. Where necessary, specific details on feature operation and capacities are requested from the bidders. Bidders should be advised to state their own nomenclature for each feature and to describe any differences in feature operation from the descriptions appearing in the Feature section of the RFP.

At the beginning of the Feature section of the RFP, the following information should be requested from the bidders, where applicable, for each required feature:

- Method of activation and deactivation
- Capacities, such as maximum number of intercom groups/codes per group, speed calling lists, etc.

- Features that preclude or limit the use of other features, together with a statement of the degree of impact

- Features requiring a station line assignment

- Features that cannot be provided to all station users at a particular system size (cutover configuration versus wired capacity)

It is important to request that bidders state whether a particular feature is standard or optional in the proposed system, and if optional, whether it requires hardware, software, labor, and/or other chargeable cost elements. Optional costs should be identified by the bidder, and itemized in the Pricing section of the proposal, but not incorporated into the total system price.

Where a bidder suggests additional features which are not specified in the RFP, but which the bidder deems appropriate to the system, such features should be described and associated prices quoted as above, but not incorporated into the total system price.

Lessons Learned from Recent Procurements

This section examines the telecommunications product/service selection process by analyzing proposals from a large PBX manufacturer and a midwest RBOC, who submitted bids for a fourth-generation PBX and a digital city-wide Centrex system, respectively. The telecommunications project involved a new high-technology industrial park. The project was scheduled to begin with a single building served by 120 lines and to grow to an eventual 5000-line, 25-building campus. Vendors were instructed to depict per-line proposal costs as a function of project growth. To attract the broadest spectrum of tenants, particularly those engaged in the provision of high-technology products and services, the park operator wanted to evaluate a range of modern telecommunications capabilities, together with various levels of park management involvement in providing shared tenant services. The discussions that follow, focus on the PDS and switching equipment portions of the project.

Overall, the objective of competitive procurements is to identify and select alternatives optimized to meet functional requirements and to minimize costs. Comparison and ranking approaches must therefore evaluate technical performance against specified criteria, employing discriminators that support a rational, repeatable selection making process. Because today's technology equips PBX vendors and RBOC/LECs alike with the ability to satisfy technical requirements (rarely do proposals take exception to any major performance stipulation), discrimination among bid alternatives comes down to quantitative cost differences and usually more qualitative technical management and risk factors.

The PDS bids for the above project are a case in point. The PBX vendor and the RBOC submitted nearly identical PDS price quotes, which are summarized in Fig. 6.4 on page 113. The RFP specified unshielded twisted-pair (UTP) wiring conforming to standards described in Chapter 6. In addition, the RFP included a mandatory optional requirement for FDDI-capable fiber optic horizontal wiring direct to desktops. The request for fiber quotes was intended to assist the park manager in deciding whether to make an additional initial investment to increase the likelihood of attracting tenants with needs for 100 Mbps desktop connections (EIA/TIA TSB-36 had not yet been published). The results of the PDS bid and evaluation process are discussed below.

The upper portion of Fig. 6.4 reflects costs associated with four different PDS cabling approaches, labeled A, B, C, and D, in perspective with those of switching equipment. Note that the switching cost per station decreases as system size increases, while the PDS cost per station remains constant.

Illustrated in the lower portion of Fig. 6.4, PDS alternatives A and B provide unshielded twisted-pair (UTP) copper wire only, conforming to IEEE 10BaseT and EIA/TIA standards. Alternative B implements the EIA/TIA standard of two UTP wiring runs terminating in separate RJ-45 jacks at each workstation outlet. Each jack is capable of supporting voice or data/LAN operations at a 10 Mbps rate, with developing standards potentially raising the rate to 100 Mbps. Alternative C adds two strands of multimode fiber to each workstation outlet, but leaves them *dark,* or unterminated, for future use. Alternative D terminates them for immediate use.

Figure 6.4 furnishes some useful insights. First, PDS costs did not materially change with growth, since they are linear, and driven by per-foot labor and material costs not subject to economies of scale. Also, the economy of UTP horizontal wiring is underscored, in light of the large-cost "step function" presented by fiber-to-the-desktop. Note that a hidden technical danger in alternative C is that without terminating the fiber, it cannot be tested, and therefore cannot be certified for satisfactory performance.

In the interbuilding PDS backbone cable application, however, fiber cable became extremely cost-effective. Potentially high per-building station counts were involved in this campus wiring scenario. The costs of installing long runs of multiple UTP backbone cables consisting of up to 1200 pairs each, together with the costs of constructing ductbank (typically around $100 per linear foot) to house the cables, made fiber very attractive. Used for linking remote switching modules (one per building) with a single two- or four-pair multimode fiber cable, the PDS cost-per-station using fiber backbone cable was a small fraction of the UTP copper alternative. For reasons explained in Chapter 4, PBXs in remote switch module configurations are better equipped than Centrex to capitalize on such benefits.

In the end, the PDS selection decision was not based on technical or cost differences in bid responses, but rather on options to:

1. Install dark or terminated fiber

2. Not plan for 100 Mbps desktop service

3. Bank on the success of FDDI-over-UTP technologies now under development for 100 Mbps desktop service

As noted in Chapters 6 and 13, a number of reputable companies—united in a UTP Forum—are confident that UTP can effectively and economically support 100 Mbps rates in horizontal PDS wiring. Yet developments had not progressed to the point where the park manager could rely on fixed price quotes to assist him in his selection decision. As indicated at the start of this section, selection could not be based on technical or quantitative cost differences among competing proposals, but rather had to be based on harder-to-quantify performance risks associated with an emerging technology.

Similar results surrounded the choice between PBX and Centrex for premises switched services. A 60-month, one-dollar buyout lease approach for the PBX alternative had been specified in the RFP to facilitate proposal comparison with the Centrex bid, and to normalize requirements for initial capital investment. On a technical, cost, and implementation schedule basis, the Centrex and PBX approaches proved nearly identical. Figures 11.3 and 11.4 summarize categories of PBX and Centrex bid costs as the project grows to 5000-line capacity. The RFP for this procurement requested bidders to provide proposal information in the formats illustrated in Figs. 11.3 and 11.4, in order to facilitate a thorough, service-by-service complete life cycle (initial, recurring, operations, maintenance, administration, etc.) cost comparison. Note that the total cost per line is nearly equal over the expansion period.

Depending on the individual system, life cycle costs are influenced by the following:

- *System management.* Traditionally less costly for Centrex, this is now changing with more CPE management provided by the LEC under tariff or contract.

- *Switching equipment maintenance and repair.* These costs are included in the Centrex tariff but must be added for the PBX after the warranty period expires.

- *Moves, adds, and changes.* Incurred for both Centrex and PBX, but may be lessening for Centrex because of trend toward "total package" LEC support.

In the campus system described above, system management and maintenance support provided at no additional charge by the LEC had to be added to the PBX costs, resulting in nearly equal total costs. For this reason, system selection rationale

Date	BASIC COSTS			O/M&A COSTS		TOTAL COSTS	
	#Additional Stations/ Total Stations	Subtotal Cost/Month	Subtotal Cost/ Station/Month	Subtotal Cost/Month	Subtotal Cost/ Station/Month	Total Cost/Month	Total Cost/ Station/Month
Start Date	120/120	$3478	$28.98	$2675	$22.29	$6153	$51.27
3 Months	425/545	$12670	$23.25	$10745	$19.72	$23415	$42.96
6 Months	800/1345	$21488	$15.98	$18360	$13.65	$39848	$29.63
Year 1	445/1800	$26463	$14.70	$19216	$10.68	$45679	$25.38
Year 2	3200/5000	$61118	$12.22	$40096	$8.02	$101214	$20.24
Year 3	0/5000	$61118	$12.22	$54871	$10.97	$115989	$23.20
Year 4	0/5000	$61118	$12.22	$56160	$11.23	$117278	$23.46
Year 5	0/5000	$61118	$12.22	$57834	$11.57	$118953	$23.79
Year 6	0/5000	$11935	$2.39	$59503	$11.90	$71439	$14.29

Fig. 11.3 PBX life cycle costs (60 month lease with $1 buyout).

| Date | BASIC COSTS | | | O/M&A COSTS | | TOTAL COSTS | |
	#Additional Stations/ Total Stations	Subtotal Cost/Month	Subtotal Cost/ Station/Month	Subtotal Cost/Month	Subtotal Cost/ Station/Month	Total Cost/Month	Total Cost/ Station/Month
Start Date	120/120	$4900	$40.84	$290	$2.42	$5190	$43.25
3 Months	425/545	$16515	$30.30	$1323	$2.43	$17838	$32.73
6 Months	800/1345	$36010	$26.77	$3269	$2.43	$39279	$29.20
Year 1	445/1800	$46401	$25.78	$4378	$2.43	$50779	$28.21
Year 2	3200/5000	$119777	$23.96	$12163	$2.43	$131939	$26.39
Year 3	0/5000	$119777	$23.96	$12163	$2.43	$131939	$26.39
Year 4	0/5000	$119777	$23.96	$12163	$2.43	$131939	$26.39
Year 5	0/5000	$119777	$23.96	$12163	$2.43	$131939	$26.39
Year 6	0/5000	$119777	$23.96	$12163	$2.43	$131939	$26.39

Fig. 11.4. Centrex life cycle costs.

shifted toward assessment of two risk factors. The first was the added client burden of managing the PBX and third party support contractors. The second was effectiveness and supportability of the PBX after 60 months of use.

The key decision driver with respect to the first factor was that the LEC offered significant, second-year-and-out support services that were bundled into the Centrex line rate. The key decision driver with respect to the second factor was that significant cost savings could be achieved in the outyears, if the PBX's useful life could be extended past the 60-month buyout point.

As in the PDS example, the selection decision could not be based on technical or cost differences among competing bids. It became dependent on assessment of:

1. Residual risks associated with operating and managing a large telephone system (even after competent third-party support contractors had been identified and related costs had been taken into account)

2. Risks involved with the predictions of technological sufficiency and continued supportability of five-plus-year-old PBXs

PBX vendors make strong arguments that large telephone system operations present business opportunities, not risks, and that their modular, software-driven PBX designs ensure seamless expansion far into the future. LECs counter with equally convincing arguments that economics of scale favor using shared public facilities, that customer operation of large telephone systems is prone to failure and that Centrex service provides the ultimate insurance against technical obsolescence.

Although differences in proposed technical and management approaches will continue to challenge the evaluator on a project-by-project basis, the above examples demonstrate the important role that RFP content and format instructions play in the telecommunications product/service evaluation and selection process.

Data Telecommunications Fundamentals

Chapter 12. Basic Concepts and Techniques

Chapter

12

Basic Concepts & Techniques

This chapter presents fundamentals of data communications in preparation for a discussion of data services in forthcoming chapters. It relies on the introductory material presented in Part I for terms of reference and background. In particular, it builds on the definitions and descriptions of digital electrical signals, binary bits, error detection and correction, data terminal equipment (DTE), digital carrier, time division multiplexing (TDM), and digital circuit switching presented in Chapters 1 through 4.

While both voice and data services can be supported on common network elements, significant differences in requirements, combined with the traditionally analog nature of voice networks has resulted in the development of largely separate voice and data facilities at private, public, local, and long-distance network levels.

A significant difference between voice and data service is the extent to which human intervention is required to ensure end-to-end communications integrity, including diagnosis and recovery under failed or inadequate service conditions. For example, if an American places a telephone call to Japan that is answered by someone who cannot speak English, human intelligence is relied upon to seek an interpreter or to take alternative action. Similarly if a call cannot be completed due to a network failure, a human determines the problem and takes corrective steps.

By contrast, data services are provided with minimal human intervention. As a consequence, more elaborate mechanisms are required to ensure that transmitting and receiving DTEs "speak the same language," and that service restoral actions are promptly taken under network failure conditions. This generally requires higher levels of hardware and software compatibility among DTEs and intervening data network elements, than is required in voice networks.

For private data networks, it might be feasible to specify hardware and software from a single source, achieving compatibility through proprietary design. For public networks relying on universal connectivity among different companies, and supported by multiple vendors, standards and protocols defined by U.S. and worldwide

209

organizations must be used. *Protocols* are strict procedures for the initiation, maintenance and termination of data communications, as described later in this chapter.

Traffic characteristics impose different requirements on voice versus data network design. For circuit switched voice communications, a nominal post-dial delay (*call setup*) interval of upwards of 20 seconds is acceptable. However, data traffic often occurs in short bursts, resulting in long inactive periods interspersed with high-speed information exchange. So a dedicated non-switched channel would result in inefficient network utilization.

Nonetheless, set-up time to establish a circuit switched call would result in unacceptable response times for on-line data transactions, where terminal-operator requests for data must be responded to in fractions or, at most, very few seconds. Developed in the 1970s for long-distance data communications, *packet switching,* an alternative to circuit switching, drastically reduces or eliminates call set-up time and is therefore well-suited to bursty data traffic.

Additionally, although most digital PBXs have data capabilities, high-port termination costs and sub-optimal performance characteristics have caused voice/data PBX applications to be passed over in favor of separate local area networks (LANs) for on-premises data communications. (Data service capabilities of LANs, *metropolitan area networks* (MANs) and *wide area networks* (WANs) are discussed in Chapters 13 through 15.)

Although the majority of the traffic carried on today's public networks is voice, the annual growth rate in voice traffic is currently only 3–5% versus 15–20% for data communications, reflecting new emphasis on integrated services digital networks (ISDN) and other advanced technologies such as fast packet switching, on which they rely. Developments in ISDN, broadband integrated services digital networks (BISDN) and underlying technologies are treated in Chapters 16 through 19.

Finally, unlike voice communications, which has evolved solely from the telecommunications industry, standards-setting and academic community influences, data communications evolution has been driven by both the telecommunications and data processing segments of information systems ventures. Beyond real differences in technical requirements, some of which have been noted above, this dual influence has resulted in the need to reconcile dissimilar telecommunications and data processing system designs and standards, which in a perfect world could have been avoided—or minimized—with an integrated approach.

While appreciating this perspective doesn't reduce the added complexity or penalties involved with dual path developments, it helps put in proper context the data technologies, services, standards, and business segments described in this chapter and in the remaining parts of the book. Furthermore, it reveals why integrated voice and data telecommunications trends, although motivated by operational convenience and economy, have nevertheless been restrained by the pace of technology and standards development.

Packet switching

A *packet* is a group of bits that is switched as a unit. A packet contains user data, destination and source information, control information, and error detection bits, arranged in a particular format. A typical packet is shown in Fig. 12.1. Packets are formed by the segmentation of user message information or data (which may be any length, any number of bits or bytes) into packets of limited length by *packet assembler-disassemblers* (PADs), as shown in the figure. Packetization is used in virtually all data communications systems.

Sufficient information is embedded in packets to enable packet switches to route them through networks. A *packet header*, which precedes user data, may contain destination address, source address, link numbers, packet numbers, and other information. Specifically, a header is control information appended to a segment of user data for synchronization, routing, and sequencing of a transmitted data packet or frame. Among adjacent and connected switching nodes, packets are encapsulated in frames which themselves include headers and trailers (codes), usually hardware generated, to indicate start-of-message and end-of-message events. Packets can vary in length but are usually limited to 128 bytes (1024 bits).

A *packet switching network* is designed to switch and transport information in packet form. Figure 12.2 illustrates how packet switching works and the differences between packet and circuit switching. In the figure, user messages, represented by the rectangles labeled A, B, and C are shown as DTE inputs and outputs. Message length is indicated by the length of the rectangles.

For circuit switching, illustrated in the upper half of Fig. 12.2, channels between switches are used exclusively for individual message transmissions A, B, and C, as-

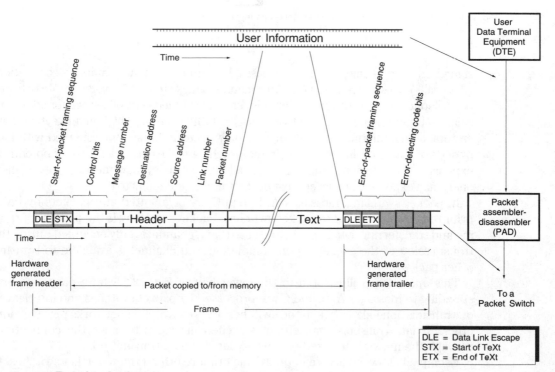

DLE = Data Link Escape
STX = Start of TeXt
ETX = End of TeXt

Fig. 12.1 Typical packet format.

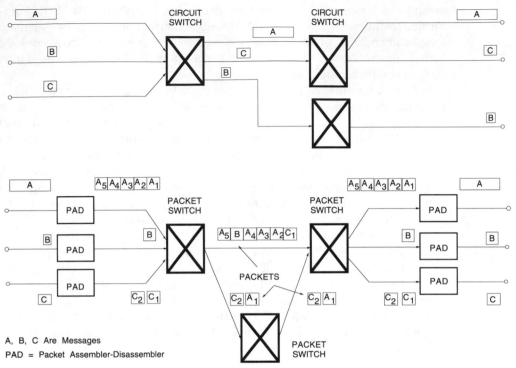

Fig. 12.2 Example of circuit and packet switched connections.

signed on a first-come, first-served basis. The circuit switches establish connections (data calls) between remote DTEs in a manner similar to that for voice traffic between two telephones. As with voice, channels remain occupied (out-of service for additional calls) until released. For interactive message traffic generated by human keyboard operators, actual information transfer may only occur in bursts, interspersed with long periods of inactivity. In this case, circuit switching makes inefficient use of potentially expensive transmission resources. Once all links are busy, new messages, even short ones, may experience unacceptable delay waiting for circuits to clear.

In packet switching, messages A, B, and C are segmented into packets by PADs prior to being connected to packet switches. Packet switching is the process of routing and transferring data in packet form so that a channel is only occupied during the transmission of the packet. Upon completion, the channel is available to transmit other packets.

This operation is illustrated in the lower half of Fig. 12.2 where packets corresponding to messages A, B, and C are processed by packet switches and interleaved on transmission links between the switches. Packet switched networks provide more efficient data transport than circuit switched networks because the connections through the network are used only while data is being transmitted.

Each packet switch is connected to one or more other remote packet switches. In transmitting message A, note in the figure that some packets are directly routed between transmitting and receiving end-users, whereas others are routed through an intermediate switch. This type of routing is typical in some packet switching networks, resulting in the possibility that all packets may not arrive at the terminating switch in the same order or sequence that they were transmitted. With *connection-oriented service,* the terminating switch stores and reorders packets, reassembles them into original message formats, and provides error control.

Packet switching operations produce various end-to-end message delays. For data applications this is normally not a problem, but it would degrade voice and video communications. For this, and other reasons, until recently, packet switching has been used exclusively for data communications. Advanced, *fast packet switching* technologies that support voice, data, video, and other services are currently under development to overcome these limitations.

Packet switch functions and capabilities

A packet switch consists of the following functional entities:

- Input and output buffering (memory elements to temporarily store packets)
- Processing for decoding header address, routing, and other information; error detection and switch and network control
- Internal switching to connect input and output buffers

The transmission of packets through a network requires three types of packet control procedures:

1. *Routing control.*   to determine the routes over which packets are transmitted
2. *Flow control.*   to prevent congestion in the network and lock-ups or traffic jams
3. *Error control.*   to deal with any transmission errors that occur

In contrast to circuit switched voice networks where signaling is invoked once to establish call connections for the duration of the entire call or transaction, in packet networks, each packet is examined for source and destination address information and acted upon accordingly. While this operation does result in efficient utilization of transmission resources, overall routing, error and flow control impose significant processing requirements on packet switches. In fact, the throughput of packet networks is limited primarily by the processing capabilities of the packet switches.

Typically, packets have a maximum length of around 1024 bits. In the U.S. Defense Data Network (DDN), the C30E packet switches have a capacity of 300 packets per second. New fast packet switch technologies are projected to handle up to 50,000 packets per second and to offer relief from the current practical limitations of 255 switches per network, and eight host processor connections per switch. Telenet, now called Sprint Net, was the first U.S. commercial packet switched network. It presently handles in excess of 55 million packets per day and has 500 access centers across the U.S.

Access & Transport Services

As noted above, once a call is established in circuit switched networks, a dedicated, physical connection is established between telephones or other user station equipment, to be torn down at the end of the call. Analogously, a connection-oriented packet switched network transport service establishes logical connections in response to station equipment (DTE) requests. All packets entering the networks are delivered to terminating DTEs in the order in which they were received. As with voice calls, connection-oriented data services use separate procedures for connection establishment and end-to-end *information transfer* (connection establishment must take place prior to information transfer).

This service is referred to as *virtual circuit service* since, in the absence of degradation, message transmission is logically identical to transport over ideal circuit switched facilities. Note, however, that circuit switching inefficiencies are avoided since packets from multiple sources can be interleaved over the same physical transmission paths.

Two different internal packet switched network designs can be implemented to provide users with virtual-circuit service. In the first method, following a request for service, a single route or path through the network is established with all packets following that route and arriving in sequence. Thus, a virtual circuit is established internally.

In the second approach, packets presented to the network are numbered and can be routed through the network on different paths, determined on a packet-by-packet (*connectionless*) basis. This *datagram* operation handles each packet independently and may not deliver packets to the terminating switch in order and reliably (without errors). To deliver virtual-circuit or connection-oriented service to users using datagram internal operation, the terminating switch must store, reorder and perform error recovery functions prior to delivering data to the terminating DTE. One advantage of permitting switch nodes to determine routes as a function of instantaneous traffic requirements is that in the event of line failure or consistent congestion, the network can dynamically find the best available route.

ARPANET, the first operational packet switched network is a datagram network; TYMNET and most other packet networks today operate on a virtual-circuit basis.

Although most packet switched networks used for wide area or long distance data communications offer users virtual-circuit service, they can be designed to offer users connectionless datagram service. This permits networks to be designed with simple switches and control procedures. It eliminates connection set-up, lowers overhead, and results in faster transmission times. Connection-less modes are widely used in local area networks to increase throughput and to reduce complexity and cost. Figure 12.3 illustrates connection-oriented and connectionless network service.

A *permanent virtual circuit* (PVC) is a virtual circuit resembling a leased line in that invariant logical numbers identifying PVCs are dedicated to a single user. Thus, at a particular interface point a network service provider assigns a fixed number of virtual circuits to a user, each of which connects specific network/user interface points. Alternatively, switched virtual circuit service permits users to establish vir-

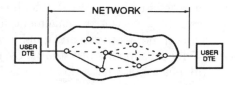 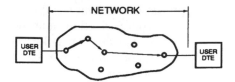

CONNECTION-LESS
NETWORK SERVICE (CLNS)

- DATAGRAM

CHARACTERISTICS:

- DYNAMIC SWITCH DETERMINED
  PATHS THROUGH THE NETWORK

- NON-SEQUENTIAL PACKET ARRIVAL

CONNECTION-ORIENTED
NETWORK SERVICE (CONS)

- PERMANENT VIRTUAL CIRCUIT SERVICE

- SWITCHED VIRTUAL CIRCUIT SERVICE

CHARACTERISTICS:

- SINGLE PATH THROUGH THE
  NETWORK

- SEQUENTIAL PACKET ARRIVAL

Fig. 12.3 Connection-oriented and connectionless service comparison.

tual circuits between arbitrary network interface points, much like direct distance dialing in circuit switched voice networks.

Figure 12.4 shows two methods for physically accessing packet switched networks. On the right-hand side of the figure is a host computer attached to a communications _front-end processor_ (FEP) with integral PAD functional capabilities. The FEP is connected directly to a packet switch, which is either located on a customer's premises, or connected via digital access facilities. In the latter case, the user pays a monthly fee for the dedicated line and a monthly access or port charge, plus a charge proportional to the amount of data actually transmitted.

FEPs, also called _stored program communications controllers,_ are dedicated computers or systems of computers that control data communications between host

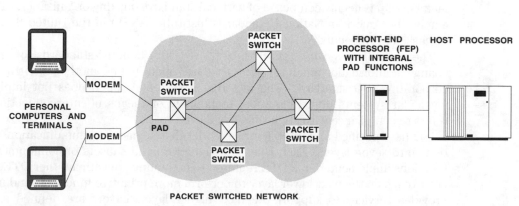

Fig. 12.4 Access to packet switched networks.

processors and various types of data communications networks. FEP functions include route selection, multi-host access, data switching, network management, message sequencing, and flow control. FEPs support both private and public data network operations. For example, the IBM 3745 model supports private IBM System Network Architecture (SNA) networks, but with network packet switching interface programs, it can connect with public packet switched networks.

The left side of the figure shows how modems are used to connect terminals or other DTEs to remote packet switches and PADs. In this case either dedicated (leased) or public switched dial-up voice network services can be used to access packet switched network services using modems. With *dial-up service,* log-on to the packet network is required each time a user wishes to obtain service. Charges are normally made on the basis of connect time, the quantity of data transmitted, or both.

Conventional packet switching incorporates a rich set of protocols capable of rendering high-quality, error-free data communications services using relatively marginal interswitch transmission extant at the time of its introduction. Some packet switched services cannot be extended via off-network access arrangements such as the modem configurations shown in Fig. 12.4. In such arrangements, for example, if error control is needed on an end-to-end basis it must be separately implemented on network ingress and egress access links.

Protocols

From the foregoing, it is evident that data communications networks require a high degree of compatibility and interoperability among DTEs and network elements, particularly with respect to physical and logical interfaces and controls. A challenge is presented by different vendor equipment and/or even different models from the same vendor, all of which must be interconnected.

In 1977, the International Organization for Standardization (ISO) established a subcommittee to develop a standards architecture to achieve the long-term goal of *open systems interconnection* (OSI). ISO is a voluntary international body concerned with developing standards for a variety of subjects. Data communications standards are developed through the workings of its Technical Committee 97. ISO membership is mainly composed of national standards-making organizations, for example, the American National Standards Institute (ANSI) in the United States, as discussed in Appendix A.

The term *open systems interconnection* (OSI) denotes standards for the exchange of information among systems that are "open" to one another by virtue of incorporating ISO standards. The fact that a system is open does not imply any particular system's implementation, technology or means of interconnection but refers to compliance with applicable standards.

ISO has specified an OSI Reference Model that segments communications functions into seven layers. Each layer is assigned related subsets of communications functions implemented in a DTE required to communicate with another DTE. Each layer relies on the next lower layer to perform more primitive functions, and in turn provides services to support the next higher layer. Layers are defined so that changes in one layer do not affect other layers.

Information exchange occurs when corresponding (peer) layers in two systems communicate by means of a set of rules known as protocols. Protocols define the *syntax* (arrangements, formats, and patterns of bits and bytes) and the *semantics* (system control, information context or meaning of patterns of bits or bytes) of exchanged data, as well as numerous other characteristics such as data rates, timing, etc.

Defining the details of seven layers of protocols for data communications is an enormously complex task. Understanding the concepts and objectives of layering is simplified with an example used by Andrew S. Tannenbaum in his book *Computer Networks*, published by Prentice-Hall, Inc., 1981. Figure 12.5 illustrates three layers of communications between two philosophers, one in Africa and one in India. The exchange of ideas between the two philosophers represents layer-three, peer-to-peer communications.

Since the philosophers have no common language, they each engage the services of a translator. In the figure, message exchange in French represents layer-two peer-to-peer communications. Each translator must engage the services of an engineer for transmission by letter, telegram, telephone, computer network, or some other means. Just as the translators had to agree on a common language, layer-one peer-to-peer communications requires that agreements regarding physical transmission media must be made between the two engineers.

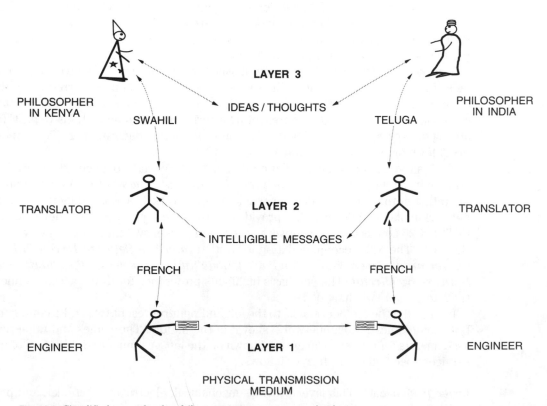

Fig. 12.5 Simplified example of multilayer peer-to-peer communications.

Note that the translators could change to English to speak to each other without affecting either the layer-one or layer-three process. Similarly, neither message integrity at layer three nor the translation process of layer two is affected should the engineers change layer-one physical media choices. However, performance aspects such as message delivery time could be drastically different if, for example, postal service was substituted for a real-time form of communications.

Note in the figure that physical information transfer between the remote locations takes place only at level one. This last characteristic and the others noted conform to the seven-layer OSI reference model, shown in more detail in the next figure.

ISO Reference Model for OSI

Figure 12.6 illustrates the ISO Reference Model for OSI, the objective of which is to solve the problem of heterogeneous DTE and data network communications. However, the OSI model is not a product blueprint. Two companies can therefore build computers consistent with the model, but unable to exchange information. The model is only a framework—meant to be implemented with standards developed for each layer.

Standards must define services provided by each layer, as well as the protocols between layers. Standards do not dictate how the functions and services are implemented in either hardware or software, so these may differ from system to system.

The International Consultative Committee for Telegraph and Telephone (CCITT) is the international standards body of the International Telecommunications Union (ITU), a UN treaty organization that considers all technical, operational, and tariff matters for telecommunications worldwide.

The results of the CCITT work are published every four years (following a plenary meeting) as "recommendations" in a series of books commonly referred to by the color of their covers (such as "orange book"). CCITT recommendations are denoted by $A.nn$, where A is a letter representing a series of recommendations (e.g., V for analog networks, X for digital networks), and nn is a two-digit number. (See Appendix A for more standards-setting information.)

X.25 is the CCITT recommendation defining the interface between data terminal equipment (DTE) and *data circuit terminating equipment* (DCE) for terminals operating in the packet mode over public data networks. DCE is a generic term for network-embedded devices that provide an attachment point for user devices. In a CCITT X.25 connection, the network access and packet switching node is viewed as the DCE. The X.25 recommendation is entitled *Interface Between DTE and DCE for Terminals Operating in the Packet Mode and Connected to the Public Data Network **by Circuit***. The emphasis highlights preference for direct access connections to packet switching nodes.

In Fig. 12.6 the protocol stacks to the left and right represent two DTEs connected by a communications subnetwork, shown in the middle. The names and numerical assignments for the seven layers are shown on the left. A summary description of the services specified for each layer follows.

Layer 1: Physical.   This layer provides mechanical, electrical, functional, and procedural characteristics to activate, maintain, and deactivate connections for the

Layer

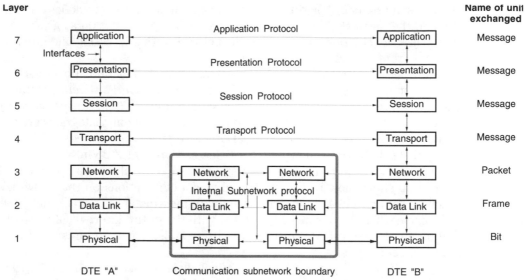

Fig. 12.6 ISO reference model for open system interconnection.

transmission of unstructured bitstreams over a physical link. The physical link can be connectors and wiring between the DTE and the DCE at a network access point, and fiber optic cable within a network. Layer 1 involves such parameters as signal levels and bit duration. X.21 is an option for layer 1 within the CCITT X.25 recommendation. In the U.S., the RS-232 C standard is generally used at layer 1, and bits are the data units exchanged.

Layer 2: Data Link.   The data link layer provides for reliable transfer of data across the physical link. It provides for mapping data units, from the next higher (network) layer to frames of data for transmission. Figure 12.1 presents the format for a typical data link frame. The data link provides necessary synchronization, error control and flow control. Layer 2 provides for multiplexing one data link onto several physical links when necessary. Also, a single physical link can support multiple data links.

Link Access Protocol-B (LAP-B).   LAP-B is an option for layer 2 in the CCITT X.25 recommendation. It is a subset of the ISO-developed high-level data link control (HDLC) asynchronous balanced mode (ABM) protocols.

Layer 3: Network.   Layer 3 provides higher-level layers with independence from routing and switching associated with establishing a network connection. Functions include addressing, end-point identification, and service selection when different services are available. An example of a level 3 protocol is the CCITT X.25 recommendation.

Layers 1 and 2 can be described as *local* DTE (station) to DCE (network node) protocols. Most of the layer 3 dialogue is between stations and nodes. For example, stations send requests for virtual circuits, and address packets to nodes for delivery through the network. There is also, however, a station-to-station aspect of layer-3 protocols. The stations must provide the network with addressing and other information to route data to another station.

Figure 12.7 illustrates station-to-node, station-to-station, and subnetwork or internal protocol categories. As described above, the *internal subnetwork* protocols may be different from those presented at station-to-node (user-to-subnetwork) interfaces. That is, even if the internal protocols implement only datagram service, the network can still provide reliable virtual circuit service.

Figure 12.8 illustrates how network DCEs can present X.25 interface protocols to attached DTEs and still support different internal or subnetwork protocols. The figure also shows how DCE layer-1 local media connections on the network side (such as copper wire) can be interfaced with long distance media (such as fiber optic cable). These conversions, together with the entire internal subnetwork operation are accomplished transparent to user DTEs, which are presented with an X.25 interface.

The *media access control* (MAC) terminology is consistent with IEEE LAN standards discussed below. In general, this protocol discussion applies to local area, metropolitan area, wide area, or any specialized network that uses layered protocols conforming to the OSI Reference Model.

When two or more separate networks connect end-point DTEs, an *internet protocol* (IP), sometimes referred to as a *layer 3.5 protocol,* is needed. The IP protocol sits on top of a network protocol and relies on it for internetwork routing and delivery.

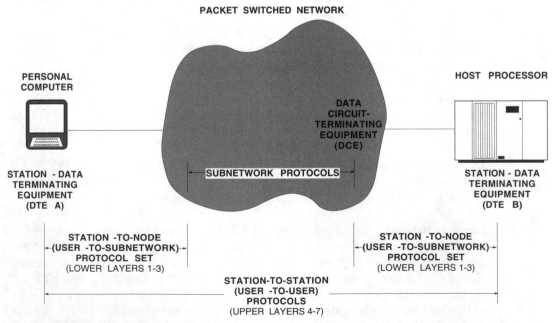

Fig. 12.7. Protocol categories.

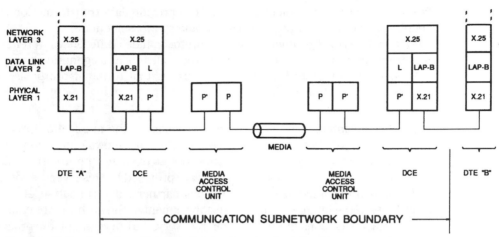

Fig. 12.8 Station-to-node and internal subnetwork protocol relationships.

Layer 4: Transport. In conjunction with the underlying network, data link and physical layers, the transport layer provides end-to-end (station-to-station) control of transmitted data and optimizes use of network resources. This layer exists to provide transparent data transfer between layer 5 session entities. In ISO terminology, an *entity* is the network processing capability (hardware, software, or both) that implements functions in a particular layer. Thus entities are identified for each layer, e.g., the layer 5 session.

Transport-layer services are provided to upper layers in order to establish, maintain, and release transparent data connections over two-way, simultaneous data transmission paths between pairs of transport addresses. The transport protocol capabilities needed depend upon the quality of the underlying layer services.

When used with reliable, error-free virtual circuit network service, a minimal transport layer is required. If the lower layers provide unreliable, datagram service, then the transport protocol must implement error detection and recovery, and other functions. ISO has defined five transport protocol classes, each consistent with a different underlying service.

Layer 5: Session. A session is a connection between stations that allows them to communicate. For example, a host processor may need to establish multiple sessions simultaneously with remote terminals to accomplish file transfers with each. *File transfer access and management* (FTAM) is an example of an ISO application-layer standard for network file transfer and remote file access.

The purpose of the session layer is to enable two presentation entities at remote stations to establish and use transport connections by organizing and synchronizing their dialogue and managing the data exchange.

Layer 6: Presentation. The presentation layer delivers information to communicating application entities in a way that preserves meaning while resolving syntax differences. Towards this objective, layer 6 can provide data transformation (e.g., data compression or encryption), formatting, and syntax selection.

Virtual terminal protocol, a layer 6 protocol, hides differences in remote terminals from application entities by making them all appear as generic or virtual terminals. When two remote host processors use virtual terminal protocols, terminals appear as locally attached to either host.

Layer 7: Application. The application layer enables the application process to access the OSI environment. It serves as the passageway between application processes using open systems interconnection to exchange information. All services directly usable by the application process are provided by this layer. Services include identification of intended communications partners, determination of the current availability of the intended partners, establishment of the authority to communicate, agreement on responsibility for error recovery, and agreement on procedures to maintain data integrity.

Protocols currently being defined or enhanced for this layer include: file transfer (FTAM), *transaction processing* (TP), directory services (ISO 9595, CCITT X.500), and *job transfer and manipulation* (JTM). The CCITT X.400 *message handling system* (MHS) protocols were first issued in 1984, and were then substantially revised in 1988 as an application layer protocol.

OSI Summary

Work on the OSI standards began in 1978. The basic family of standards is still not yet complete. OSI standards continue to evolve, meeting new requirements and supporting new environments and applications. Because multiple standards are available within a given protocol layer, and because each standard permits significant parameter variations and options, the fact that vendors design to standards is not sufficient to ensure ISO OSI objectives of open system interconnection and interoperability among heterogeneous hardware and software products.

Two approaches have been developed to address this problem. First, user groups with common business interests have been formed to develop a complete and precisely defined set of services, and a detailed specification for the *suite* of protocols (a *profile* of applicable standards) that are used to implement those services.

For example, the National Institute of Standards and Technology (NIST) is responsible for the Government Open Systems Interconnection Profile (GOSIP), contained in the Federal Information Processing Standard (FIPS) 146, published in August, 1988. GOSIP is a functional standard—a precisely defined, specific subset of other standards, together with rules for implementing and using the specified subset suite of protocols. GOSIP specifies operational configurations, selects the particular protocol suite, and identifies options and values.

The effective date for GOSIP Version 1 was February 15, 1989, and it became mandatory for all government solicitations and contracts for new network products and services August 15, 1990.

Manufacturing Automation Protocol (MAP), and Technical and Office Protocols (TOP), the primary standards efforts by the manufacturing, technical, and office communities are two other examples of suites established for specific industry applications.

The second approach to resolving practical problems associated with OSI is testing and certification to ensure compatibility among vendor equipment and data communications networks. GOSIP, MAP, and TOP all include efforts to build a set of conformance and interoperability testing tools and facilities for protocols and suites.

The Corporation for Open Systems (COS), is a nonprofit organization composed of manufacturing, service and user organizations in the computer communications area. COS seeks to facilitate the development of the international, multivendor marketplace through the development, introduction, and verification of OSI and ISDN standards and by ensuring vendor equipment interoperability. COS does this by developing and distributing products designed to test for conformance to international standards and by providing impartial verification of the standards-compatibility of OSI and ISDN products and services.

Fast Packet Switching

Earlier circuit switching, packet switching, and layered protocol discussions provide the foundation for explaining fast packet switching, a new technology encompassing both voice and data services. In Chapter 4, circuit switching was defined as equipment arranged to establish connections between lines, between lines and trunks, or between trunks. This definition, consistent with AT&T Reference 9, "Engineering and Operations of the Bell System," principally applies to voice services. More generally, *switching* has been defined as the routing of signals in circuits to transmit data between specific points in a network. Recalling earlier circuit, channel and packet switching concepts gives meaning to these somewhat abstract terms, underscoring the definition's application to both voice circuit switching and data packet switching.

To have gone beyond voice switching definitions in Chapter 4 would have involved introduction of even more abstract terminology, creating serious pedagogical difficulties. At this point, however, subtle distinctions among advanced switching techniques can be presented.

Figure 12.9 illustrates a taxonomy of telecommunications switching technologies. Key distinctions between circuit and packet switching are shown. In circuit switching, a user is assigned a dedicated circuit with a fixed bandwidth for some period of time. With packet switching, users share network ports, compete for shared transmission resources, and are provided variable bandwidth capacity (i.e., they only use resources when they have information to transmit).

As noted earlier, packet switching efficiently serves users with "bursty" traffic—long periods of inactivity punctuated by short transmissions. On the other hand, circuit switching is ideal for a users with constant levels of traffic over extended periods, especially when the traffic demands a fixed bandwidth equal to the allocated resource's capability, e.g., voice and video.

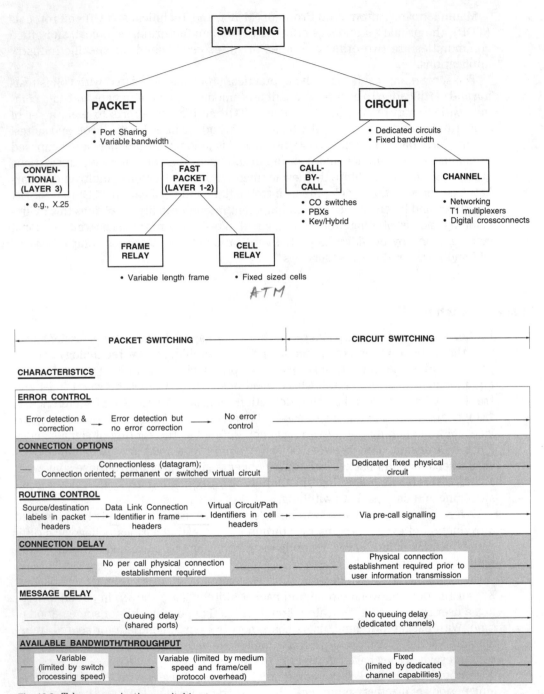

Fig. 12.9 Telecommunications switching taxonomy.

Circuit switching subcategories include *call-by-call* and *channel* switching. In call-by-call circuit switching, a resource is selected and allocated upon each request for service. Channel switching provides the ability to remotely modify multiplexers and network configurations in response to outages or time-of-day traffic variations, or to accommodate growth and organizational changes. In yesterday's networks, channel switching was accomplished manually via patching or physical setup (*provisioning*). Today, programmable T1 multiplexers and digital cross connect systems permit *soft provisioning,* and support the trend away from point-to-point private networks toward flexible network topologies.

Packet switching subcategories include *conventional* and *fast packet* techniques. In conventional packet switching, communications networks provide user interfaces which furnish services associated with the bottom three ISO protocol layers. These networks have afforded years of reliable data communications, being designed to sustain error-free service using older, poorer quality, lower bandwidth analog circuits. Networks designed for X.25 typify this category, where attached DTE devices lack processing capabilities (*dumb terminals*) to support error control and other higher-level protocol services, so the networks themselves provide those functions.

Fast packet is a term referring to a number of broadband switching and networking paradigms. Implicit in the fast packet technology is the assumption of an operating environment that includes reliable, digital, broadband, nearly error-free transmission, and intelligent end-user equipment. Because intelligent end-user systems have sufficient processing power to operate sophisticated protocol suites, it is now possible to perform error detection, correction, and requests for retransmission on an end-to-end basis. In combination with virtually error-free fiber optic transmission systems (bit error rates of better than one error in a billion [10^9] bits without error control), network designs can be streamlined to emphasize speed of service and minimum message delay.

Frame relay and *cell relay* are two fast packet technologies. Frame relay is defined in CCITT Recommendation I.122 "Framework for additional packet mode bearer services," as a packet mode service. In effect, it combines the statistical multiplexing and port sharing of X.25 packet switching with the high-speed and low delay characteristics of time division multiplexing and circuit switching. Unlike X.25, however, frame relay implements no layer 3 protocols and only the so-called "core" layer 2 functions. With these simplifications, frame relay makes practical use of interswitch transmission rates of 1 to 2 Mbps (e.g., DS1 rate services), considerably greater than the 56/64 kbps channels ordinarily used in X.25 networks. While initial offerings are at the DS1 rate or less, DS3 rate frame relay operation is considered feasible.

The basic units of information transferred are variable length frames, using only two bytes for header information. Delay for frame relay is lower than for X.25, but it is variable and larger than that experienced in circuit switched networks. This means that currently frame relay is not suitable for voice and video applications where excessive and variable delays are unacceptable.

Cell relay is the process of transferring data in the form of fixed length packets (cells). Cell relay is used in high-bandwidth, low-delay, packet-like switching and multiplexing techniques. The objective is to develop a single multiplexing/switching mechanism for dividing up usable capacity (bandwidth) in a manner that supports its allocation to both voice and video traffic and packet data communications services. Standards groups have debated the optimum cell size. Small cells favor low-delay for these applications but involve a higher header-to-user information overhead penalty than would be needed for most data applications. The current specification is for a 53-byte cell which includes a 5-byte header and a 48-byte payload.

In addition to the difference in units of information transferred, frame and cell relay differ in terms of how and where the techniques are used in a network. Frame relay, like X.25, is a network interface specification. Cell relay, used at network cores, implements common switching/multiplexing for all types of information (including frame relay). The bottom of Fig. 12.9 lists other important technical characteristics that distinguish the switching technologies described above. The structure of the figure is such that entries should be read from left to right, and the spatial location of the entries roughly corresponds to the switching technologies with the same left-to-right justification.

For example, in the error-control row, conventional packet switching networks (like X.25) provide comprehensive error detection and correction service; frame relay networks provide error detection but no error correction; and networks using cell relay typically provide no error control, a characteristic shared by networks offering circuit switching service. In terms of connection options, any of the packet switching techniques can theoretically support connectionless (datagram) or permanent or switched virtual circuit connections. Circuit switching provides only fixed, physical circuit connections. Early frame relay offerings are limited to permanent virtual circuit service.

For routing control, conventional packet switching includes explicit source and destination address information in each packet header. This permits networks to route each packet independently from source to destination. With frame relay, headers include only a *data link connection identifier* (DLCI), which permits networks to assign a fixed route for all frames between network end-points by simple "table look-up" procedures. Cell relay uses *virtual circuit identifiers* (VCIs) and *virtual path indicators* (VPIs) in much the same way that frame relay uses DLCIs. In circuit switching, networks select a fixed connection path prior to information exchange in response to user provided called-party address information.

Because all packet switching techniques employ port sharing, delays associated with establishing connections in circuit switched networks simply do not exist. On the other hand, message delays might occur in port-shared packet networks when many sources seek to use common switch and transmission resources simultaneously. In these cases, queues may develop subjecting individual messages to queuing delays, a phenomenon described in Chapter 7. In circuit-switched networks, once a connection is established, the transmitting terminal has dedicated use of the resource and hence experiences no queuing delay.

In terms of bandwidth or throughput, packet switching provides users with variable bandwidth capabilities, ideal for bursty traffic. In contrast, circuit switching provides users with fixed bandwidth limited by the physical characteristics of connecting channels. Whereas useful bandwidth in conventional packet switching networks is limited by switch processing speed, in frame and cell relay, useful payload bandwidth is limited by transmission medium speed and protocol overhead.

Although concise, the above discussion reveals the major architectural factors being used to design tomorrow's integrated digital networks. As a result of the influence of high-quality fiber optic transmission and high-speed microprocessor technologies, it can be concluded that many of the circumstances that justified separate voice and data networks in the past are no longer valid. The rate of transition appears to be limited only by market economics and the deliberations of standards setting bodies. Succeeding chapters discuss existing and emerging premises, metropolitan and wide area network services that incorporate the technologies described above. Accordingly, those chapters present additional telecommunications switching insights, viewed from the perspective of network carrier and equipment service applications.

Part

4

Data Services

Premises (Local Area) Services

This chapter describes approaches for planning and implementing premises data services. Emphasis is on terminal-to-terminal (peer), terminal-to-server, and terminal-to-host processor connections. Accordingly, some previous definitions are expanded, and some new terminology is introduced. The chapter briefly traces early technologies, then treats the broad category of local area networks (LANs), which in the last decade have emerged as the premises data services approach of choice, having become a focal point of IEEE and other standards setting bodies.

IBM defines a *terminal* as a device, usually equipped with a keyboard, often with a display, capable of sending and receiving data over a communications link. *"Dumb" terminals* are limited to low-speed operation, do not incorporate local processing, and transmit characters one at a time as they are typed by an operator. Normally hard-wired (via dedicated nonswitched connections) to a host processor, these terminals cannot support line sharing, polling, or addressing. Polling occurs when a host processor signals a terminal to determine if it is ready to receive or send data.

Workstations are input/output devices used by operators which can process data independent of host processors and can be configured to exchange data with other workstations, host processors, or servers. Workstations vary in complexity, ranging from simple personal computers (PCs) to devices with host processor-like capabilities supporting text, database, graphics, imagery, and other data-intensive applications. Workstations can implement multilayered protocols, as previously described, and therefore can support error control, addressing, and other capabilities to achieve reliable data communications.

Point-to-Point Premises Networks

Figure 13.1 illustrates a point-to-point approach for connecting terminals to host processors. *Cluster controllers* or *multiplexers* can be installed in telecommunica-

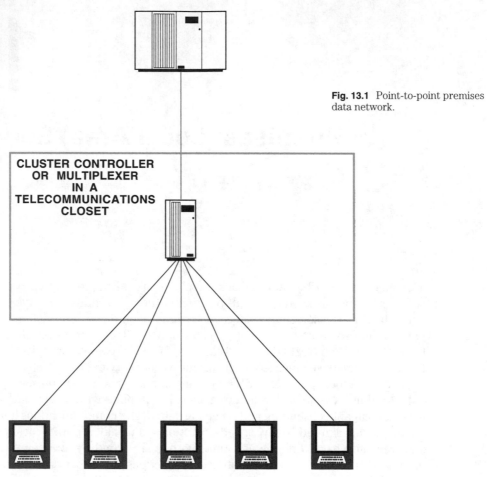

HOST PROCESSOR IN A COMPUTER ROOM

CLUSTER CONTROLLER OR MULTIPLEXER IN A TELECOMMUNICATIONS CLOSET

TERMINALS IN OFFICES

Fig. 13.1 Point-to-point premises data network.

tions closets to reduce the amount of cabling required between terminals in offices and hosts in central computer rooms. Also, communications front-end processors can be used between hosts and *remote cluster controllers* (cluster controllers in other premises), wherein the front-end processor and cluster controllers are connected by an external network.

This approach is typically used with terminals without local processing capability. However, PCs and other workstations can be programmed for emulation, for example, of IBM's 3270 driven by a centralized IBM host system.

The traditional cabling method for terminals such as the IBM 3270 has been coaxial cable. Premises networks using coaxial cable historically require completely discrete and separately administered premises distribution systems. These installations require make and model compatibility between hosts and terminals. In the past, different equipment required different cable types, so that there was no possibility of standardized, universal wiring. As we saw in Chapter 6, this is no longer the case.

Data PBX-Based Premises Networks

A second approach for connecting premises terminals, hosts and servers, shown in Fig. 13.2, employs a data PBX. A *data PBX* supports data applications in the same way that PBXs support voice applications. The data PBX is a circuit switch designed to handle data traffic among various types of DTE equipment. Data PBXs can establish data calls to different host computers, to different ports on the same host, or to shared servers. Data PBXs can also be configured to provide connections to modem pools and other gateway devices which access external networks.

Data PBXs are manufactured by various companies—such as Gandalf, Develcon, and Infotron—and support a variety of DTE vendors. Data PBX performance is limited by the number of physical circuits available. For example, if a system has only ten circuits to a host, when they are all in use, subsequent users are blocked from completing calls to the host and are placed in a queue.

In 1988, the cost per data PBX port was in the range of $100 to $300, which was less expensive at that time than other premises network alternatives. Another data PBX advantage was that it could use twisted-pair telephone wiring and consequently could be supported by a universal (*structured*) wiring environment.

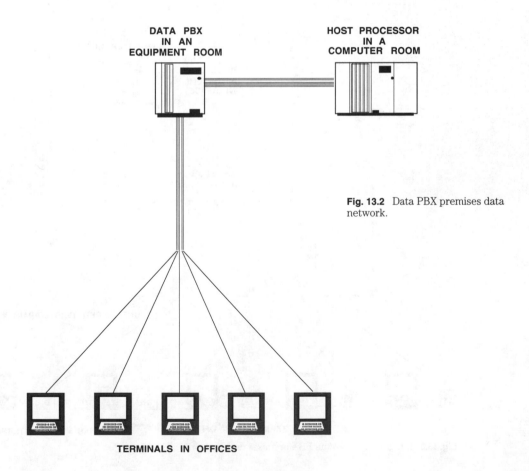

Fig. 13.2 Data PBX premises data network.

Integrated Voice/Data PBX-Based Premises Networks

Figure 13.3 illustrates an integrated voice/data PBX, a third alternative for connecting terminals, hosts, servers within a premise. For data applications, integrated voice/data PBX operation is similar to that of data PBXs. Additional discussion appears in the Circuit Switching Fundamentals and Private Branch Exchange sections of Chapter 4.

Integrated voice/data PBX products use unshielded twisted pairs (UTP) for both voice and data signals, and are, therefore, totally compatible with universal wiring. Each port typically supports data communications at 64 kbps rates. As described in Chapter 4, DS1 (1.544 Mbps) connections between the PBXs and host processors (using either CPI or DMI designs) are available. This greatly simplifies PBX-to-host wiring since each DS1 connection replaces 24 separate wiring connections. Integrated voice/data PBXs establish connections using circuit switching.

Integrated voice/data PBX approaches have been overshadowed by packet-switching-based local area networks, described next, primarily because they incur the highest cost per data port connection and because LANs typically offer higher

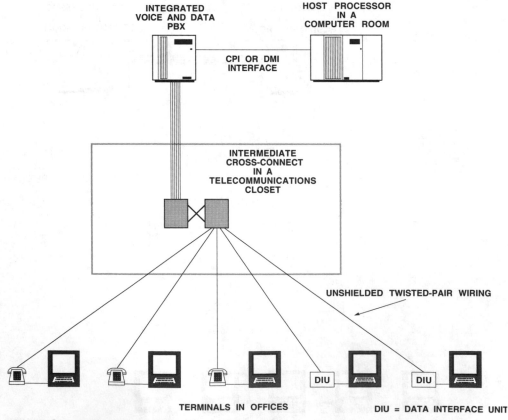

Fig. 13.3 Integrated voice/data PBX premises data network.

terminal-to-terminal data communications rates (up to 200 kbps under some circumstances). However, PBXs can be configured to accommodate thousands of 64 kbps data ports, resulting in aggregate throughput rates in the order of 100 Mbps, whereas single-packet-switching-based LAN segments are normally limited to aggregate rates somewhat less than 10 Mbps.

CSMA/CD LAN-Based Premises Networks

One of the most popular premises data network approaches is the *CSMA/CD LAN*, which employs Carrier Sense Multiple Access with Collision Detection. A *LAN* is a high-speed (typically in excess of 2.0 Mbps) data communications system wherein all segments of the transmission medium (typically coaxial cable, twisted-pair or optical fiber) are in an office or campus environment.

Today's CSMA/CD LANs use *network interface cards* (NICs)—printed circuit cards plugged into workstations or PCs—which implement LAN protocols and provide a medium interface, a transmission medium to which all NICs are connected, and optionally, file servers, print servers, gateway servers, and for larger networks, a network management system, as shown in Fig. 13.4.

Servers are devices that provide multiple attached workstations with various host-like services. Examples include storage of and access to centralized data files (file servers), printers, facsimile, and communications gateway services. Servers permit workstations to share resources such as computer mass storage devices, multiplexers, modems and data communications devices for connecting premises networks to external networks.

A *gateway* is a server which permits client terminal/station access to external communications networks and/or information systems. A gateway is usually a protocol-translating device that connects a local area network to an external network or any two networks that use different protocols and operating environments.

Figure 13.4 illustrates a CSMA/CD LAN that uses coaxial cable as the transmission medium. Attachments to the cable are made using transceivers. *Transceiver* is a generic term describing a device that can both transmit and receive. In IEEE 802 LAN standards, a transceiver consists of a transmitter, receiver, power converter and—for CSMA/CD LANs—collision detector and jabber detector capabilities. The transmitter receives signals from an attached terminal's NIC and transmits them over the LAN cable medium. The receiver receives signals from the medium and transmits them via the transceiver cable and NIC to the attached terminal. The *jabber detector* is a timer circuit that protects the LAN from a continuously transmitting terminal.

Originally, the cable followed a bus or serpentine route from one office to the next. Transceivers located in the ceiling or some other convenient location were connected to NICs by transceiver drop cables (limited to about 150 feet in length). Such wiring plans deviated greatly from telephone twisted-pair universal wiring approaches and have since given way to designs that use standard unshielded twisted-pair cabling in conformity with standards described in the Premises Distributions System section of Chapter 6.

CSMA/CD is a local area network contention-based access-control protocol technique by which all devices attached to the network "listen" for transmissions in progress before attempting to transmit themselves and, if two or more begin transmission simultaneously, are able to detect the "collision." In that case each backs off (defers) for a variable period of time (determined by a preset algorithm) before again attempting to transmit.

The network shown in Fig. 13.4 uses a *bus* topology. A bus is a transmission path or channel using a medium with one or more conductors such as used in Ethernet, wherein all network nodes listen to all transmissions, selecting certain ones based on address identification. Bus networks employ some sort of contention-control mechanism for accessing the bus transmission medium, such as the CSMA/CD method.

An alternative contention-control method is used in a *token bus,* a LAN access mechanism and topology in which all stations actively attached to the bus listen for a broadcast token or supervisory frame. Stations wishing to transmit must receive the token before doing so; however, the next logical station to transmit might not be the next physical station on the bus. Access is controlled by preassigned priority algorithms.

In a *token ring* LAN, a token or supervisory frame is passed from station to adjacent station sequentially. Stations wishing to transmit must wait for the token to arrive before transmitting data. In a token ring LAN, the start and end points of the medium are physically connected, leading to the ring terminology. An example of a token ring LAN is presented below.

Ethernet, a LAN design trademarked by the Xerox Corporation, is characterized by 10 Mbps transmission employing CSMA/CD as the access-control mechanism. This architecture has been adapted by the Institute of Electrical and Electronic Engineers (IEEE) and incorporated into the *ANSI/IEEE 802.3,* one of a family of stan-

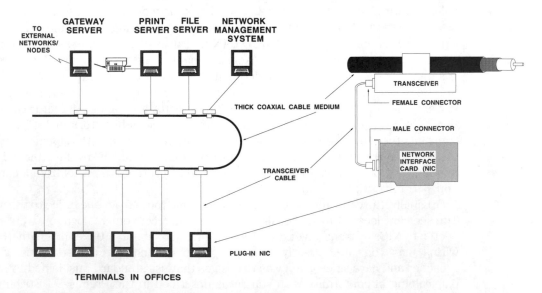

Fig. 13.4 CSMA/CD LAN-based premises network.

dards that deals with physical and data link protocol layers, as defined by the ISO Open System Interconnection Reference Model (OSI). The relationship between the OSI and the IEEE physical partitioning is shown in Fig. 13.5.

ANSI/IEEE has defined two 802.3 coaxial cable CSMA/CD subcategories, namely *10Base5* (known as *thick Ethernet*), and *10Base2* (known as *thin Ethernet* or *cheapernet*). Physically the difference between the two specifications is the diameter of the coaxial cable and the connectors. The thin coaxial cable is much easier to install and less expensive, but sacrifices segment length and supports fewer connections (taps) per segment, as indicated in Fig. 13.6. A third 802.3 CSMA/CD standard, *10BaseT*, recently issued, defines LANs using unshielded twisted-pair cable (UTP), described below.

Although the signaling rate for CSMA/CD LANs is 10 Mbps, actual throughput is less, since as traffic increases, there are more chances for collisions which reduce effective throughput. Under some busy conditions a 10 Mbps LAN may only be able to support about a 4 Mbps aggregate throughput. CSMA/CD LANs work best (encounter fewer collisions) under lightly loaded conditions. Maximum throughput between two attached DTEs is normally limited by other factors and can be as high as 200 kbps using available NICs and network software. Token passing LANs perform best under high traffic conditions, providing users with aggregate throughput close to 100% of the signaling rate.

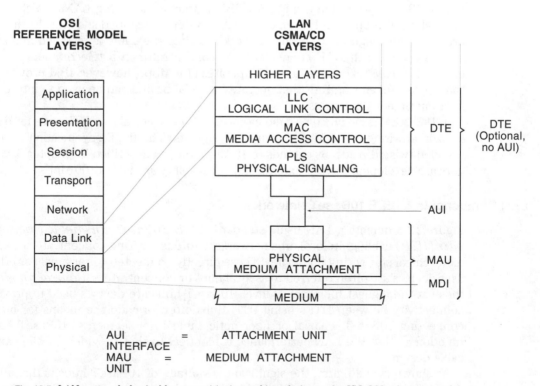

Fig. 13.5 LAN protocol physical layer partitioning and its relation to the ISO OSI reference model.

	THICK	THIN
IEEE 802.3	10Base5	10Base2
CABLE TYPE	802.3 coax	RG-58 coax
MAX. SEGMENT LENGTH	500m (1640 ft.)	185m (607 ft.)
NETWORK SPAN	2500m (8700 ft.)	925m (3035 ft.)
TAP SPACING	2.5m (8 ft. 3 in.)	.5m (1 ft. 8 in.)
MAX. TAPS/SEGMENT	100	30
MAX. STATIONS/NETWORK	1024	1024

Fig. 13.6 ANSI/IEEE 802.3 specifications.

Token Ring LAN-Based Premises Networks

Figure 13.7 shows a token ring LAN. IBM markets token ring LANs which were originally based upon an IBM shielded cable system developed specifically for IBM products. Although logically the network is a ring, the cable is installed in a star topology with each DTE connected to a concentrator in a telecommunications closet. The IBM token ring network operates at 4 Mbps, however, IBM provides a "backbone" token ring LAN that operates at 16 Mbps and can be used to interconnect multiple 4 Mbps rings.

In 1986, IBM issued its type 3 media specification, which allows UTP to be used for horizontal wiring in 4.0 Mbps token ring segments. The IEEE specification for unshielded twisted-pair now extends to 16 Mbps, but, as described in Chapter 6, vendors are developing ways of extending that capability as high as 100 Mbps.

LAN Components & IEEE 10BaseT Networks

Figure 13.8 depicts a LAN using standard, *IEEE 10BaseT unshielded twisted-pair (UTP)* cabling. In local area networks, a *hub* is a wiring concentrator used in hierarchical star wiring topologies. Those directly connected to terminals or other user devices are often referred to as local hubs or concentrators. *Central hubs* are those at the highest hierarchical level. 10BaseT hubs are devices used to provide connectivity between DTEs using UTP. Hubs often provide the means for interconnecting 10BaseT, coaxial or fiber optic cable LAN segments. 10BaseT hubs function so that the LAN performance is identical to that provided by all coaxial cable designs.

As shown in the figure, the significant advantage of 10BaseT hubs is that all of the horizontal wiring can be identical to that used for voice services. This includes the modular phone jacks and UTP cable as described in the Premises Distribution System section of Chapter 6. Also eliminated are the coaxial transceivers and bulky transceiver drop cables resulting in simpler, less costly installation, administration, and maintenance. Recent vendor trends incorporate multiport bridging and network management functions within *intelligent hub* products, functions which are discussed below. Intelligent hubs are available from Cabletron, Synoptics, Bytex, 3 Comm, and others.

A hierarchical *hub/concentrator* arrangement, shown in Fig. 13.9, illustrates the flexibility afforded by modern multimedia products. UTP can be used where it is most efficient, such as for terminal to telecommunications closet runs, and fiber optic cable used for longer runs to central equipment rooms. If needed to support existing terminal equipment and wiring, hub coaxial cable, and fiber cable modules are available, as are repeater modules for interconnection to coaxial cable LAN segments.

In IEEE 802 LAN standards, *bridges* are devices that connect LANs, or LAN segments, providing the means to extend the LAN environment in terms of the number of stations, performance and reliability. An example bridge connection arrangement

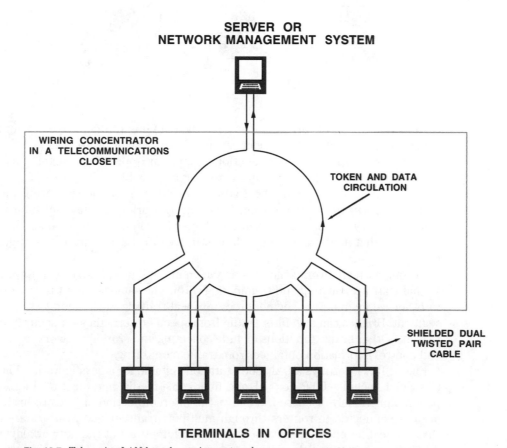

Fig. 13.7 Token ring LAN-based premises network.

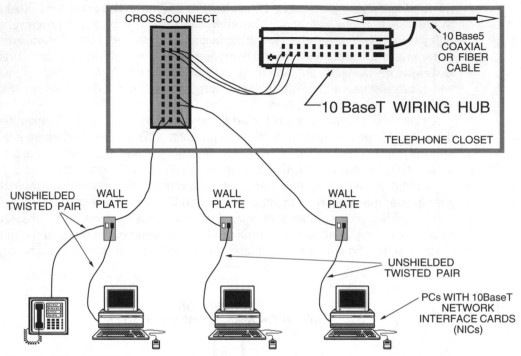

Fig. 13.8 10BaseT LAN hub arrangement.

is shown in Fig. 13.10. Bridges operate at the data link protocol layer (layer 2), and consequently pass packets without regard to higher layer protocols.

Bridges perform three basic functions: frame (as opposed to packet) forwarding; learning of station addresses; and resolving of possible loops in the topology (as might result if bridge 3 in the figure were added) by incorporating an IEEE-defined *spanning tree* algorithm. *Self-learning* bridges construct tables of network addresses by listening to source address information contained in data signal frames, so that no installation or administrative effort is required to configure LAN nodes.

In connecting network segments as shown in the figure, bridges can filter packets so that traffic originating and terminating within the same segment is not forwarded to other segments. This segmentation of local traffic reduces congestion on each segment. Bridges can also filter traffic from specified terminals for security or other reasons. Other bridge functions include collecting and storing network management and control information obtained from traffic monitoring.

Figure 13.11 portrays a LAN configuration suitable for a large, multibuilding campus which includes UTP, coaxial, and fiber cable media, and most of the LAN components previously discussed. It shows the interconnection of hubs to bridges, and bridges connected to routers through multiport transceivers. *Routers* are devices that connect autonomous networks of like architectures at the network layer (layer 3). Unlike a bridge, which operates transparently to communicating end-terminals at the data link layer (layer 2), a router reacts only to packets addressed to it by either a terminal or another router. Routers perform packet (as opposed to frame) routing and forwarding functions; they can select one of many potential paths based on transit delay, network congestion or other criteria. How routers perform their functions is largely determined by the protocols implemented in the networks they interconnect.

Routers are often used to interconnect autonomous network segments in multiple buildings or within a campus as shown in Fig. 13.11. For these applications routers are configured to interface with optical fiber to interconnect buildings, as well as intrabuilding 802.3 LANs. Routers can also be configured to provide gateway services to external networks. The router market is experiencing rapid growth with 1990 revenues of $155 million projected to increase to $565 million by 1992. Vendors include Wellfleet, Cisco, and Proteon. *Brouters,* devices that combine the functions of bridges and routers, are also available.

Figure 13.12 shows the protocol levels involved with repeater, bridge, router and gateway devices in terms of the OSI reference model. In digital transmission, a *repeater* is equipment that receives a pulse train, amplifies it, retimes it, and then reconstructs the signal for retransmission. In IEEE 802 local area network (LAN) standards, a repeater is essentially two transceivers joined back-to-back and attached to two adjacent LAN segments.

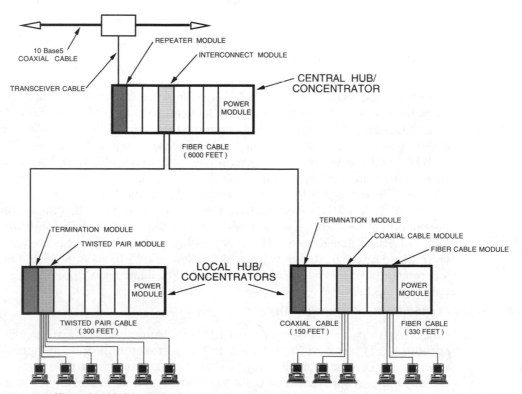

Fig. 13.9 Hierarchical hub/concentrator arrangement.

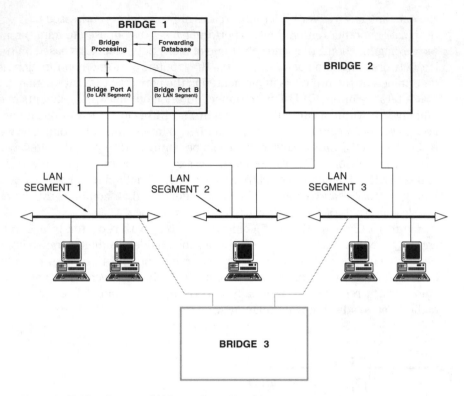

Fig. 13.10 Bridges between LAN segments.

To draw conclusions about future LAN trends, it is informative to compare market share in various types of LAN products. Between 1989 and 1991, the ratio of token ring to Ethernet CSMA/CD products increased from about 60% to 80%. However, earlier predictions that token ring LANs would catch and surpass Ethernet CSMA/CD sales have apparently been revised due to the growth of 10BaseT products. In the same period, the ratio of 10BaseT to coaxial Ethernet LANs rose from 25% to 60%, and by 1994 the ratio is projected by some to reach 165%. It now appears that 10BaseT UTP designs have resolved many of the difficulties caused by nonuniform and expensive coaxial cable Ethernet systems. In 1990, approximately 30% of LANs installed used UTP cabling and growth to 50% is expected by 1992. Overall, the LAN industry grew from $2.4 billion in 1987 to about $6 billion in 1991, and some predict that over 50% of the extant 100 million personal computers will be connected to LANs.

Fiber-Distributed Data Interface (FDDI)

Fiber-distributed data interface (FDDI) is the American National Standards Institute (ANSI) standard X3T9.5 for a 100 Mbps token ring using an optical fiber medium. FDDI was originally proposed as a packet switching network with two primary areas of application: first, high performance interconnection among mainframes, and among mainframes and associated mass storage/peripheral equipment; and second, as a backbone network for interconnecting lower-speed (e.g., 10 Mbps Ethernet) LANs. The latter application, previously discussed, is shown in Fig. 13.11.

The FDDI standard, which basically specifies the two lower protocol layers, is written in four parts: *physical-layer media-dependent* (PMD), *physical-layer protocol* (PHY), *media access control* (MAC), and *station management technology* (SMT). SMT defines functions involved with how the network handles emergencies and how workstations enter or exit the network. Relative to IEEE 802.5 (on which FDDI was originally based), FDDI uses an enhanced *token passing* access protocol which, under light conditions, results in negligible access delay, a property shared with lightly loaded LANs using CSMA/CD (Ethernet) IEEE 802.3 protocols. However, whereas CSMA/CD data throughput is limited to only 37% of the LAN bit rate under heavy loads, FDDI efficiency at heavy loads approaches 100%, a feature similar to IEEE 802.5 token ring protocols.

Architecturally, an FDDI network employs a dual counter-rotating "ring" design, depicted in Fig. 13.13. Note that when a node is not transmitting its own data frames, signals received from a neighboring node are passed to the next neighboring node. Although connected in a ring topology, physical wiring resembles a star or hub topology. The figure indicates that class A nodes are actually connected to two rings, such that if a cable break (illustrated as an example) or other failure occurs, a back-up ring sustains operation. Class B nodes have only one connection to a wiring concen-

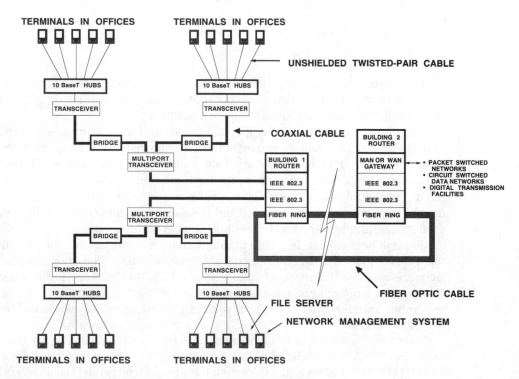

Fig. 13.11 A 10BaseT, coaxial, and optical fiber-based premises network.

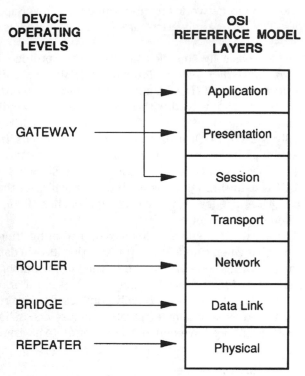

DEVICE OPERATING LEVELS

GATEWAY

ROUTER

BRIDGE

REPEATER

OSI REFERENCE MODEL LAYERS

Application

Presentation

Session

Transport

Network

Data Link

Physical

Fig. 13.12 Device relationships to the ISO OSI reference model.

trator or hub. Concentrators are equipped to detect failed cable segments or nodes and simply bypass the failed connection at the hub itself. The dual ring design greatly enhances service reliability and availability.

In addition to 100 Mbps throughput and the above-noted advantages, since it uses optical fiber, FDDI is superior to metallic cable alternatives in terms of lower noise susceptibility from *radio frequency interference* (RFI) and *electromagnetic interference* (ELI), lower attenuation (signal loss), and greater cost-efficiency. Fiber has a considerably longer span distance (2 kilometers or 1.24 miles using multimode fiber) before repeaters are required and is easier to install because of its flexibility, smaller size, and lighter weight (about 30 pounds per kilometer).

Although not completely secure from intrusion, it is substantially more difficult to tap than copper twisted-pair or coaxial cable, and intrusion detection fiber cable designs are available that meet requirements of U.S. government intelligence agencies. Fiber cables can easily be sourced to meet government TEMPEST (a reference to compromising emanations that may result in unauthorized disclosure of information) specifications.

Government agencies and military services have invested heavily in programs such as the Navy's SAFENET (Survivable Adaptable Fiber-Optic Embedded Network) and MFOTS (Military Fiber-Optic Transmission System) to integrate FDDI networks into government communication systems on bases, reservations and various shipborne, airborne and space vehicle environments. In addition, the FDDI standard is included in the U.S. GOSIP specification, which outlines networks and protocols approved for government system installations. FDDI and FDDI precursor equipment is available from vendors such as FIBRONICS, Proteon, Canoga Perkins, and Raycom Systems.

As FDDI nears ratification, with only SMT, Version 6.2 awaiting final approval, ANSI is already preparing two additions. The first includes a *single-mode fiber option* that extends internode distances from 2 to 40 kilometers (about 25 miles). The second initiative, known as *FDDI II,* adds *isochronous* data (nonpacket data) transmission capability, enabling support of voice and video as well as data traffic. Isochronous signals are periodic signals in which the time interval that separates any two corresponding significant occurrences or level transitions, is equal to some unit interval or a multiple of that unit interval. For example, in digitized voice signals, voice samples ideally occur at precisely the sampling interval or frame rate. Originally intended to be compatible with FDDI, it now appears that nodes/workstations will require both packet and isochronous media access controllers (P-MACs and I-MACs), and that on any given ring, transitioning to an FDDI II mode will demand that all nodes be FDDI II compatible.

Despite considerable debate regarding details of the FDDI standard, interest in the technology is high. At the Interop 91 show, over 40 vendors interoperated on the same ring, including some from Japan, England and Germany. Perhaps the principal technical disadvantage of FDDI is the fact that nodes must have operational capabilities to intercept and process signals at 100 Mbps. When used as a lower-speed LAN interconnection backbone medium, only routers need the high-speed capability. As a competitor for horizontal wiring to the desktop, FDDI imposes high-speed transmission requirements on every workstation or attached device.

Practical FDDI horizontal wiring disadvantages include high unit termination costs, traceable to advanced craftsperson skill levels needed to polish fiber ends and attach connectors, and expensive fiber network interface cards that have yet to benefit from economies of large scale production. The magnitude of these trade-offs and the role that FDDI over UTP cable for horizontal wiring are treated in Chapters 6 and 11.

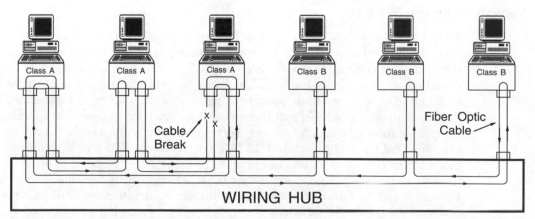

Fig. 13.13 Fiber distributed data interface (FDDI) network approach.

Other Premises Services Options

Apart from the mechanisms described above, this section identifies additional premises data service options. ARCNET, popular during the early 1980s, is a 2.5 Mbps token-passing LAN developed by Datapoint Corporation. Originally designed for RG62 A/U coaxial cable but later adapted for UTP cabling, ARCNET is relatively inexpensive, easy to install, expand, and modify, and is a proven technology that supports up to 255 nodes. Although now giving way to IEEE 802.3 CSMA/CD LANs for new installations, many current users are hesitant to upgrade to different technologies in the absence of operational requirements that ARCNET cannot meet.

Another product, AT&T's Information Systems Network (ISN), uses packet switching in a hierarchical star topology. This product uses concentrators, secondary connection points for groups of devices, to statistically multiplex data from a number of sources on a single link to a *packet controller*. A packet controller is the network's central switch and management device. All messages between sending and receiving endpoints pass over a short high-speed bus, which offers technical advantages relative to contention-based CSMA/CD designs. Optical fiber or private line connections through carrier networks permit the creation of large networks. Although the ISN design has merit, its relationship to AT&T's integrated voice/data PBX and CSMA/CD LAN products has not been fully articulated.

Finally, some LECs offer central office-based services to compete with CPE LAN and data PBX approaches. LEC LAN offerings, known as *CO-LANs*, are proposed as alternatives to CPE LANs using any of the IEEE 802 CSMA/CD, token-ring, or token-bus approaches. LECs also offer virtual circuit switched services which essentially duplicate on-premises data PBX functions described earlier. The factors which differentiate on-premises from LEC offerings include the need for both terminal-to-LEC CO access lines, plus, in the case of terminal-to-host computer traffic, an equal number of host-to-terminal return links. Evidence to date indicates that LEC offerings are constrained by both the physical delay and cost disadvantages associated with remote local area network arrangements.

Network Operating Systems

To this point, our LAN discussion has focused on equipment and operational functions. Embedded or installed in the equipment, and implementing most of the functions are *network operating systems* (NOS), i.e., software that controls the execution of network programs and modules. Structurally, networking software is made up of multiple modules, most residing in network servers, but some installed in each terminal/station accessing network resources. Peer-to-peer NOSs permit any terminal/station to act as a resource server or a client, and can be based on Microsoft's Disk Operating System (MS-DOS) designed for IBM and compatible PCs. Since MS-DOS is not designed to run multiple programs or respond to many simultaneous users, most NOSs designed for large networks with dedicated servers/superservers have a *multitasking* and *multiuser* architecture. Advanced NOS products support network management, diagnostics, and administration, as well as primary server, client, device, and external network driver functions.

Novell's *Netware* product has captured 70% of the NOS market, which amounts to 44 million terminations and 400,000 LANs. Netware supports a heterogeneous hardware environment including 30 different LANs and 100 different network adapters. The product supports true file sharing and incorporates system-fault tolerant/transaction-tracking software which enhances data integrity and implements disk mirroring and duplexing. Netware 3.X is a 32-bit NOS supporting 32 terabytes of disk space, 250 users per server, and 100,000 open files.

Microsoft's NOS product, called *LAN Manager,* is backed by AT&T, DEC, HP, IBM, NCR, 3 Comm, and others. LAN Manager uses the multitasking operating system, which is required on equipment used for servers but not for client terminals/stations. One of the product's unique capabilities is its support of distributed processing. Other NOS vendors include Banyon's *Vines,* Artisoft's *LANtastic,* and NOSs from AT&T, DEC and others.

LAN Performance Considerations

The operation and maintenance of Local Area Networks (LANs) is often characterized by recurring and unpredictable failures that demand "fire-fighting" corrective actions by highly qualified staff. One reason for this is that LANs typically incorporate and support heterogeneous vendor hardware and software, the interoperation of which is not specified or guaranteed by any single source. As a consequence of complex device interactions, problem diagnosis and isolation can pose a significant challenge.

The result is that there are a large number of LANs in existence today which exhibit chronic failures, and for which LAN operators have little insight into design sensitivities, weaknesses, or performance margins in terms of user traffic and utilization. Based on industry reports, 25 two- to six-hour outages per year have been reported by large network users, and administration costs per 100 nodes range from $18 to $32,000. The ramifications of this situation have been documented in a 1989 Infonetics study *The Cost of LAN Downtime*, which concluded that the impact on business revenues that results from networks being unavailable dwarfs any other cause.

Another problem is that LANs typically evolve over time, are implemented with minimal front-end requirements analysis, and little formal product development, test and evaluation. As in any data communications network, response time and throughput are key performance metrics in large heterogeneous networks, but these parameters are difficult to predict based on manufacturer's specifications. Complicating the problem further is the fact that bridge and router performance varies with traffic statistics in different ways for different vendors. Thus the problem reflects the complexity of LAN components and differences in performance characteristics among vendors of apparently plug-compatible components.

For example, Fig. 13.14 illustrates the percentage of packets dropped by Cisco and Wellfleet routers as a function of packet byte size under heavy traffic conditions. Note that not only do the vendor products differ for each packet size, but that the trend from large to small packet performance is exactly opposite. Furthermore, the industry has not yet developed benchmark testing standards, so predicting theoretical and actual performance using traffic estimates and manufacturer's specifications, although highly important as a point of departure for new networks, is difficult.

Part of the difficulty stems from the use of multi-vendor "open system" products in configuring LANs and the fact that neither user organizations, nor third-party integrators, can afford to duplicate the large product engineering investments normally required to develop integrated products of a similar scope, such as large PBX or proprietary mainframe systems. It is unlikely that anyone could undertake to implement a 10,000-line PBX from an assortment of different vendor products, including computers, line adapter cards, and multiplexers, with software from yet another source, without expecting anything other than chaos.

Although $5 to $10 million is not considered an unreasonable cost for a large-scale, fully engineered and product-developed PBX, LAN projects can represent equivalent or greater complexity. Yet LAN-project budgets of that magnitude would be unthinkable. Nevertheless, the open-system, multivendor trend in the evolution of LANs is irrevocable, and so the long-term solution lies in emerging standards work and improved open network management techniques, systems, and products.

Network Management

A growing realization among business leaders is that continued viability of their corporations and institutions, is inextricably linked to telecommunications operations. For some, the cost impact of service outages is measured in tens, or even hundreds of thousands of dollars per hour.

Beyond a quest for higher intrinsic system reliability, the high-cost impact of outages places a premium on rapid fault detection and isolation, performance and traffic monitoring techniques that can detect deteriorating network conditions and alert operators to action before catastrophic failures occur. Centrally controlled reconfiguration and restoration capabilities constitute important aspects of *performance management,* a subset of network management.

Information service users often require ready access to information resources distributed throughout buildings, among buildings on a campus, and at remote locations spanning the nation or the globe. Consequently, the requirement for interconnecting local area, metropolitan, and wide area telecommunications networks—public and private—has become commonplace. Once users employ interconnected networks, they must monitor and manage both individual subnetwork and end-to-end network domains. This task was made more complex by divestiture, which in most instances, necessitates the use of multiple carrier networks.

The operational need, and the market for products that respond to that need, has prompted vigorous efforts from standards setting bodies and network management product vendors. It is reported that at one time, IBM had assigned over one thousand people to NetView, SystemView, and other projects related to its network management products. Digital Equipment Corporation has stated that Enterprise Management Architecture is the third largest development project in the company's history. AT&T with its Unified Network Management Architecture

Percentage of Packets Dropped

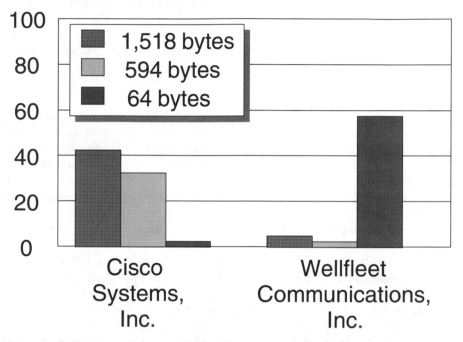

Fig. 13.14 LAN router performance (*Business Systems Group, Inc.*).

(UNMA), Hewlett-Packard, Sun Microsystems, and others are making similar levels of investment.

While these efforts center around homogeneous vendor products and systems—such as IBM's NetView as the management resource for an IBM Systems Network Architecture (SNA) network—one of today's key challenges is achieving integrated, centralized control of distributed systems composed of dissimilar subnetworks and numerous heterogeneous vendor products within those subnetworks. Without integrated management, a service interruption, such as the loss of a main communications line, could trigger a deluge of failure alerts from divers network components, each measuring a different parameter, encoded differently, and transmitted and/or displayed using different protocols and formats.

To address this problem, ISO is developing a *common management information protocol* (CMIP)—an OSI protocol for network management, designated ISO/IEC (International Electrotechnical Commission) 9596. Building on OSI Reference Model work, the scope of CMIP activities addresses peer-to-peer subnet as well as hierarchical management, and includes protocols applicable to all seven OSI layers.

In CMIP parlance, a *management domain* may be decomposed into one or more *managed systems* (sub-domains) and managed systems further decomposed into managed objects. Managed objects are resources (e.g., modems, T1 multiplexers, bridges, routers, LAN hubs, etc.) to be monitored and controlled by one or more management systems. A *management agent* is hardware and/or software in an object which exchanges management information with a *management station.* In networks, management stations run by an operations staff constitute a network operations center (NOC). CMIP establishes a structure for formatting messages and transmitting management information between reporting devices (agents in managed objects) and data collection programs residing in management stations.

Network management operations require an extensive *management information base* (MIB) to store system/network configuration, performance, traffic, maintenance, security, accounting, technical support, and other data. The MIB contains information about managed objects including the attributes, nominal and threshold alarm values, management actions that may be invoked (e.g., initiation of equipment test modes), events that may be reported (error conditions exceeding defined thresholds), and other data associated with each object. Naturally, to be universally (internationally) applicable in heterogeneous network and vendor product environments, there must be explicit agreement on strict object definitions and names, as well as associated values, actions and events.

While CMIP theory and the structure are sound, its full definition and penetration of traditional local and wide area networks has yet to take place. Filling the gap is *simple network management protocol* (SNMP), a network management protocol originally produced for use within the Internet—a scaled-down approach which resembles the CMIP structure and functionality outlined above. Internet is a large collection of connected networks, primarily in the U.S., running the Internet Suite of protocols sometimes referred to as the DARPA (Defense Advanced Research Projects Agency) Internet, or the National Science Foundation (NSF/DARPA) Internet or the Federal Research Network.

The Internet protocol suite is a collection of protocols produced under DARPA sponsorship, which is currently the *de facto* industry standard for open system networking. Included are the *transmission control protocol* (TCP), a transport protocol offering *connection-oriented* transport service, and the *Internet protocol* (IP), a network protocol offering a *connectionless-mode* network service.

Developed jointly by the Department of Defense, industry and the academic community, SNMP was ratified as an Internet standard in request for comment (RFC) 1098, in April of 1989. SNMP is currently supported by a large contingent of LAN and WAN manufacturers, with over 100 companies shipping devices that support SNMP agents. The earliest implementations having been in bridge and router equipment. Figure 13.15 depicts the generic SNMP architecture in a local area network application.

Although LAN management addresses only a part of enterprise system management, in no other segment has the incidence of heterogeneous equipment interconnection been higher. While no single product can represent a complete solution, Novell's recent announcement of its NetWare management system (NMS) product may represent significant progress in open systems management. Since success in a heterogeneous environment is strictly dependent on compatible multivendor design and operation, the product's promise is in no small measure due to the fact that more

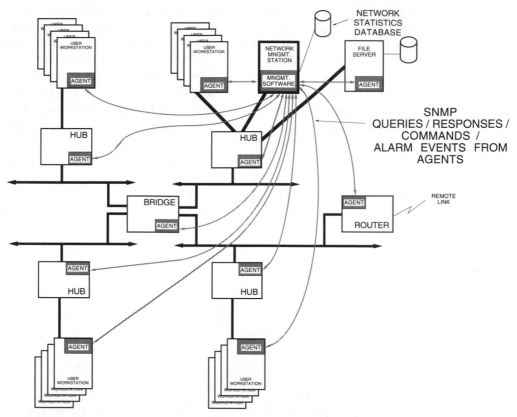

Fig. 13.15 Simple network management protocol (SNMP) LAN management concept.

than 50 computer and network industry vendors have pledged NetWare management system support.

NMS consists of three NetWare-loadable modules: NetWare and SNMP management agents, a services manager, and a management map. Agents, which must be installed in each server controlled by NMS, collect statistics like memory usage, available disk space, and data from attached devices such as hubs or client workstations. Agents also report attribute values and alarm thresholds related to traffic congestion or failure conditions. Management information is exchanged between agents and management stations using Novell's inter-network packet exchange (IPX) or the TCP/IP protocols.

The NetWare services manager is an application that runs on a management station which can be either a dedicated or nondedicated personal computer using Microsoft Windows or IBM OS/2 Presentation Manager software. Using a data base, the NetWare services manager processes and stores agent information enabling an operator to monitor and control network components, to initiate packet test transmissions, and reconfigure the network to bypass failed components or ports. NetWare's management map can "discover and learn" servers, routers, bridges, cable segments and workstations, and also provides graphical network representations.

An SNMP limitation is that the network being managed is also the network used to exchange agent-manager information. This means that during network installation or under severe failures (cable breaks, etc.) the SNMP process itself is disabled. Some proprietary network management systems employ dial-up modems through switched voice circuits to maintain critical remote monitoring and control functions. Also the application of SNMP in large networks involving a hierarchy of management domains may not be adequate due to the need to manually configure agents to address particular managers. As SNMP functionality evolves, many predict that SNMP and CMIP approaches will converge.

Metropolitan & Wide-Area Network Services

Today's LANs frequently begin as stand-alone networks serving small work groups with related job assignments and homogeneous equipment. In large enterprises, however, user requirements for access to company-wide data bases and other information services often lead to incorporation of separate LANs into large multibuilding enterprise networks, connecting heterogeneous host and workstation equipment. When individual LANs are unable to meet mounting capacity demands, they typically are split into two or more segments connected by bridges. As described in the last chapter, bridges isolate LAN segments by keeping traffic within individual segments, unless it's addressed to terminals on other segments.

Enterprise data communications requirements normally extend beyond single building or campus locations, necessitating interconnection of local area networks by metropolitan and wide area networks, using LAN routers or gateways. IEEE defines a *metropolitan area network* (MAN) as a network in which communications occur within a geographic area of diameter up to 50 kilometers (about 30 miles) at selected data rates at or above 1 Mbps consistent with public network transmission rates. In this book, MAN means any network in which communications cross public rights-of-way and occur within a geographic area of diameter up to 50 kilometers, independent of data rates. *Wide-area networks* (WANs) provide services beyond the distance limitation of MANs. As a context for subsequent discussions of current and emerging MAN and WAN services and technologies, this chapter summarizes how user requirements and the telecommunications environment has evolved over the past two decades. Later chapters build on this framework to portray how future integrated digital networks are being developed to support the gamut of voice, data, and video services.

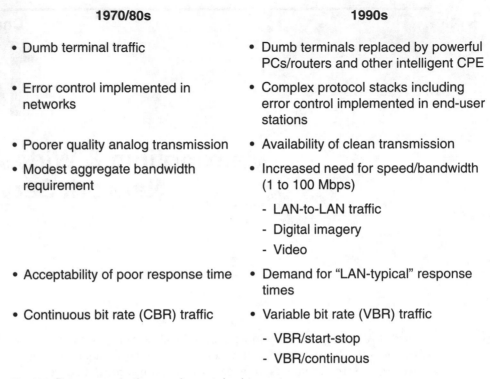

1970/80s	**1990s**
• Dumb terminal traffic	• Dumb terminals replaced by powerful PCs/routers and other intelligent CPE
• Error control implemented in networks	• Complex protocol stacks including error control implemented in end-user stations
• Poorer quality analog transmission	• Availability of clean transmission
• Modest aggregate bandwidth requirement	• Increased need for speed/bandwidth (1 to 100 Mbps)
	- LAN-to-LAN traffic
	- Digital imagery
	- Video
• Acceptability of poor response time	• Demand for "LAN-typical" response times
• Continuous bit rate (CBR) traffic	• Variable bit rate (VBR) traffic
	- VBR/start-stop
	- VBR/continuous

Fig. 14.1 Data communications requirements/environments.

Requirements & Environmental Trends

To understand data communications system design trends, it is instructive to contrast data service requirements/environments of the 1970s and 80s with today's situation, and make projections for the balance of the 90s. Figure 14.1 lists key factors shaping developments. In the early 70s, communications between so called "dumb terminals" and central host computers dominated data network traffic. As noted in Chapter 12, these terminals could support only simple interface protocols so to achieve reliable transport service, networks had to supply error control, addressing, connection establishment, and other high-level protocol services.

As a result, existing wide area data communications networks are either SNA, X.25 or circuit-switched architectures. *SNA (Systems Network Architecture)* is IBM's proprietary description of the logical structure, formats, protocols, and operational sequences for transmitting information units (packets) and controlling network configuration and operations. The purpose of SNA's layered structure is to permit end user CPE to be independent of and unaffected by the way in which the specific network services and facilities are provided. While many criticize it as being antiquated and inefficient, in its day SNA had no peer. Moreover, its multilayer protocol served as a forerunner for ISO's reference model.

Unlike yesterday's dumb terminals, today's powerful PCs, workstations, routers, and other intelligent CPE are capable of supporting complex protocol stacks and assuming end-to-end responsibility for reliable transport. These capabilities make simpler, higher-speed networks not only possible but desirable. Also, in the 1970s, data traffic was carried on analog voice transmission facilities. Voice traffic is relatively unperturbed by transmission errors (one error in one thousand bits is tolerable), but data communications demands near error-free transport.

Powerful error-detection and correction techniques used in SNA and X.25 packet networks furnish end-users with reliable, error-free transport, masking poorer-quality transmission facilities. The price paid, however, is low throughput and high message delay due to extensive processing within networks. As noted in Chapter 12, the combination of virtually error-free fiber optic transmission (bit error rates of less than one error in a billion (10^9) bits without error control) and intelligent end-user CPE devices now permits streamlined network designs which significantly enhance speed and reduce message delay.

Another important factor distinguishing the current situation from the 70s and 80s is an increased amount of data traffic, and higher bandwidth requirements associated with that traffic. While the demand for voice service is increasing at 3% to 5% annually, demand for data service is expanding at a 15% to 20% annual rate. Furthermore, in contrast to earlier interactive applications with modest bandwidth requirements (2.4 to 9.6 kbps), some current and future data communications applications require orders-of-magnitude more bandwidth (1 to 100 Mbps). Examples include LAN-to-LAN, digital imagery, high-speed computer aided design (CAD), and video applications.

Expanding bandwidth requirements need faster MAN and WAN response times. In the past, interactive service users had to be content with slow screen refresh times and transaction response characteristics. Today's users are accustomed to near-instantaneous response available on high-speed LANs. When 10 Mbps LANs are interconnected by 2.4 kbps MAN or WAN lines, users become frustrated with response time degradation occurring when logged-on to remote LANs.

A final development driving future network designs is the need for variable versus continuous bit rate services. Traditional voice and data transmission requirements are best satisfied by *continuous bit rate* (CBR) services. For example, without digital speech interpolation (DSI) techniques as described in Chapter 2, conventional PCM voice encoders produce an output bitstream at a fixed 64 kbps bit rate, independent of pauses in speech activity. CBR services are similarly suited for file transfers at constant rates over entire sessions.

Some new applications, however, can be more economically supported with *variable bit rate* (VBR) services. For example, users may need to scan files or documents, briefly viewing pages until they find what they are seeking. While in the *scan mode,* average bandwidth must be high enough to build display screens rapidly. Once desired objects are located, the user pauses or enters an *editing mode,* during which time, greatly reduced average bit rates are sufficient (actual rates being driven by how fast users think and type). Such applications are said to exhibit *VBR/Start-Stop* characteristics. That is, during a data communications session, periods of intense activity may be interspersed with periods of less and/or no activity.

Another VBR example involves new forms of digital video encoding. Without compression, encoding standard National Television System Committee (NTSC) television analog signals produces CBR digital signals at bit rates of about 140 Mbps. Encoders employing conventional compression algorithms produce digital output signals at constant bit rates less than TV's peak 140 Mbps bandwidth requirement. Such encoders take advantage of the fact that during still periods (little image motion) as low as 1 Mbps may be adequate to transmit all essential image information. Since these encoders generate constant bit rates too low to preserve full 140 Mbps NTSC digital quality, the algorithms used are termed *lossy*. During rapid image motion intervals, lossy algorithms compromise image resolution, smooth image motion tracking, or both. The motivation behind the use of such encoders, and the lower than 140 Mbps output rates, is to provide acceptable quality while conserving expensive transmission bandwidth capacity.

The David Sarnoff Laboratory is currently experimenting with a new class of video encoding algorithms that generate *VBR/Continuous* outputs. These new algorithms are *lossless*; they involve no loss in resolution quality or motion tracking capabilities. During periods of rapid image motion, the new encoders can produce peak bit rates enabling full image quality. During periods of little motion, the Sarnoff algorithms produce lossless encoded video outputs at as low as 1 Mbps.

For networks to support VBR traffic with maximum economy (equivalent to handling the greatest number of VBR signals possible within a fixed maximum bandwidth and a specified grade of service), they must employ switching and multiplexing designs that provide users with VBR or bandwidth-on-demand services. Today's circuit switching networks cannot provide bandwidth-on-demand services. Rather, as noted earlier, following call establishment, *fixed bandwidth channels* are dedicated to single users for entire call durations. In circuit switched networks, the fixed bandwidth would have to be at least as great as the peak VBR bit rate. This means that for those intervals where actual VBR rates are less than the peak rate, some or most of the fixed circuit switched channel capacity is wasted. That is, the channel isn't fully used by the connected parties, and the network has no way of allocating the unused capacity to other users.

Packet switched networks can be designed to provide bandwidth on demand for users with VBR traffic. But for the network to determine how many VBR users can be simultaneously accommodated within a transmission facility of fixed bandwidth, it needs to know both the peak and the average bit rate statistics of each user's traffic sharing the facility. The problem of designing VBR capable networks to ensure grade of service levels is not unlike the circuit switched process described in Chapter 7—except, of course, that it is a good deal more complicated. Blocking probability is not simply a function of the traffic intensity expressible in numbers of fixed bandwidth call arrivals and numbers of available channels, but must take into account both the number of calls, and peak and average bit rates involved with each call.

Today's packet switched networks are not equipped to provide true bandwidth-on-demand services for users with VBR traffic. CCITT's Study Group XVIII is currently developing a paradigm which defines fundamental concepts, rules, and procedures to be used in such a network. The study group defines *asynchronous transfer mode* (ATM) as a broad-bandwidth, low-delay, packet-like (cell relay) switching and multiplexing technique. It is essentially connection-oriented, although it is envisioned as supporting all services. ATM networks will accept or reject connections based on traffic already being carried by the network, and the new user's average and peak bandwidth requirements. ATM networks also implement policing mechanisms to ensure that users don't violate peak or average bit rate commitments. When available, ATM networks will provide flexible and efficient service for LAN-to-LAN, compressed video, and other VBR bandwidth-on-demand applications.

Service providers will benefit from ATM since their networks will be able to carry more traffic than circuit or non-ATM packet switched alternatives. Users will benefit from ATM services since it is anticipated that service providers will pass on some of the savings and charge only for actual quantities of information transferred.

As stated at the beginning of this chapter, this introductory review of evolving user requirements and telecommunications environments provides a context for evaluating the merits of existing and future MAN and WAN services. We next present a review of these services, reserving Chapters 16 and 17 for additional discussion focusing on narrowband integrated services digital network (ISDN), future broadband ISDN (BISDN) developments, and underlying ATM and advanced networking technologies.

Figure 14.2 depicts the array of MAN and WAN network and service options and indicates their availability. It should be noted that some options—frame relay, for example—may be offered by both LECs and IXCs (or implemented as private networks). In fact, the technical and legal ramifications of MAN and WAN boundaries in terms of intra-LATA and inter-LATA jurisdictions are yet to be fully explored. The presentation that follows emphasizes technical characteristics without attempting to resolve precisely how the regulatory and judicial actions will ultimately shape U.S. data communications.

Data Communications Service Options	Availability
• PSTN Modem Networks	Now
• Dedicated Digital Private Lines	Now
• Circuit Switched Digital Capability (CSDS)/ Virtual Circuit Switch Services	Now
• Conventional Packet Switched Data Networks	Now
• Narrowband Integrated Services Digital Network (ISDN)	Limited
• FDDI MAN Offerings	Limited
• Frame Relay	Limited
• Switched Multimegabit Data Service	Trials
• Broadband ISDN (B-ISDN)	Future

Fig. 14.2 MAN/WAN solutions.

PSTN Modem Networks

The data communications service with the greatest potential for universal, world-wide connectivity still involves the use of modems and public switched voice telephone networks. Chapter 2 describes how modems facilitate data communications over the ubiquitous PSTN, or over dedicated analog voice grade private line networks. Estimates of the percentage of the 100 million PCs that are connected to modems range to 25%. Even large corporations that employ SNA or packet switched data networks often use modems over PSTN facilities as back-up or to perform "out-of-band" data network diagnostics during failure recovery. Reflecting the infusion of modern large scale integrated-circuit (LSI) technology, reliable and efficient modems are available at costs as low as $100 (Chapters 7 and 10 present design and tariff guidelines pertinent to modem-based data network services).

Dedicated Digital Private Lines

A second and popular MAN and WAN data communications service option involves dedicated digital private lines. LEC and IXC offerings range from high quality but relatively expensive 9.6 and 56 kbps services provided over special carrier facilities, to fractional T1, DS1 (T1) and higher rate services, as described in Chapters 8 and 9. Chapters 7 and 10 present network design and tariff guidelines for digital private line data services. The frame relay section of this chapter contrasts private line and frame relay LAN-to-LAN interconnection alternatives.

Circuit-Switched Digital Capability (CSDC)
Virtual Circuit Switch (VCS)

CSDC service provided by some LECs affords customers with dial-up access to 56 kbps data service. CSDC is an economical alternative to dedicated lines for users with occasional needs to transmit large blocks of data for a few hours per day or a few days per month. Virtual circuit switched (virtual circuit switching is defined in Chapter 12) networks, in local and wide area applications, are intended to provide host-to-host and terminal-to-host data communications. LEC VCS offerings include bridging of remote customer LANs, access to wide area and packet networks, and central office-based LAN (CO LAN) services. CSDC and VCS networks employ circuit switching, and involve operations similar to those described for data PBXs in Chapter 13. Perhaps the most popular circuit switched digital application is for dedicated private line service back-up.

Conventional Packet Switched Networks

Conventional and fast packet switched network concepts, operations, and technologies are discussed in Chapter 12. Some LECs offer conventional packet switched public data network (PSPDNs) services. Nationwide packet switched networks currently provide WAN services offered by AT&T (ACCUNET Packet Service), British Telecom's Tymnet, CompuServe Inc.'s CompuServe Network Services, General Electric's GE Information Network Services, IBM's IBM Information Network, and Sprint International's Sprintnet (formerly Telenet).

As pointed out in Chapter 12, to fully capitalize on network capabilities requires direct digital connections to packet switches. Connections via modems and PADs adjacent to the packet switches located on carrier premises, do not extend error control and other services to customer premises CPE. For high-volume users, technical and economical factors favor collocating packet switches or PADs on customer premises. As a consequence, many large users operate private packet switched networks. Examples include the Treasury Department which had 10 nationwide networks prior to the implementation of its single packet switched Consolidated Data Network (CDN), and the Federal Aviation Administration which currently operates two X.25-based packet switching networks, one for administration, and one for operations support.

Chapter 15 characterizes tariff structures for basic and value added networks offered by service providers.

Integrated Services Digital Network (ISDN)

The Integrated Services Digital Network (ISDN) is a set of standards being developed by the CCITT and various domestic standards organizations. The first set of CCITT recommendations, adopted in October, 1984, defines ISDN officially this way:

> An ISDN is a network, in general evolving from a telephony integrated digital network, that provides end-to-end digital connectivity to support a wide range of services, including voice and non-voice, to which users will have access by a limited set of standard multipurpose user-network interfaces.

ISDN is now available on a limited basis in the U.S. Appendix A explains why progress towards ISDN in the U.S. has lagged behind the rapid deployment in Europe and Japan. Chapter 16 explains interface definitions and discusses ISDN basic and primary rate services.

Fiber Distributed Data Interface (FDDI) MAN Offerings

Although FDDI was originally proposed as a high-performance interconnection among computer mainframes and mass storage peripherals, and as a high-speed backbone network interconnecting lower speed LANs within building or campus locations, it has been adapted for MAN use. In 1991, Metropolitan Fiber System Inc. had installed FDDI networks in Houston, Texas, effectively creating a crosstown MAN service for interconnecting LANs and other data applications. Metropolitan plans to expand its service to 10 additional cities. FDDI's technical characteristics are presented in Chapter 13.

Frame Relay

Current X.25 packet switches with typical 300 packet-per-second capacities (approximately 300 kbps) are no match for multimegabit LAN-to-LAN connectivity applications. More powerful microprocessors can increase conventional packet switching throughputs, but hardware improvements alone are not adequate.

As it stands, the computing complexity imposed by X.25's overhead-intensive protocol, designed for individual packet-by-packet routing and link-by-link error control, while appropriate for yesterday's poorer-quality analog-derived transmission systems, stands as an insurmountable obstacle to multimegabit MAN and WAN applications.

see page 225

Frame relay is a high-speed switching technology that achieves 10 times the packet throughput of existing X.25 networks by eliminating two-thirds of the X.25 protocol complexity and adding out-of-band signaling. The frame relay concept as one branch of fast packet switching is discussed in Chapter 12. Frame relay service is based on the transparent delivery of frames defined at the data link layer 2 of the OSI reference model. As noted in Chapter 12, the frame relay standard is described by CCITT under its I-Series recommendations, I.122 "Framework for Additional Packet Mode Bearer Services," and within ANSI T1S1/88-2242 as "Frame Relay Bearer Service—Architectural Framework and Service Description."

Frame relay is derived from CCITT's link access protocol (Link Access Procedure-D, LAP-D) and has begun to emerge as a keystone for future fast packet networks using high quality fiber optic transmission. Figure 14.3 compares X.25 and frame relay protocol stacks. Where X.25 uses the bottom three layers, frame relay performs its switching functions using only the physical layer 1 and the "core aspects" of the LAP-D data link layer 2. The frame relay frame format, in the middle of the figure, shows that only two bytes are used for header information. Frame relay provides *connection-oriented* service and, although either switched or permanent virtual circuit service can be supported, in 1992 carrier offerings were limited to *permanent virtual circuit* (PVC) service. (See Chapter 12 for definitions of connection-oriented and PVC terminology.)

The *data link connection identifier* (DLCI) part of the header specifies PVC paths connecting pairs of network end points. PVC paths are allocated by service providers to customers, with up to about 1000 PVCs available for each network interface. Including DLCI routing information in the layer 2 header eliminates the need for a layer 3 protocol. Customer CPE provides DLCIs to frame relay network switches in much the same way that humans provide manually inputted telephone numbers to circuit switches to identify called parties. Bit patterns representing DLCIs have local significance only, and can be reused at every network entry point.

Other parts of the header defined on Fig. 14.3 are used for congestion control and other network operations. The lower part of the figure illustrates that X.25, SNA, and data signals in other protocols' formats can be encapsulated as information in frame relay frames. In X.25 networks, transmitting switches store packets until an acknowledgment is received that each packet has arrived without errors at the destination switch. When destination switches receive packets with errors, they return a negative acknowledgment to the transmitting switch, which then retransmits the packet. In frame relay, a frame check sequence appended after the information portion of the frame enables detection of errors within frames. Unlike X.25, however, errored frames are simply discarded, and it is the responsibility of higher layer protocols implemented in intelligent CPE to take whatever corrective action is required.

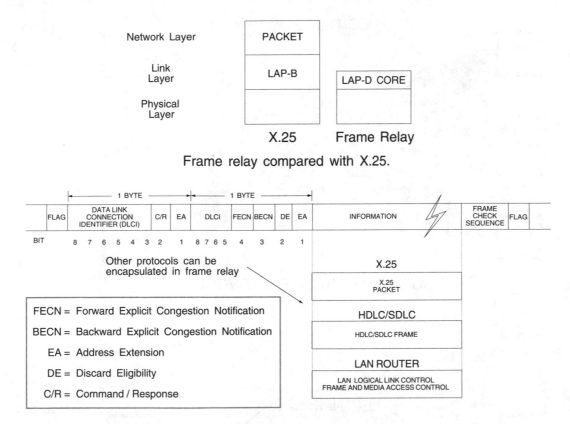

Frame relay compared with X.25.

Q.922 (Annex A) LAP-D frame format.

Fig. 14.3 Frame relay frame structure.

Figure 14.4 depicts another important frame relay advantage. The top of the figure shows four LANs, each connected to every other LAN using dedicated private lines. Since there are no intervening switches, full-mesh connectivity is necessary, not unlike the telephone example shown in the first figure of this book, i.e., Fig. 1.1. As in Chapter 1's telephony example, switching can significantly reduce the number of separate transmission channels required. Because of the throughput and delay problems of X.25 networks, prior to the appearance of fast packet frame relay networks, many users had no viable choice other than full mesh networks.

For the purposes of an example, assume that each node in the top of the figure uses three 56 kbps lines to every other node. This results in the need for 12 LEC private lines (local channels in the tariff parlance introduced in Chapter 10), 6 IXC private lines (interoffice channels in tariff parlance) and 12 router ports. With frame relay, shown in the bottom part of the figure, 4 LEC private lines, 4 router ports, and an IXC frame relay switched service is required. It is interesting to extrapolate these comparisons by increasing the number of customer premises nodes. As network size grows the case for switched services becomes even more compelling.

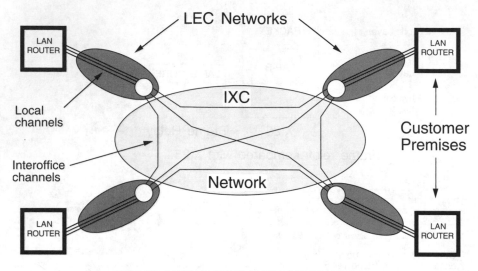

FOUR NODE PRIVATE NETWORK

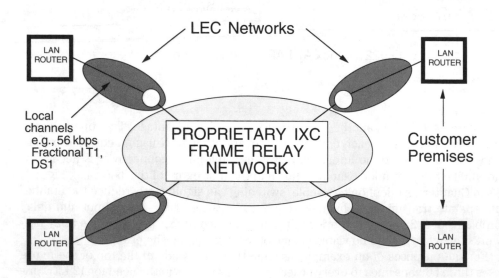

FOUR NODE FRAME RELAY NETWORK

Fig. 14.4 Frame relay/private line network comparison.

At the present time, frame relay pricing varies considerably among service providers, but ballpark savings relative to private line alternatives of 30% to 40% can reasonably be expected. Frame relay service providers typically price their service on a per port (e.g., a 56/64 kbps, fractional T1, DS1 port speeds) basis, independent of the type of access service employed. For large users, DS1 (T1 lines) local channel LEC access lines may be justified. Users under the impression that with DS1 access, higher than 56 kbps connections could be made between two company locations have discovered that service providers (Sprint is an exception) do not permit higher than agreed upon end-to-end bit rates, even when such traffic would not cause network congestion. To date frame relay services are not suited for voice, video or the variable bit rate traffic described above.

With LAN-to-LAN connectivity emerging as a major issue today, both LAN bridge/router and T1 equipment vendors are striving to address the frame relay market. T1 vendors who bring the strength and advantages of telecommunications carrier experience to bear include AT&T, Network Equipment Technologies (NET), Newbridge Networks, Netrix, Northern Telecom, StrataCom, and Timeplex. Strata-Com offers a proprietary, fast packet switching product now being tested in several field trials. DEC, Cisco, Vitalink, Wellfleet, and RAD Data Communications Inc. are LAN equipment vendors modifying existing customer premises products for frame relay application.

In late March 1991, Williams Telecommunications Group, Inc. (WilTel) became the first carrier to announce general availability of a public frame relay service. WilTel's frame relay service is based on StrataCom's IPX-32 fast packet multiplexers, which have been installed in the carrier's nationwide 11,000-route-mile digital microwave and fiber optic backbone network. US Sprint introduced its public frame relay service later in 1991. Sprint will continue to support its standard X.25 based packet switching service, formerly known as Telenet, as well as the new frame relay service. AT&T's Interspan frame relay service, scheduled for general availability in 1992, will employ the StrataCom IPX-32 product as the access node of its high-speed network. BT North America, CompuServe and Graphnet are other announced or current frame relay service providers. BT North America offers service in 116 cities and three countries.

It is interesting to note that although frame relay was to be a new ISDN service (recall its CCITT title,"Framework for Additional Packet Mode Bearer Services," I.122, part of the I series which addresses ISDN), no U.S. ISDN service provider has announced frame relay as a *call-by-call* additional mode for its existing ISDN service. To date, each frame relay service provider's network is proprietary. Like X.25, frame relay is merely an interface specification leaving internal network design issues to service providers. Consequently, interoperability among different frame relay networks has yet to be addressed. Moreover, with the 1992 permanent virtual circuit limitation, no switched virtual circuit mechanism is available to support "public-telephone-like" connections to randomly selected frame relay subscribers. Thus as of 1992, carrier-based frame relay services permit subscribers to share common facilities and benefit from lower costs than would be incurred using private frame relay networks, but those offerings cannot be characterized as true "public fast packet switched" service.

In spite of these limitations, frame relay service could be offered by LECs in addition to switched multimegabit data service (SMDS), to be discussed next. For example, PacBell and NYNEX have indicated that they may provide frame relay service in 1992. Because of constraints on bit rates between pairs of end points and the growing need for flexibility as driven by evolving user requirements, some pundits believe that private networks may dominate the near term frame relay scene.

Switched Multimegabit Data Service (SMDS)

With an eye toward the future, IEEE and ANSI are presently working on advanced MANs that are better suited for interconnection of high-speed traffic from today's LANs. IEEE has prepared a draft MAN 802.6 standard. The proposed standard defines a high-speed shared-access protocol for use over a dual unidirectional bus network. Interim versions supporting data rates of 45 Mbps (DS3) are being developed by Bellcore and might be available in the 1992 timeframe.

Switched Multimegabit Data Service (SMDS) is a Bellcore-proposed LEC service offering public, connectionless, high-speed, fast packet-switched (cell relay—see Chapter 12 for definitions) data service throughout metropolitan areas. When introduced, SMDS will provide 1.544 Mbps (DS1) or 43.736 Mbps (DS3) access to fiber optic-based switched networks. The use of MAN technology now being defined in IEEE 802.6, will likely be the means of attaining early SMDS availability in carrier networks. Every regional holding company (RBHC) is currently running or planning a MAN trial, and all are working through Bellcore to set national standards.

The need for switched carrier-based versus private MAN service stems from:

1. User traffic characteristics

2. The potential need for ad hoc "public" connections between different user organizations

3. The requirement to cross public rights-of-way

Regarding user traffic considerations, a key MAN application is the interconnection of LANs supporting bursty, broadband traffic with short response times.

To be successful, MAN interconnection must be transparent to LAN users. Ideally, these users would enjoy the same throughput and response time performance, whether connected through a single LAN, or via two LANs and an interconnecting MAN. Typically, average LAN-to-MAN-to-LAN traffic may be only 0.5 Mbps, but critical applications may require peak 10 Mbps capacities. It is hard to imagine that a user organization would opt to lease a DS3 (44.7 Mbps) service for occasional use. Thus, along with items 2 and 3 above, traffic demand and throughput economics increasingly justify switched public service.

MAN standards are based on a three-level SMDS interface protocol (SIP) stack defined in IEEE 802.6, (covering layer 1 and parts of layer 2 in the OSI reference model). As depicted in Fig. 14.5, the proposed standard calls for a distributed queue dual bus (DQDB) architecture, which defines a high-speed, shared medium access protocol for use over a dual, unidirectional, fiber optic bus network. This connectionless service is similar to other IEEE 802 LANs. (Connection-oriented and con-

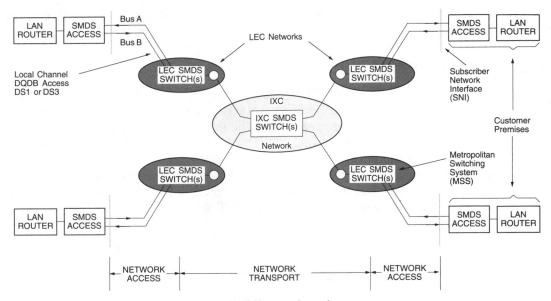

Fig. 14.5 Switched multimegabit data service (SMDS) network topology.

nectionless service are defined in Chapter 12.) As illustrated in the figure, in the literature an *SMDS switch* is termed a *metropolitan switching system* (MSS) and the *network interface* is called a *subscriber network interface* (SNI).

Distributed queueing is a media access protocol that, like FDDI, delivers near-perfect access characteristics. In particular, the protocol enables all of the theoretical "payload" bandwidth (100% efficiency), to be used and the average slot access delay approximates that of a perfect scheduler at all traffic loading levels. *Payload bandwidth,* the bandwidth available for user information bearing signals, is the channel bit rate—less switching/multiplexing signaling and control overhead.

Security is a high-profile MAN issue. User concern is twofold. First, no organization wants its confidential data passing through a competitor's building. Second, there must be a means to prevent one customer from disrupting the service of another, either accidentally or maliciously. Current designs segment MANs into *access* and *transport* network elements, with the access element the only one entering the customer's premises. Bridges are used to connect the segments and to screen data, reportedly making the transport element invisible to customer equipment. Authoritative verification of security issues will likely be required before shared medium SMDS/MAN service is universally accepted.

Cooperation among IEEE 802.6 working groups and CCITT's study group XVII is paving the way toward compatibility between SMDS MAN developments and BISDN's asynchronous transfer mode (ATM), with agreement already reached on a common cell size. Consequently, while initial SMDS work is focused on MAN applications, eventually extension is planned to encompass wide area, inter-LATA applications, as implied in Fig. 14.5. SMDS is incorporating CCITT E Series, E.164

addressing so, unlike frame relay, SMDS will provide true public high-speed switched data service.

Broadband ISDN (BISDN)

In one of the first-draft CCITT documents, *Broadband integrated services digital network* (BISDN) is simply defined as "a service requiring transmission channels capable of supporting rates greater than the primary rate." Primary rate in the U.S. refers to ISDN's 1.544 Mbps rate. With this definition of BISDN, some literature now refers to conventional ISDN as narrowband ISDN or NISDN. Chapter 17 describes BISDN services and the ATM packet switching and multiplexing and *synchronous optical network* (SONET) transmission facilities to be incorporated into the integrated digital networks that will support BISDN.

15

Selecting Data Network Services

MAN and WAN data network design objectives are similar to those of voice networks—to provide specified service and quality levels using the least expensive approaches. Additionally, the principles of traffic aggregation to achieve economies of scale, explained in Chapter 7, also apply to data networks. For large users, the main difference between voice and data network design reflects the current lack of maturity in public data service offerings. Keen competition among the IXCs for shares of the large total voice business revenue has led to aggressive pricing. As noted in Chapter 10, the evolution of public IXC infrastructures towards digital, software-driven architectures has made virtual private network service the IXC's most economical offering. This phenomenon has all but eclipsed facilities-based private voice networks (so prevalent in the 1960's and 70's) from the large user network scene. However, the migration from private to public data networks has not been so dramatic because large data users can still obtain more flexible and lower-priced service from dedicated facilities.

The main reasons for the persistence of private data networks has been the state of switching technology and the profile of user demand. As noted in the last chapter, switch processor limitations and the complexity of the X.25 protocol have made it infeasible to build large central office packet switches capable of aggregating and handling the quantity of traffic that would make public facilities more economical than private facilities. Consequently, today's large X.25 users operate private networks with packet switches located on their premises, an arrangement with both technical and cost advantages. In SNA-type networks, the user's own host computers normally perform switching functions, using private line transmission to interconnect front-end communications processors, as described in Chapter 13. Reports indicate that there may be as many as 20,000 SNA networks operating today.

Moreover, business demands for data services alone have not justified the development of high-speed CO-based fast packet switches, essential to the provision of public data service at prices low enough to wean large users from private networks.

Nevertheless, as noted in the last chapter, technology has improved to the point where several carriers provide packet switched public network services, largely under the banner of *value-added networks* (VANs). In addition to transporting data, VANs provide enhanced services such as protocol conversion (several different protocols operating simultaneously on one network) and electronic mail (E-mail). *E-mail* is the generic term for *non-interactive communication* of text, data, image, or voice messages between a sender and designated recipients using telecommunications. Note that E-mail is tolerant of the unpredictable and relatively long message delay characteristics of conventional packet switched networks.

The pricing structure of such public packet switched services is usage-based, so traditional trade-offs between shared solutions and dedicated solutions apply. Pricing currently favors dedicated networks for medium to large users, but this may change as IXCs upgrade their switching technology to fast-packet based architectures for voice, data, video and all other services. (See Chapter 17.)

The same jurisdictional separation exists for data networks as for voice networks; LECs provide intra-LATA (MAN) service and access to IXC POPs, and IXCs provide inter-LATA (WAN) service. As indicated earlier, the ramifications of jurisdictional regulations on MAN/WAN technical performance is not fully understood. The following sections describe LEC and IXC packet switched service rate structures and include public versus private network comparisons where appropriate.

LEC Packet Switched Services

LEC *packet switched public data network* (PSPDN) services are a recent development with most services introduced after divestiture. These services are described variously as *public packet switched networks* (PPSN), *packet switched networks* (PSN), or *public data networks* (PDN), depending on the particular LEC providing the service. Service is provided using LEC-owned packet switches in selected central offices within LATAs. The switches may be accessed via dedicated private lines between user premises and a CO equipped with a packet switch. Alternatively, users can access PSPDNs using dial-up service via public switched voice networks. In the latter case, connections to packet switches are made through a *pool* of shared CO-trunk terminations.

Figure 15.1 depicts access arrangements and network elements, related to monthly tariff charges. Note that as in the case of LEC and IXC voice networks, a unique tariff terminology is used. For example, *transport* is generally defined as network switching, transmission, and related services that support information transfer capabilities between originating and terminating access facilities. In data communications tariff parlance, *packet transport facilities* refer to transmission facilities between customer premises and a CO with a packet switch. Two possibilities exist. If a customer premises is directly homed on a CO with a packet switch, then the packet transport facility involves only *dedicated local loop* transmission. When a customer premises is homed on a CO without a packet switch, both dedicated local loop and *interoffice transport facilities* are involved.

Figure 15.2 illustrates rate elements for a typical PSPDN service. *Basic service* supports X.25 interfaces. When conversion from other interface protocols is re-

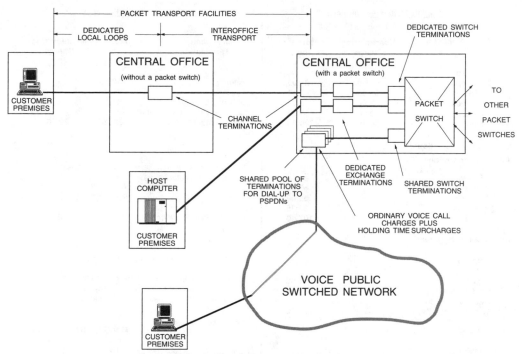

Fig. 15.1 LEC Intra-LATA packet switched public data network tariff elements.

quired at dedicated access connections, additional charges apply. Protocol conversion services are not regulated and vary from LEC to LEC. For dedicated service, the user procures an *exchange termination,* a *switch termination,* and *packet transport facilities* for each access line to the PSPDN switch. As shown in Fig. 15.2, fixed monthly charges are assessed for these services. Dial-in users pay normal voice charges plus a *holding-time surcharge* to access the network.

Usage charges recover costs for the use of LEC network facilities connecting packet switches in different central offices. Most LEC and IXC usage charges are based on the number of *kilosegments* of data sent per month. One kilosegment is one thousand segments of data, and a *segment* is defined as a unit of data with a maximum length of 64 bytes (characters). Where carriers use kilosegments as the billing quantity, care is needed to not confuse billed kilosegments with the number of packets of data transmitted per month. Carriers offer customers the ability to specify *maximum packet size,* a selection made on the basis of traffic types and network operation, as noted below. The default packet size for most networks is 128 bytes. In some circumstances segments are not completely filled with user data, but are billed, nonetheless, as full segments.

The relationship between the number of bytes of user data transmitted and the number of segments billed depends on both the type of data and the protocol being utilized. Typically a packet is generated when protocols determine that data must be transmitted or when accumulated user data reaches a defined maximum

PSPDN Dedicated Services

Switch Terminations:

Speed	Nonrecurring Charge ($)	Monthly Rate ($)
1.2K bps	69.50	21.00
2.4K bps	69.50	25.00
4.8K bps	69.50	29.00
9.6K bps	69.50	33.00
56K bps	69.50	220.00

Exchange Terminations:

Analog Synchronous

Speed		
1.2K bps	---	65.00
2.4K bps	---	65.00
4.8K bps	---	83.35
9.6K bps	---	107.60

Analog Asynchronous

Speed 0.3K bps		
1.2K bps	---	51.00
0.3/1.2K/2.4K bps	---	51.00
(Autoband Exchange Termination)*, Each	---	51.00

Digital

Speed		
2.4K bps	---	56.25
4.8K bps	---	56.25
9.6K bps	---	56.25
56K bps	---	64.00

Packet Transport Facilities

Channel Terminations:

	Monthly Rate ($)	Nonrecurring Charge ($)
Analog	13.50	---
Digital*	45.90	---
56K bps Digital*	54.50	---
56K bps Extended Distance Arrangement**	225.00	1,280.00

Local Loops (2 Each Required) — **Access Area**

	A ($)	B ($)	C ($)
Per Pair, Per Month	5.16	9.00	12.53

* Does not apply for PSPDN dedicated service lines when customer's serving CO is a PSPDN serving CO.

** Some 56K bps packet facilities need more than the standard arrangement and may require case-by-case construction and pricing.

Interoffice Transport

Up to and including 9.6K bps:	Monthly Rate ($)	Nonrecurring Charge ($)
Per Mile	0.71	---
Analog CO Equipment	101.25	---
Digital CO Equipment#	50.40	---
56K bps Digital:		
Per Mile	0.71	---
CO Equipment	50.40	---

Usage Charges

	Normal Packet Kilosegments W/O Protocol Conversion ($)	Priority Packet Kilosegements W/O Protocol Conversion ($)
Day Rate: 8 am-9 pm	0.120	0.155
Night Rate:	0.080	0.105

PSPDN Public Dial-Up Service

Usage Charges: Packet kilosegment charges are the same as dedicated rates

	Each 1/10 Minute ($)
Holding Time Charges:	0.0005

Fig. 15.2 Intra-LATA packet switched public data network rate example (SOURCE: *CCMI*).

packet size. For interactive traffic and some protocols, pressing the enter key on a terminal may send a single line of data to a host or, worse yet, each character may be sent and echoed back to the terminal. In the former case, average packets would contain 40 bytes. In the latter case, packets contain only one byte. With file transfer traffic, virtually all packets are of maximum length, or for the default case, 128 bytes.

For file transfers, *packet fill* is 100%. Normally there are two segments per packet, and the segment user data fill rate is also 100%. For the interactive traffic with line buffering example, the packet fill rate is about 31% (40 bytes—an average typed line worth of data—divided by 128—the packet size). Now there is one segment per packet, and the segment fill rate is 40 user data bytes divided by 64—the segment size, or about 63%. To estimate the number of billable segments per month for any traffic type, the following formula can be applied:

$$\text{segments/month} = \frac{\text{characters/month}}{(\text{packet size} \times \text{packet fill})} \times \text{segments/packet} \qquad (15.1)$$

Using this formula, and the usage and fixed charges in Fig. 15.2, it is relatively easy to compute traffic levels at which private lines are less expensive that PSPDN service between two locations. At 56 kbps line speeds, the break-even point may occur at as low as 10 to 60 minutes of traffic per month, illustrating the relatively high pricing profile of PSPDN service. However, for configurations involving large numbers of nodes, each with small amounts of traffic, the PSPDN may prove more economical than large numbers of private lines. A central bank data center serving numerous automatic teller machines, is such an application.

In reviewing Fig. 15.2, under the Packet Transport Facilities part, access area classes A, B, and C refer to LEC rates related to normal urban versus suburban access charge differences. In the Usage part, *priority packet kilosegment* charges refer to a class of service that gives preferential treatment to packets under periods of network congestion, relative to *normal packet kilosegments*, which may incur longer delays.

IXC Packet Switched Services

MFJ directives and rulings, cited earlier in reference to IXC voice operations, apply as well to IXC inter-LATA packet switched service operations. As with voice, LEC or other bypass carrier facilities provide local channel access from customer premises to IXC Points-of-Presence (POP). As two examples likely to be encountered in the inter-LATA arena, this section considers selected Sprint International and AT&T dedicated and dial-up service and rate structures. Sprint supports a large number of protocols and features. On the other hand, under its ACCUNET Packet Service with dedicated access, AT&T supports only X.25 interface protocols.

Figure 15.3 illustrates Sprint's VAN access, transport, and special feature rate structures, which are comparable to the LEC tariff elements and structures described above. As in LEC PSPDN services, access can be either dedicated, using private lines, or dial-up. Both access methods are offered at various line speeds

Dedicated Access Services

	Type	Speed (bps)	Monthly Rate ($)	Nonrecurring Charge ($)
Asynchronous	---	300-1200	600.	1200.
IBM	Terminal	4800	425.	1200.
Synchronous	Terminal	9600	650.	1200.
and Bisync	Host	4800	1200.	1200.
	Host	9600	1525.	1200.
X.25	---	2400	1000.	1200.
	---	4800	1200.	1200.
	---	9600	1525.	1200.
	---	14400	2000.	1500.
	---	19200	2550.	1500.
	---	56000	Quotable	Quotable

Other Services and Features

Access Management Services

	Monthly Rate ($)	Nonrecurring Charge ($)
Mnemonic Host Codes		
Each Host Code		
Reserve Host Code	50.	50.
Change Address of Host Code	25.	50.
	N/A	100.
ID/Passwords		
0-500 (Each)	4.	5.
Next 2000 (Each)	3.	5.
Each Additional	2.	5.
Change ID Parameters (Each)	N/A	5.

Dial Access Services

Datacall Plus- Asynchronous Service (Traffic Charges Included)	Location	Speed (bps)	Hourly Rate Standard ($)	Hourly Rate 1-Year Term ($)
	Class A	300-9600	7.50	6.00
	Class B & C	300-9600	9.00	7.50
	In-WATS	300-9600	15.50	14.00

Datacall Plus- X.25 Service (Traffic Charges Included)	Location	Speed (bps)	Hourly Rate Standard ($)	Hourly Rate 1-Year Term ($)
	Class A	1200-9600	7.50	6.00
	Class B & C	1200-9600	9.00	7.50
	In-WATS	1200-9600	15.50	14.00

Usage Charges

Traffic charges apply for network access unless otherwise specified. Traffic is charged at $1.40 per kilosegment. A segment has a billable length of up to 64 characters.

Fig. 15.3 Sprint Inter-LATA packet switched public data network rate (SOURCE: *CCMI*).

and protocols, including dial-up X.25. Popular business applications for the user include connection to public information services and intra-company terminal-to-host data communications.

Recurring and non-recurring dedicated access charges include costs for private lines between customer premises and IXC POPs, packet switch ports and associated modem and termination arrangements. Although private line access facilities are paid for as part of the service, they are actually provided by LEC or bypass carriers. In addition to basic services, a multitude of access management services are offered such as host mnemonic codes and ID/passwords.

Dial-up access services are available at selected line speeds at standard and discounted extended term hourly rates. Dial-up charge rates are dependent on the city in which the customer premises is located. Sprint publishes lists in which cities are categorized as classes A, B, or C. For cities which fall outside of the LATA in which the packet switch is located, an INWATS rate category is provided.

Usage charges of $1.40 per kilosegment are applicable to dedicated access virtual calls, but not to dial-up calls.

Figure 15.4 illustrates parts of AT&T's ACCUNET Packet Service rate schedule. The rates cover the IXC services only, i.e., switch ports and packet transport between AT&T's switches. Dedicated access lines can be ordered as local channels under AT&T's Tariff #11. In determining total charges, Tariff #11 and *central office connection* (COC) charges need to be added to applicable charges shown in Fig. 15.4. AT&T Tariff #11 and COC charges are discussed in the IXC Private Line Services section of chapter 10.

AT&T's dial-up services can be obtained using Interspan Access Service, one of several offerings made by AT&T under its unregulated Interspan Services Family of Data Connectivity Options. AT&T's dial-up offering includes protocol conversion for IBM synchronous and other asynchronous protocols.

Usage charges shown in Fig. 15.4 are applicable to both dedicated and dial-up services. Note that business day rates apply from 6:00 A.M. to 9:00 P.M., which is much longer than the standard 8:00 A.M. to 5:00 P.M. interval used for voice rates. Note also that AT&T usage charges are based on kilopackets rather than kilosegments as used by Sprint and most LECs.

As indicated in Fig. 15.4, AT&T discounts usage rates based on total monthly usage. Since AT&T is regulated, its packet switched service offering is published as a tariff. In contrast, rates for most packet switched services are not filed as tariffs with the FCC. Standard non-regulated rates can be obtained directly from service providers, but discounted rates can often only be determined through competitive bid processes. For large users, this represents an opportunity to negotiate rates below advertised rates, but it makes a priori service price comparison difficult.

While packet switched services are most popular for the general mix of corporate traffic, circuit switched digital services are becoming increasingly popular for backup and business applications like desktop video conferencing. Since network transport for switched voice traffic is provided by 56 kbps circuits, IXCs are offering switched digital service at the same speed. Although separate subnetworks without echo cancellation and separate network management are involved, competition has kept prices close to switched voice levels. Moreover, newly developed customer

AT&T Accunet Packet Service Functions

Monthly	Speed (bps)	Maximum Number of Calls	Monthly ($)	Installation ($)
Digital:	2.4K	7	500.00	500.00
	4.8K	127	500.00	500.00
	9.6K	127	500.00	500.00
	56K	511	1000.00	500.00
Analog:	2.4K	7	500.00	500.00
	4.8K	127	650.00	500.00
	9.6K	127	800.00	500.00

AT&T Accunet Packet Service Usage Charges

	Business Day (*) ($)	Non Business Day (*) ($)
Per Kilopacket	0.69	0.35

* Business Day: 7 am - 6 pm Monday through Friday excluding holidays
Non Business Day: 7 am - 6 pm Monday through Friday on New Year's Day, July 4, Labor Day Thanksgiving Day and Christmas Day.

Usage Charge Discounts

Volume Usage Discount Plan - Applies to the total combined usage charges in a billing month for packet charges on Accunet Packet Service network calls.

Total Monthly Usage Charge ($)	Usage Charge Discount (%)
0.00 - 5,000.00	0
5,000.01 - 8,000.00	5
8,000.01 - 10,000.00	7
10,000.01 - 15,000.00	10
15,000.01 - 25,000.00	12
25,000.01 - 50,000.00	15
50,000.01 - 100,000.00	20
100,000.01 +	25

Fig. 15.4 AT&T's ACCUNET Packet Service rate example (SOURCE: *CCMI*).

premises equipment (CPE), known as *inverse multiplexers,* make multiple switched 56 kbps circuits appear as a single, switched, synchronous high speed channel. This provides the equivalent of switched 112 kbps and 384 kbps service, suitable for signals generated by video codecs. As a consequence, corporate-wide video conferencing, wherein telecommunications costs are incurred only during conferences, is now possible, avoiding the considerable cost involved with large numbers of wideband private lines previously needed.

Figure 15.5 shows a switched 56 kbps rate schedule implemented as part of an IXC's virtual private network offering.

Miles	Initial 18 Seconds			Each Additional 6 Seconds		
	Day ($)	Evening ($)	Night ($)	Day ($)	Evening ($)	Night ($)
55	0.1049	0.0998	0.0998	0.0083	0.0066	0.0066
292	0.1163	0.1091	0.1091	0.0121	0.0097	0.0097
430	0.1274	0.1178	0.1178	0.0158	0.0126	0.0126
925	0.1346	0.1238	0.1238	0.0182	0.0146	0.0146
1910	0.1346	0.1238	0.1238	0.0182	0.0146	0.0146
3000	0.1367	0.1253	0.1253	0.0189	0.0151	0.0151
4250	0.1478	0.1343	0.1343	0.0226	0.0181	0.0181
Over 4250	0.1478	0.1343	0.1343	0.0226	0.0181	0.0181

Fig. 15.5 Inter-LATA switched 56 kbps rate example (SOURCE: *CCMI*).

IXC Special Contract Services

As with LEC PSPDN services, the point at which private packet switched networks become more economical than public IXC services occurs at relatively low traffic levels. As an example, a 56 kbps private line can support approximately 125 packets per second of interactive traffic. If the circuit costs $1,500 a month, it becomes economical at a usage level of only 10 minutes per day (assuming an average cost of $1.00 per kilo-packet for public service). Understanding this, the IXCs prime business strategy is to offer *special contract proposals* to large users. This approach is consistent with the fact that provisioning data communications services for large corporations is a complex, multidimensional problem, and that each large user's configuration is unique.

In many cases, IXCs propose solutions placing packet switches on user premises, with private lines connecting the switches to form a network. A well designed private network must always be a large user consideration, in which all private network life cycle cost elements are evaluated, e.g., equipment investment, installation, maintenance, network management, and administration.

Network Design Summary

The network design tools and objectives discussed in Chapters 7 and 10 apply to packet switched networks. The application is, however, more complex. Using hub switches to aggregate traffic, making more efficient use of transmission, applies to data as well as voice communications. The discussions in this chapter make it clear that packetizing data considerably increases the complexity of design optimization. Relatively simple voice traffic requirements are displaced by the need to consider not only traffic intensity, but also the impact of various protocols and the impact of packet size on throughput performance. Private hub switches in metropolitan areas can not only provide packet switched services to company locations in that area, but can also serve to aggregate and forward traffic to hub switches in other cities, resulting in more economical long distance operations.

Optimum placement of switches and the design of private line networks connecting switches is a non-trivial undertaking. The general problem has been a research topic for the past twenty years at major universities throughout the country. Without highly capable in-house telecommunications staff, businesses need competent telecommunications design specialists equipped with computers and sophisticated network models to help articulate requirements and to develop private or hybrid networks. The purpose of this chapter has been to put the reader in the position to understand the principles behind vendor alternative designs so that he can have confidence that proposed solutions are complete, meet established technical performance specifications, and satisfy business needs.

Part
5

Integrated Digital Networks

16

Integrated Services Digital Network (ISDN)

The concept of an ISDN was first formulated in the councils of the International Consultative Committee for Telephone and Telegraph (CCITT) before the 1976 plenary session. While the concept incorporated significant contributions from the U.S., it can fairly be said to have reflected a European view of telecommunications. In the early 1970s, European telecommunications administrations, and to some extent Japan, viewed evolving digitization of their telecommunication infrastructure, initiated by U.S.-based activities—as needing a standards framework to ensure future interoperability.

In those days of analog technology, switching and transmission systems were viewed as independent network segments. Digitization of the switching and transmission plants, allowed the integration of many functions, particularly multiplexing, leading to the *"Integrated Digital Network"* (IDN) concept. But this technological improvement could not reach customers until systematic provisions for access were established. Thus, the concept of user access to an existing IDN underlies the Integrated Service Digital Network (ISDN).

ISDN consists of a set of standards being developed by the CCITT and various U.S. standards-setting organizations. The CCITT formal recommendations, adopted in October, 1984, first defined ISDN as:

> . . . a network, in general evolving from a telephony integrated digital network, that provides end-to-end digital connectivity to support a wide range of services, including voice and non-voice, to which users will have access by a limited set of standard multipurpose user-network interfaces.

Note that the CCITT concept of ISDN was never meant to be a product or service specification, but rather a guideline for development of such offerings in a manner that would promote international connectivity and interoperability. ISDN characteristics include:

- *End-to-end digital connectivity.* All signals are transmitted in digital form from terminal to terminal.

- *Common channel signaling.* As described in Chapter 5, ISDN uses out-of-band signaling, transmitted independently over the network in the form of messages containing addresses, information, and protocol elements in standard formats. As CPE and network equipment adopt the same set of standards, function and feature transparency across both private and public networks will no longer be dependent on the use of proprietary products from a single vendor (a problem discussed in Chapter 4). (Note: The concept of CPE, customer premises equipment has U.S. but no CCITT relevance.)

- *Multipurpose user network interfaces (UNI).* ISDN permits the user to connect to voice, data, video, and other services by a single access mechanism, in contrast with the separate arrangements now required.

The U.S. version of ISDN as propounded by Bellcore is intended to support a wide variety of new products and services in evolutionary fashion, delivering them via two standardized user network interfaces. It is also meant to establish a platform for future telecommunications utilities, such as AT&T's Universal Information Services (UIS). ISDN is expected to reduce service costs through more efficient use of network resources, improve network management on the part of service providers as well as customers, and create a better-coordinated network environment.

History

- 1968—The CCITT Study Group D is formed to evaluate digital transmission systems.

- 1972—CCITT Study Group D issued discussion of CCITT functions and publications and issues recommendations and a proposal to study a worldwide Integrated Services Digital Network.

- 1976—CCITT Study Group XVIII establishes questions for an ISDN ("Green Book").

- 1980—CCITT Study Group XVIII establishes a formal plan to define an ISDN conceptually at the next Plenary (1984).

- 1984—CCITT issues the I series recommendations, defining the framework for ISDN ("Red Book").

- 1988—CCITT issues the "Blue Book," further defining ISDN and outlining additional broadband ISDN studies.

- 1990—In the U.S., numerous field trials are under way, and working ISDN "islands" are in place; formation by the Corporation for Open Systems (COS) of the ISDN Executive Council; NIST (National Institute of Science and Technology) establishes the National ISDN Users Forum to broaden the utilization of ISDN and accelerate the development of applications.

- 1991—Publication of Bellcore-developed report covering procedures referred to as *National ISDN-1*.

Services

Figure 16.1 illustrates *access* and *transport* services provided by ISDN. Access service for smaller systems and equipment, such as voice/data workstations, is provided by the *Basic Rate Interface* (BRI). BRI consists of two 64 kbps information bearer channels (B channels) and one 16 kbps packet switched data channel (D channel) which performs signaling for the B channels, and furnishes a mechanism for packet switching user data. Note that ISDN defines new access arrangements, but that transport—in the intra-LATA and inter-LATA contexts for the U.S.—is provided by the existing IDN.

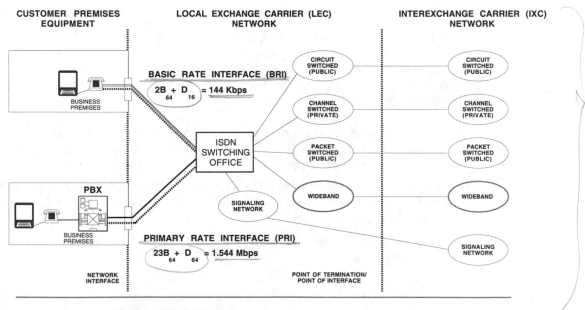

Fig. 16.1 Integrated Services Digital Network (ISDN).

ISDN CHANNEL TYPES (CCITT)

B channel	64 Kbps
D channel	16 or 64 Kbps *
H0 channel	384 Kbps
H11 channel	1536 Kbps
H12 channel	1920 Kbps
H4 channel	622.08 Mbps

* The D channel is 16 Kbps for basic rate interface and 64 Kbps for primary rate interface.

Access service for host processors and CPE switching systems (e.g., PBXs) is provided by the *Primary Rate Interface* (PRI). PRI is based on the DS1 transmission rate of 1.544 Mbps and consists of 23 B channels and one 64 kbps D channel. In addition, six B channels may be bundled together to form a 384 kbps H0 channel, or 24 B channels may be bundled to form a 1.536 Mbps H11 channel. These bundles would be used for applications requiring high data rates, such as compressed video, or host-to-host bulk data transfers.

Exactly what services will be offered by LECs in the U.S. is not yet clear. In fact, Appendix A explains why, as a result of judicial and regulatory decisions, including FCC's Network Channel Termination Equipment (NCTE) Order, MFJ intra- and inter-LATA distinctions, and stipulated IXC roles in the provision of public services to customers, ISDN developments in the U.S. cannot precisely adopt all CCITT recommendations. In particular, the NCTE Order, prohibiting carrier owned NCTE on customer premises has been cited as causing progress towards ISDN in the U.S. to seriously lag that of Europe and Japan.

Standards

The guiding principle for ISDN standards has been interoperability. While CCITT is responsible for ISDN, CCITT has adopted the International Standards Organization's (ISO), Open System Interconnection (OSI) reference model and incorporated it into ISDN data service recommendations. Three types of ISDN standards-based offerings are envisioned by CCITT:

- *Bearer services*, described above, consist of network services delivered over the B channels and encompass the first three layers of the OSI model—i.e., the physical, data link, and network layers.

- *Teleservices*, implemented in the ISDN terminal equipment, may encompass any or all of the upper four layers of the OSI model—i.e., the transport, session, presentation, and application layers.

- *Supplementary services* are provided by the network for use with bearer services and teleservices.

Interfaces

CCITT has defined V, U, R, S, and T *reference points* to facilitate access and provide interface specifications, as depicted in Fig. 16.2. Underlying the CCITT concept of ISDN as access to the IDN, a basic notion is the NT, the *network termination*. As originally conceived, NT is service provider-owned equipment on customer premises which establishes the point of telecommunications service delivery by a carrier (USA) or an administration (most of the rest of the world). However, in the U.S., the UNI must be at the U reference point, since the FCC NCTE Order mandates the separation of any and all forms of customer premise equipment from the service provided by a carrier.

In CCITT I Series parlance, basic and primary rate access are standardized at the

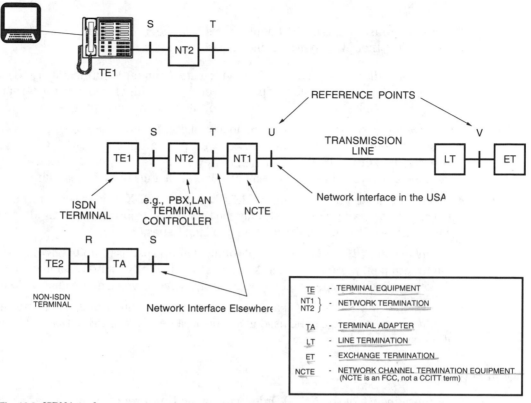

Fig. 16.2 ISDN interfaces.

customer side of the NT1, or as shown in Fig. 16.2, the S or T reference points. The inscribed "boxes," shown between the reference points on the figure, represent functional groupings defined by CCITT. For example, NT2 and LT1 encompass all of the functions needed to adapt the classic two-wire loop to a customer premises to support basic and primary rate signals. NCTE, a U.S.-only concept implements NT1 functions. While the "boxes" in Fig. 16.2 are not intended to define any particular equipment envelope, in the U.S. various manufacturers may incorporate any or all of the NT1, NT2, TE1, and TA functions in any particular equipment.

NT2 functions may be incorporated in single stand-alone terminals (telephones, data terminals, or other ISDN user devices) or digital PBX, LAN, or multiplexer equipment. ISDN-compatible terminals are designated *Terminal Equipment 1* (TE1), conforming to the S and T reference points. Non-ISDN terminals are designated *TE2*, requiring a *terminal adapter* (TA) for physical connection to the network at the R reference point. The American National Standards Institute is developing ISDN standards for the U.S. ANSI T1.601-1988 is the BRI standard in the U.S. Since 1988, several manufacturers, notably Northern Telecom Inc. (NTI) and AT&T have had BRI and PRI capability, but only between their own proprietary switchgear and telephones.

Applications

While numerous potential ISDN applications have been identified, those that have actually made their way to market include:

- *Incoming call identification.* Using Automatic Number Identification (ANI), the originating telephone number is passed over the D channel, allowing the terminating CPE to display the number, or use it to call up a database profile of the caller.

- *Call-by-call service selection.* Previously, digital facilities had to be preassigned to specific outbound or inbound services, e.g., T1 circuits for WATS and 800/900 services. Use of the ISDN D channel, however, allows dynamic B channel assignment on a real-time basis, producing trunk cost savings of up to one third.

- *Outbound station identification.* ISDN can pass a PBX user's telephone number to the network, where previously only the PBX trunk group number was identified. This feature is valuable for resolution of billing issues.

- *Switched digital bandwidth.* ISDN offers dial-up speeds of up to 64 kbps terminal to terminal, whereas transmission rates are currently limited to a maximum of 38 kbps using V.32 type adaptive modems on switched voice grade analog circuits.

- *Feature portability.* PBX features such as uniform numbering, call forwarding, and message waiting can be used in a wide area network of PBXs that perform as if they were a single system.

Terminal Equipment

As a consequence of the NCTE Order making the U reference point the network interface (NI) in the U.S., the NT functions must be implemented in CPE. The NT functions performed for physical and electrical connection of an ISDN terminal to the BRI include:

- Full-duplex, 160 kbps (2B1Q) transmission, with echo cancellation, over two wires.

- Two-wire to four-wire conversion.

- Contention resolution for multiple terminal access (passive bus).

- Environmental protection (lightning and power cross).

NT1 termination equipment is manufactured by AT&T, NTI, NEC, GTE/Fujitsu, Ericsson, Siemens, and others. The adoption of ANSI T1.601-1988 now enables this equipment to be compatible with all LEC service provision conforming to the standard, which was not previously the case.

Several manufacturers are producing equipment compatible with AT&T's 5ESS CO switch, on a PRI basis only. The AT&T Definity Generic 3 PBX possesses both PRI and BRI capability.

NTI's ISDN Meridian telephones have BRI capability, but can only be used with the DMS-100/SL-100, and other Meridian 1 family members. Interestingly, AT&T has had BRI telephones (7500 series) compatible with the 5ESS since 1988, using two-wire designs for the U reference point. These telephones are not, however, compatible with the Definity PBX, which uses four-wire T reference point telephones. Backward compatibility among AT&T PBXs is achieved by interworking BRI with AT&T's proprietary, non-ISDN, digital communications protocol (DCP).

Carrier Activities

AT&T, MCI, and Sprint have implemented ISDN on a broad scale, both in virtual private networks for nationwide corporate networks, and to a lesser degree, in government networks such as the Federal Telecommunications System 2000 (FTS2000), the Washington Interagency Telecommunications System (WITS), and the Pacific Zone Aggregated Switch Procurement (ASP) program.

PRI services are offered in several hundred cities by the three dominant IXCs. LEC offerings, however, are limited, with only Bell Atlantic providing BRI on a region-wide basis. LECs have only recently begun to file PRI tariffs for intra-LATA service, the first being Pacific Bell in late 1990.

Bell Atlantic has announced its intention to equip all digital COs (providing about two-thirds of its access lines) with ISDN by the end of 1992. Yet, corresponding numbers for Pacific Bell and Nynex are only 50% and 25%, respectively. In fact, the FCC forecasts that only about 20 percent of all CO switches will have ISDN capability by 1994. Of course, this reflects the relatively slow rate of retirement of the analog but feature-rich 1AESS CO switch.

Data applications for ISDN currently tariffed by Bell Atlantic include printer sharing services, automated software update and installation, backup of remote PCs, and electronic transfer of order data.

Impediments to ISDN Growth

For years, widespread rollout of ISDN has been delayed by the "connectivity without compatibility" issue, together with the lack of portability of terminal equipment. Prior to the ANSI T1.601-1988, deployment of ISDN had been impeded by lack of uniform standards, which caused manufacturing delays and market confusion.

Interface issues have also caused problems, with IXCs providing only PRI service and LECs providing more BRI than PRI service. This is the result of PRI being a long-haul network tool, and BRI being a CPE or local loop-related capability. Thus, ISDN as a concept is alien to the bifurcated intra-/inter-LATA carriage concept adopted in the U.S. as a consequence of divestiture and deregulation.

Use of CCITT Signaling System No. 7 (SS #7) in the underlying IDN is absolutely essential for the provision of ISDN and its service offerings. The AT&T and Bellcore versions of SS #7 are commonly referred to as CCS 7 and SS 7 respectively. Although these versions are functionally similar, gateways will be required between LECs and IXCs to account for differences, another penalty of divestiture. Nevertheless, implementation of these signaling systems is gradually providing a foundation to link the ISDN islands that have grown up inside urban areas creating compatibility between inter-LATA and local loop trunking facilities. NTI has released version BCS-31 for their DMS-100 CO switch, and AT&T will soon release generic 5E7 for their 5ESS CO switch. Most field testing has been completed, and by year end 1994, some 60 percent of all CO switches will have implemented some version of SS #7.

A major impediment centers around the concept of interworking. *Interworking* refers to the ability to maintain operational integrity when IDN networks providing ISDN access are connected to networks not supporting ISDN, or to other non-ISDN compatible systems or equipment. While possible, successful interworking is difficult to accomplish, consuming additional development and operational resources. A related issue is achieving compatibility and interoperability among PBX switches designed to be ISDN capable but produced by different manufacturers. This is a potentially significant cost item, constituting a barrier to market entry. Field trials in this arena have been conducted over the past five years, without significant increases in working ISDN implementations.

In many cases, the LECs are unsure about the ability of their existing local loop facilities many of which are forty or more years old—to technically support ISDN without large-scale, expensive rehabilitation. The T BRI reference point also demands an additional pair relative to the traditional analog one pair-per-service arrangement.

There is also the issue of ISDN traffic concentration. As an inherently efficient transmission method, it tends to place "all the eggs in one basket." Recent history has not helped the non-redundant network image; cases in point: the Hinsdale, Illinois, CO fire; the Framingham, Massachussets and New York City power failures; nationwide IXC SS #7 failures; and regional SS #7 failures.

Probably the biggest impediment to the spread of ISDN has been the phenomenal growth of the LAN. In 1984, when ISDN was being defined, there were a handful of proprietary, premises-based computer networks, led by Datapoint's Arcnet. Two years later, ISDN was still attempting to tackle 64 kbps and 1.544 Mbps transmission rates—already provided by non-ISDN services. LANs, meanwhile, had begun to proliferate around Ethernet and token ring technologies, offering high speed premises data services for a relatively small investment. Today, the promise of ISDN capable PBX or LEC provided premises services as ubiquitous, low-cost, high-speed local data pipelines has been overtaken by LAN marketplace events. As a consequence, PBX/Centrex systems for voice, and LANs for data are becoming de facto standards for premises-based business communications needs. If this trend continues, the often-touted advantages of integrated voice and data via basic rate access, will melt away leaving only a yet-unproven residential demand to fuel future growth. Worse yet, no ISDN service is appropriate for the burgeoning LAN-to-LAN and MAN/WAN requirements outlined in Chapters 13 and 14.

IDN Requirements

Actions contributing to the formation of the *critical mass* of IDN capabilities necessary to make ISDN a reality include:

- Upgrade of the PSN with all-digital CO switching
- Deployment of Signal Transfer Point (STP) packet switches throughout the PSN for SS #7 signaling, crucial to ISDN long-haul service

- Establishment of automated systems and databases for network operations, customer applications, maintenance, and administration

- Upgrade of all DS1 transmission facilities between network switches with extended superframe (ESF)—see Chapter 3—as the first step to attaining 64 kbps clear channel capability

- Continued adoption of U.S. versions of CCITT ISDN recommendations via ANSI/ECSA (Exchange Carriers Standards Association), IEEE, etc.

Opportunities

Growth in any network brings with it increased marginal revenue opportunities and lower unit costs. ISDN service marketing drivers include:

- Two channels per digital line, allowing theoretical per-line price umbrella of double that for analog and non-ISDN digital service

- New and better telecommunications features with ISDN

Historically, the telecommunications marketplace has not accepted a new product or service even those with more features at the same price level as existing products and services. The new generation offering must be not only better, but cheaper. Price has always driven marketplace acceptance of telecommunications innovation.

The key challenge is for the IXCs and the LECs to match price with services at levels attractive to potential ISDN users. Perceived value in the marketplace is critical, e.g., AT&T's Princess phone coup, as contrasted with their Picturephone fiasco.

Significant enabling factors for ISDN include:

- Orderly implementation of Bellcore's National ISDN-1, the set of standards and agreements among IXCs, LECs, and equipment manufacturers designed to accelerate ISDN rollout by establishing interoperability and ubiquity of service.

- Since ISDN is optimized for conservation of transmission facilities, redundancy can be folded into individual service offerings, thereby avoiding the stigma of vulnerability to single thread failures.

- SS #7 is the signaling technology of choice in U.S. public and private networks, and the only practical means of connecting ISDN islands. Growth and development of PRI will benefit from the pull effect exerted by continuing nationwide deployment of SS #7.

- ISDN can gain leverage from the presence of the ubiquitous LEC local loop, in many cases giving a competitive edge to Centrex BRI services over CPE switching alternatives.

- As has been the case with Centrex, ISDN market penetration is linked to the LECs making applications available to small users, as well as offering interoperability through public and private networks.

- Price levels perceived by users as advantageous in comparison with competing services. Purveyors of ISDN must avoid false starts and bad press during the maturation period.

- Specification of ISDN as part of large-scale federal government networks, e.g., FTS2000, WITS, ASP, etc.

- Economical provision of BRI services to rural customers in independent telephone company territories based on low-cost modifications to existing CO switches and subscriber line carrier (SLC) facilities.

- Growing awareness of ISDN applications fostered by efforts such as the NIST's National ISDN Users' Forum and the National ISDN Applications Center in New York City.

Coordinated ISDN Activities

To make ISDN a success, developers must simultaneously achieve the multivendor interoperability envisioned by its creators and stimulate applications that provide end users with practical business advantages. These goals are being pursued through two major programs involving users, vendors, and service providers.

Bellcore has taken the lead in specifying interoperable sets of standards and agreements among ISDN parties, to be implemented in phases. The first set, known as *National ISDN-1,* is scheduled for completion in late 1992 or early 1993. Progress is somewhat hampered by the fact that Bellcore is not a standards body and must achieve its ends by mutual agreement.

In parallel, the NIST's National ISDN Users' Forum has established procedures for turning users' requirements into working applications. Periodic ISDN Users' Workshops are conducted to translate application ideas into prioritized requirements. Requirements are then given to vendors at ISDN Implementor's Workshops where application profiles, implementation agreements and conformance test specifications are established.

To illustrate the end result of both of these efforts, a series of national demonstrations is being planned. The Corporation for Open Systems (COS) is sponsoring the first of these events in the fall of 1992. Known as the *Transcontinental ISDN Project-92* (TRIP-92), it will showcase for the first time the true potential of a unified ISDN providing real applications support.

17

Broadband Integrated Services Digital Network (BISDN)

Chapters 12, 14 and 15 describe the growing need for high-speed MAN and WAN services currently driven by LAN-to-LAN interconnection applications, with future digital video and imagery requirements promising to accelerate the trend. Those chapters also treat emerging requirements for technologies that economically support variable bit rate traffic by providing true *bandwidth-on-demand* services. This chapter addresses ongoing efforts to develop technologies and standards responding to these requirements, going beyond what some believe to be interim frame relay and SMDS solutions outlined in Chapter 14. CCITT is developing a Broadband Integrated Services Digital Network (BISDN) umbrella standard—incorporating underlying standards for integrated digital network switching, multiplexing and transmission facilities—that will be able to meet expanding requirements well into the future.

In one of the first-draft CCITT documents, Broadband Integrated Services Digital Network (BISDN) is simply defined as "a service requiring transmission channels capable of supporting rates greater than the primary rate." As noted in the last chapter, in the U.S. the primary rate for *narrowband* ISDN (as the current standard is sometimes referred to) is 1.544 Mbps. The intention behind CCITT's Recommendation I.121 BISDN definition is to create the means for providing an all-purpose network to support a wide range of voice, data, and video services. To meet this objective, BISDN will use fiber optic transmission systems with bandwidths ranging from 155.52 to 2,488.32 Mbps, with the potential of providing users with access to bandwidths hundreds of times greater than primary rate.

As with ISDN, the concept of BISDN is a network evolving from a telephony *integrated digital network* that provides end-to-end digital connectivity to support a wide range of services, to which users will have access via a limited set of standard multipurpose network interfaces (UNIs). However, while existing IDNs possess capacities and operating modes for narrowband ISDN services, advanced switching and multiplexing technologies crucial to BISDN service are not generally available. As a consequence, much BISDN-related work is focused on standards and the development of *core* switching, multiplexing, and transmission technologies for the next generation IDN. Two major initiatives, ATM/STM and SONET are treated below.

Asynchronous Transfer Mode (ATM)/Synchronous Transfer Mode (STM)

CCITT Study Group XVIII, the current focus for international BISDN, generically calls the switching and multiplexing aspects of BISDN *transfer modes*. The rationale behind selecting a terminology that merges switching and multiplexing aspects, is the fact that in today's networks, switching and multiplexing functions are often integrated in the same equipment. Two paradigms for blending switching and multiplexing techniques are *asynchronous transfer mode* (ATM) and *synchronous transfer mode* (STM). As depicted in Fig. 17.1, in the STM paradigm, circuit switching is combined with time division multiplexing. Except for higher bandwidth capacities needed to support BISDN services, STM is basically equivalent to current circuit switching and TDM techniques, as defined and described in Chapter 4.

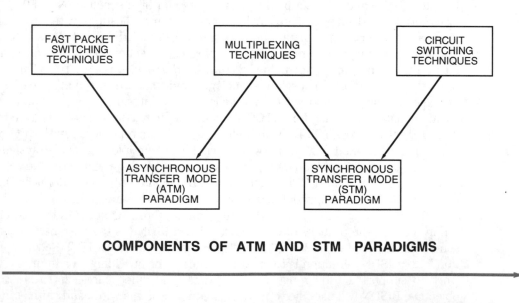

COMPONENTS OF ATM AND STM PARADIGMS

FLAG	HEADER	PAYLOAD	
1 BYTE	4 BYTES	48 BYTES	

ATM CELL STRUCTURE

Fig. 17.1 Switching/multiplexing paradigms.

Recall from Chapter 4 that modern circuit switches use time slot interchange versions of time division multiplexing to implement their matrices. Moreover, today's digital PBXs and CO switches can eliminate the need for separate multiplexers by directly producing DS"N" compatible multiplexed signals. Both constitute existing product examples of STM, and both have amply demonstrated the efficiency and economic benefit of merging aspects of switching and multiplexing.

Using STM, when a user requests service, circuit switches provide routing through *fixed-bandwidth* multiplexers. If network loading prevents the allocation of needed resources on an end-to-end basis, the call is blocked. Also, for the duration of completed calls, or *sessions*, if the assigned user doesn't fully utilize the circuit bandwidth, the unused capacity cannot be made available to other users. For example, with variable bit rate traffic, an STM derived circuit able to support peak bit rates must be assigned. Then during intervals when less than peak bit rates are generated, unused capacity goes wasted. Chapter 14 contains a broader discussion of variable bit rate traffic and its impact on switching techniques.

Figure 17.1 shows that the asynchronous transfer mode paradigm combines fast packet switching and multiplexing techniques. ATM is defined as a broad-bandwidth, low delay, packet-like switching and multiplexing technique. The name selected by CCITT is somewhat confusing since the transfer mode does not involve asynchronous (start-stop) transmission (defined in Chapter 2) associated with non-intelligent terminal traffic.

A significant feature of ATM is its ability to conserve network resources when carrying traffic, such as bursty LAN-to-LAN and video variable bit rate signals. The ATM advantage is that it statistically prevents wasted network capacity. When a user requests service, the network determines both the user's average and peak bandwidth requirements. Knowing the average and peak load requirements of other competing users, the network controller makes a determination, on a statistical basis, whether sufficient switching and multiplexing capacity is available to support the request.

With ATM, usable capacity can be assigned dynamically, (on demand), during a session. This permits a network engineering approach which can accommodate bursty services, while guaranteeing acceptable performance for continuous-bit-rate services such as voice and video. In contrast with the fundamental notion of a fixed bandwidth channel, long associated with circuit switched telecommunications, ATM defines new rules and procedures for dividing up bandwidth capacity and allocating it to users in support of a wide variety of services.

In ATM, bandwidth capacity is allocated to fixed-sized information-bearing units called cells. Each cell contains header and information fields. The header contains a *virtual channel identifier* (VCI) that is used like a time slot identifier in an STM approach. VCIs can be translated (much like time slots can be interchanged), at ATM interfaces, prior to being transported to another interface. Thus, ATM creates *virtual channels*, which can be manipulated like fixed STM channels, but which offer variable rather than fixed bandwidth capacities.

Cell relay is defined as the process of transferring data in the form of fixed length packets (cells). In terms of selecting a cell size, the objective is to develop a single multiplexing/switching mechanism for dividing up usable capacity (bandwidth) in a manner that supports its allocation to both isochronous (e.g., voice and video traffic, see the FDDI section of Chapter 13) and packet data communications services. Standards groups have debated the optimum cell size. Small cells favor low delay for isochronous applications but involve a higher header-to-user information overhead penalty than would be needed for most data applications. The current specification is for a 53 byte cell which includes a 5 byte header and a 48 byte payload, a design intended to enable ATM to support all user services. The cell composition is illustrated in Fig. 17.1.

It is interesting to observe that in ATM-based architectures, the significance of the word *integrated*, as in integrated digital networks, takes on the broadest possible connotation, encompassing both the ability to transparently accommodate voice, data, and video services, as well as the almost complete functional integration of switching and multiplexing within facilities.

Cell relay, unlike conventional packet relay, provides no error or flow control. As a consequence, cell relay is extremely efficient, minimizing processing requirements as a cell moves through a network. Figure 17.2 summarizes the comparison between packet switching and multiplexing techniques. As noted, ATM service is connection-oriented in that out-of-band call setup signaling is required prior to information transmission, and, once the call is established, all traffic is transmitted over the same logical connection between end users. That logical connection however, can support either connection-oriented or connectionless user services.

Evolution to ATM will involve significant investment, and will require interworking strategies to permit gradual implementation within networks, understanding that an entire network cannot be upgraded at once, and to permit interconnection with non-ATM networks. From a performance point of view, IDNs with ATM cores should be able to offer *public data communications services* able to compete on a cost basis with large-user private networks, a capability not now possible as discussed in Chapter 15. To date, potential data communications revenue alone has not justified investment in the high-capacity, fast packet switching facilities needed to enable carriers to offer *virtual private data networks* with the same economic advantages to the users as today's virtual private voice networks. Perhaps ATM plans that pro-

CHARACTERISTIC	X.25 (PACKET RELAY)	FRAME RELAY	SMDS (CELL RELAY)	ATM (CELL RELAY)
LINE RATES	56 kbps	56 kbps/1.544 mbps	T1/T3 (1.544/44.736 mbps)	UP TO 1200 mbps
CELL/PACKET LENGTH	VARIABLE (8 BYTE HEADER)	VARIABLE (2 BYTE HEADER)	FIXED 53 BYTES (5 BYTE HEADER)	FIXED 53 BYTES (5 BYTE HEADER)
ERROR CONTROL	ERROR DETECTION AND CORRECTION	ERROR DETECTION	NONE	NONE
PROTOCOL	LAYERS 1-3	LAYER 1 PLUS	LAYER 1	LAYER 1
STANDARDS	X.25	ANSI T1.606/618/617 CCITT I.122, Q.922/933	IEEE 802.6 AND BELLCORE TR-TSV-000772,3	CCITT.121/2
DELAY	LARGE	LOW	VERY LOW	VERY LOW

Fig. 17.2 Comparison of switching/multiplexing techniques.

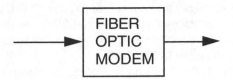

STS-N LEVEL	OC-N LEVEL	LINE RATE (Mbps)
STS -1	OC -1	51.84
STS -3	OC -3	155.52
STS -9	OC -9	466.52
STS -12	OC -12	622.08
STS -18	OC -18	933.12
STS -24	OC -24	1244.16
STS -36	OC -36	1866.24
STS -48	OC -48	2488.32

STS-N = Synchronous Transport Signal Level

OC-N   = Optical Carrier Levels

Fig. 17.3 SONET electrical and optical signals.

duce benefits for both traditional voice as well as emerging integrated services will produce acceptable return-on-investment results.

Synchronous Optical Network (SONET)

Having merged switching and multiplexing, ATM networks require wideband transmission. CCITT Recommendation I.121 states that ATM can be supported by any suitable digital transmission system and cites SONET G.707/708/709 recommendations as an example. Although flexibility in the choice of transmission within an IDN is an advantage, a standard BISDN access approach is fundamental to the provision of universal network interfaces for users. For users to capitalize on bandwidth-on-demand ATM network services, they will need wideband access to ATM cell relay switches which may involve extension of SONET-like capabilities, primarily developed for IDN internal transport, directly to user premises. We therefore present the following synopsis of SONET developments.

Synchronous Optical Network (SONET) is the name of a newly adopted standard, originally proposed by Bellcore, for a family of interfaces to be used in LEC/IXC optical networks. SONET is essentially a standard for BISDN transmission facilities. SONET defines standard optical signals, a synchronous frame structure for multiplexed digital traffic, and operations procedures so that fiber optic transmission systems from different manufacturer/carriers can be interconnected. SONET has been endorsed by the CCITT in recommendations G707, 708, and 709, and includes the definition of a *synchronous digital hierarchy* (SDH) of signals for standard interfaces at multiples of a base rate as depicted in Fig. 17.3.

The SDH is similar to the DS"N" TDM signal hierarchy described in Chapter 3. However, synchronous multiplexing allows component service signals to be combined into higher rates in a manner which simplifies multiplexing and demultiplexing of those component signals, offering easy access to SONET payloads.

For example, the OC-1 rate of 51.84 Mbps carries a DS3 payload signal of 44.736 Mbps plus overhead that includes payload pointers. Payload pointers permit direct access to lower speed components without the need to sequentially demultiplex the entire high-speed signal. This operation results in significant equipment economies at switching centers and signaling transfer points, and is an extension of today's programmable *digital cross-connect system* (DCS) capabilities illustrated at Fig. 3.13 on page 49.

Note that SONET provides standards for fiber optic transmission and the multiplexing signal structure, while permitting interconnection of competing fiber and multiplexer equipment designs.

Figure 17.4 is a conceptual block diagram of a BISDN network arranged to support voice, data and video customer premises services. The figure illustrates how the ATM and SONET portions physically relate to each other.

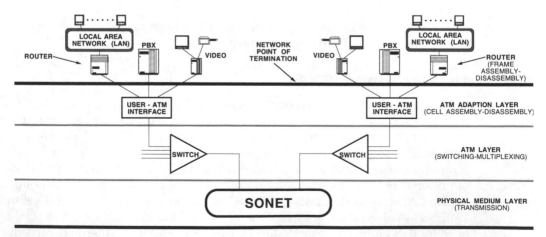

Fig. 17.4 Conceptual broadband integrated digital network.

Other Technologies, Standards, Systems & Equipment

18

Other Technologies, Standards, Systems & Equipment

This chapter describes important telecommunication developments and trends not addressed in earlier sections. Driving these developments are several factors. First, as the number and power of personal computers (PCs) continues to grow, new business applications arise involving the interconnection of PCs for work group support. Within buildings, the Ethernet (IEEE 802.3) and token ring (IEEE 802.5) LANs described previously, are currently the most popular work group interconnection methods. Such arrangements are displacing centralized mainframe computer architectures, with attached nonprocessing dumb terminals at workstation locations, in a process known as *downsizing*.

The steepness of LAN growth is one indication of the force behind this trend. As noted in Chapter 13, the LAN industry grew from $2.4 billion in 1987 to about $6.0 billion in 1991, and some predict that over 50% of 100 million personal computers will eventually be connected to LANs.

Other important developments include an extensive assortment of new products falling into the category of *personal communications,* the fastest growing telecommunications industry segment. Personal communications refer to a broad range of services, systems and equipment, e.g., facsimile machines, landline telephones, cellular telephone systems and emerging *personal communication system* (PCS) adjuncts, and a variety of radio systems including pagers, hand-held remote data entry terminals, and autonomous citizen-band like radio systems.

While no precise definition yet exists, personal communications provides at least one human operator with direct terminal access and real-time or near-real time interactive communications with a remote human operator or an information system resource. For example, the Department of Defense is transitioning from over-the-counter message service to equivalent services based upon direct PC-to-PC messaging.

Concepts and terminology associated with these developments are summarized below.

Facsimile Products & Services

Since the mid-1980s, *facsimile* (fax) growth has been exponential, with 1986 sales of 200,000 units, rising to well over one million in 1990. The growth stems largely from two factors. First, all units produced in that time frame conform to *CCITT Group 3 standards*. This means that regardless of manufacturer, all Group 3 machines are capable of interoperation. Group 3 machines work very well over dial-up lines. Thus universal connectivity between any two arbitrary users with Group 3 machines, located anywhere in the world, requires only access to standard telephone service. The second factor responsible for the dramatic growth in units produced is technology. Stand-alone Group 3 facsimile machines, now a commodity product, can be purchased for under $500.

Facsimile growth and success has not been without problems. Like LANs, most purchases to date have been made at the department level, outside the purview of centralized MIS or telecommunications planning and management. Many corporations, drowning in faxes, are beginning to take steps to get soaring fax-related telecommunications bills and overcrowded voice lines under control. The practice of routinely making plain paper copies because of the appearance and difficulty of handling *thermal paper faxes* is another of many examples of hidden manpower and material costs accompanying the fax's success.

Also, because of its universal connectivity, and because all it takes to read other people's traffic is a fax machine, the susceptibility of sensitive and private corporate data to unauthorized intentional or accidental reception, has raised significant privacy and security questions.

Fax's popularity has spawned adjunct products and services including; PC-fax cards; LAN fax servers; fax store and forward exchanges, and a variety of third-party subscription services, each of which is described below.

Stand-alone facsimile machines

Apart from voice telephone service, fax is the preeminent example of personal communications, permitting end-users to directly interact. With declining unit prices, it has become standard practice to equip individual offices and departments with their own stand-alone units so that individuals have convenient access. The stand-alone fax market has developed in two segments. A *low-end segment,* in the under $1000 price range, continues to incorporate features previously associated with more expensive units, benefiting from cost/performance advantages of newer design and manufacturing technologies.

Faxes in this segment now offer features such as single button automatic dialing, automatic redial, last number redial, automatic multiple sheet original feeding, and automatic paper cutting. Machines in this level can also combine standard telephone service as well as automatic voice answering and fax answering capabilities.

The *high-end segment* offers additional capabilities such as delayed transmission (capitalizing on low overnight telephone service rates), the ability to broadcast the same fax to multiple recipients, relay broadcast (the ability to receive a fax and later relay it to one or more other designated recipients), support for individual, department or program identification/authorization codes to limit and track fax utilization and provide extensive management reports, plain paper fax, local copying, and support for CCITT Group 4 digital transmission capabilities.

The term *Group 3* refers to faxes using standard, analog, voice telephone service and transmitting full pages in 12 to 25 seconds. Group 4 refers to faxes using digital transmission at speeds of 64 kbps, capable of higher-quality imagery, and transmitting pages in approximately 3 seconds. Some high-end units include 20 megabyte hard discs for storing thousands of pages electronically. Some of these machines offer fax store and forward, that is, they can act as hubs in a network of fax machines, and can receive and store multiple-page fax documents for later relay or broadcast.

High-end fax machines from Ricoh, Cannon, Murata, NEC, Sharp, Toshiba, and others are priced in the $1000 to $6000 range, according to features. Some units are more expensive, for example, Canon's fax L4600 Group 4 digital laser machine with the company's proprietary digital processing technology, originally sold for about $16,000.

PC-facsimile cards

Generally, a *fax card* is a printed circuit card designed to be inserted in a personal computer. Among other functions, the plug-in card performs standard Group 3 fax modem functions and is fitted with modular telephone connectors for attachment to telephone lines. With appropriate software, text and graphic images can be converted to facsimile formats and directly transmitted via telephone networks to any Group 3 capable stand-alone fax, or to another similarly equipped PC.

Fax cards eliminate the steps of making a hard-paper copy and physically placing it on a conventional fax machine. Since the vast majority of text and graphics material is generated electronically on PCs or workstations, the potential for time and material savings is great. In addition to eliminating the need for an operator to leave his PC, make the hard copy and take the time to walk to the nearest fax machine, it reduces congestion at department fax machines and the need to make repetitive visits during busy periods. Further, as a side effect, eliminating the need for optical scanning of hard copies results in higher image quality than can be otherwise achieved.

Received faxes can be directed to PC-fax cards where they can be viewed on a screen and stored electronically within the PC's hard disc. Should a hard copy of a received fax be needed, it can be printed on the PC's dedicated or shared printer, also providing the advantage of plain paper, in contrast with most stand-alone fax machine's thermal paper copies.

PC-fax cards do have several drawbacks. First, if there is a need to fax material not available electronically within the PCs, then either a stand-alone fax machine or a scanner for the PC must be purchased. Secondly, whereas stand-alone fax machines are designed to be left in a powered-on mode to automatically receive faxes around the clock, it may not be possible or desirable to operate PCs continuously for fax purposes only.

PC-fax cards are available from Gammalink, AT&T, Intel, Panasonic, IMAVOX, and others at prices from the low hundreds to $1200. External boxes that perform PC-fax card functions and connect to a PC serial port are also available.

LAN facsimile servers

LAN fax servers operate similarly to LAN file or printer servers, i.e., they provide a Facsimile capability that can be shared by any attached PCs. LAN fax servers, typically fabricated from PCs and including one or more PC-fax cards attached to one or more telephone lines, are priced in the $2000–$4000 range.

PC LAN fax servers exhibit a number of advantages over either PC fax cards or stand-alone machines, although methods of routing and notification of incoming faxes remain to be standardized. First, centralized LAN fax servers coordinate fax traffic since the servers contain sufficient storage to support delayed downloading of faxes to attached PCs until the PCs are activated. Additionally, LAN fax servers facilitate centralized access control, authentication, record keeping and management report generation.

PC LAN fax servers can support advanced, high-end stand-alone fax machine functions. Thus a single PC LAN fax server investment may be an excellent alternative to proliferating expensive high-end fax machines. Also, PC LAN fax servers can be integrated with other messaging systems like E-Mail, direct PC-to-PC message service through carrier networks, and certain PC-to-value added network (VAN) subscription services. For example, E-Mail messages can be forwarded directly to recipients with E-Mail capabilities, and the same messages can be converted to fax formats for forwarding to recipients without E-Mail, but with fax capabilities.

Since PC LAN fax servers include external telephone network interfaces as well as those internal to the LAN, they can act as hubs for metropolitan or wide area fax networks. LAN fax server products are available from OAZ Communications, Castelle, Biscom, Brooktrout, Hybrid fax, and Share Communications.

Facsimile store & forward exchanges

Facsimile store and forward exchanges (switches) are designed to receive, store, relay, broadcast and administer fax transmission. Fax store and forward exchanges cut costs by batching faxes in off-hours or by using less expensive private networks. This is particularly useful for international faxing where off-peak public tariffs or single private lines for aggregated traffic result in significantly lower costs. Intended for large government organizations or corporations who can justify the establishment of fax networks, or as the switching vehicles for third party subscription service bureaus, these machines can be configured to store 10,000 pages and transmit thousands of pages per day.

They are functionally equivalent to PC LAN fax servers, though normally larger in scale, and therefore can provide translation between E-Mail, Telex and other services. Suppliers include 3M Company, Panasonic Communications and Systems Company, Ricoh, SpectraFax, and others. Prices range from $30,000 to $300,000.

Subscription services

AT&T, MCI, and Sprint, using facsimile store and forward exchanges, offer the enhanced technical and administrative features of high-end facsimile machines, LAN fax servers, and fax store and forward exchanges described above, but on a *subscription-service* basis. Each company is attempting to position itself as a fax-capable, interexchange carrier. Users subscribing to these services have no need to purchase or install any additional on-premises equipment. To date, these enterprises have catered mainly to those customers with extremely large fax traffic volumes and extensive numbers of fax addressees.

Like fax exchanges, the current market for subscription services is small and analysts differ over its potential. As with carrier-provided SMDS LAN services, the security of sensitive and private corporate traffic is an unresolved concern, and may further limit the applicability of third party subscription services as an all-encompassing approach for fax networking and administration.

A new application-based technology called *Faxtex* is similar in concept to audiotex and *integrated voice response* (IVR), except that instead of receiving voice information, the user receives fax information. Both audiotex and IVR play a part, however, in a Faxtex network application. For example, a call to the Faxtex system initially routes the caller via audiotex, e.g., "press 1 for product A information, press 2 for product B information," etc. An IVR step next occurs with the user interactively prompted by voice to enter the number of the product for which information is desired. The Faxtex step occurs when the caller is prompted to enter his fax number.

The Faxtex system computer then searches its database, finds the product information, and using fax board technology, sends the information to the caller's fax machine, often before the caller has hung up. Faxtex has a market niche in the area of customer service, to provide distribution lists, technical data, reader response services, and other time-sensitive information on demand and over long distance. Products are supplied by Copia, Ibex, Interfax, Intervoice, and SpectraFax.

Summary

With the explosive growth in the LAN market noted earlier, it is no surprise that by 1995 LAN fax sales are projected to equal those of Group 3 plain paper machines. By 1992, plain-paper machines will overtake thermal-paper machines which initially dominated the market. Some projections show Group 4 lagging Group 3 sales until 2000. By 1995 the total fax market is expected to exceed $8 billion, with after-market sales equalling about 40% of the total. The importance of fax as a critical enterprise resource will continue into the 21st century, and integration of fax with other computer-based management information systems will continue to grow.

Cellular, Personal & Wireless Communications Systems

Cellular systems

Mobile cellular telephone, first marketed in the early 1980s, has proven to be the impetus for an ever-expanding class of wireless mobile, transportable, and personal communications products and services. The simple idea behind cellular service, which makes it and the other wireless devices practicable, is that the same frequency channels can be systematically reused. Unlike previous radiotelephone designs which used a single set of frequencies to cover an entire city, severely limiting the maximum number of simultaneous users, cellular uses low-powered transmitters covering geographic cells (usually less than 8–10 miles in radius) as shown in Fig. 18.1a. Frequency assignments are made in patterns that minimize interference (i.e., the same frequencies—f_n in the figure—are not assigned to adjacent cells).

Because the frequency spectrum is a limited resource, without a systematic way to reuse frequencies, there would be no way to accommodate ever-increasing user requirements. Implicit in the cellular concept is the capability to shrink cell sizes to accommodate growth in the number of subscribers. A cellular system may begin with 10 cells, but faced with increased demand, find that number inadequate. If the initial cell pattern has been properly laid out, it is relatively simple to split existing cells in high usage areas, such as downtown business areas, into between three or six new cells. Various cell sizes can thus be used in metropolitan areas in accordance with geographical traffic patterns.

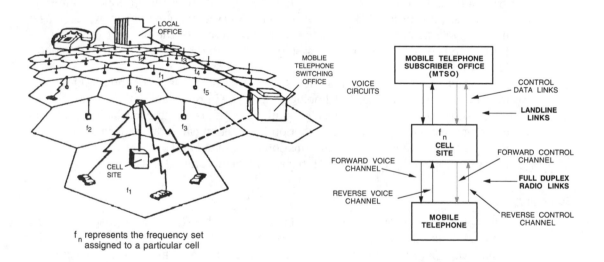

f_n represents the frequency set assigned to a particular cell

(a) CELLULAR MOBILE TELEPHONE ELEMENTS   (b) CELLULAR MOBILE TELEPHONE OPERATION

BAND	FREQUENCY (MHZ)	CHANNEL NUMBER	SIGNALING CHANNEL
BAND A BASE TRANSMIT	870.030 TO 879.990	1 TO 333	313 TO 333
BAND B BASE TRANSMIT	880.020 TO 889.880	334 TO 666	334 TO 354
BAND A MOBILE TRANSMIT	825.030 TO 834.990	1 TO 333	313 TO 333
BAND B MOBILE TRANSMIT	835.020 TO 844.880	334 TO 666	334 TO 354

(c) 666 CHANNEL FREQUENCY ALLOCATION PLAN

Fig. 18.1 Cellular telephone system design.

Cellular design imposes the requirement that as an active mobile subscriber moves through an area, his call has to be handed-off between cell site transmitters. This creates a control problem which had to wait until microcomputer and digital switch technology made it practical and affordable to achieve hand-off without losing the call connection, or being audibly imperceptible to subscribers.

Handing-off involves computers at three levels—microprocessors in the *mobile transceiver, the mobile control unit* (MCU) at the cell site, and in the electronic switches in the *mobile telephone switching office* (MTSO). The MTSO that controls cell sites, normally provides connectivity to nationwide telephone networks. Figure 18.1b shows that separate transmit and receive channels are assigned for both voice and control functions. L.M. Ericsson, AT&T, and a newly formed alliance between Motorola and Northern Telecom are the leading cellular switch vendors.

Figure 18.1c identifies cellular spectrum allocations made by the FCC for 666 channel system designs. The FCC has approved 832 channel designs, but not all service areas and not all mobile telephones are so equipped. The FCC grants two separate licenses in each area, one to the wireline LEC, and the other to independent enterprises—originally selected by lottery. In Figure 18.1c, bands A and B represent allocations to the two service providers.

In the mid-1980s, unit prices for cellular telephones hovered in the $2000–$3000 range, and monthly subscription charges averaged $125–$150, resulting in lackluster growth in demand for units and services. By 1987, unit prices broke the $1000 barrier and with steadily declining usage costs touched off an annual growth rate of 40% as depicted in Fig. 18.2. Today, mobile telephones can be purchased in three different formats: *mobile* (permanently installed in vehicles), *transportable* (bag phones or hard case models fitting in briefcases) and *handheld* (some as small as wallets). The latter two styles have self-contained batteries. Some bag phones can be purchased for under $100 with subscription service registration. While some models are as feature-laden as modern desk sets, *stairstep pricing* provides real incentives for opting for other than the lowest cost. For example, the $100 models may or may not include battery chargers. Some models can be charged from automobile cigarette lighters, but not while in use, and finally some models can be charged in automobiles, while in use, or from home or office battery chargers. Handsfree speakerphone operation and lapel microphones are other example features that can add to basic costs. By the time one pays for the base unit and ancillary cables and accessories, what started as $100 phones can easily migrate to $500 price levels.

Nevertheless, at $500, it is now possible to purchase a top-of-the-line hard-case portable telephone that fits easily in a standard briefcase. Handheld models, particularly the ultra-small models like Motorola's Micro Tac and Fujitsu's Pocket Commander (10.2 ounces and only slightly larger than a pack of cigarettes) are priced in the $900–$1500 range. With extra talk-time batteries, these units support up to 80 minutes continuous conversation and 13 hour standby operation between charges.

Even these tiny models incorporate features such as alphabetic directory and scroll, 100-number memory, 40 name alphanumeric memory, last number redial, silent alert mode, voice and ring volume control, individual call timer, accumulated call timer, desk charger, battery level monitor, and electronic locking.

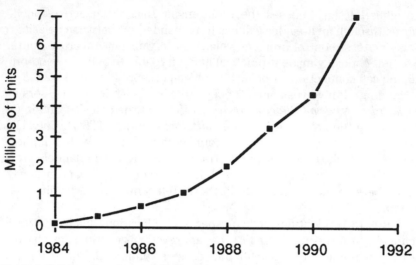

Fig. 18.2 Cellular subscriber growth.

As the 1990s began, over 3 million analog cellular telephones were in use across the U.S. Research indicates that in the years ahead, one in four cellular phones sold will be for personal use and by 1993, one out of three will be for non-business communications. By the end of 1994, the installed base of personal and consumer cellular should number at least 4 million out of a total subscriber base of nearly 12 million cellular users. These figures indicate continued strong growth in personal communications products and services. Estimates indicate the total size of the cellular industry to have reached $3 billion in 1991.

In this decade, the emergence of new mobile and personal communications technologies is expected to accelerate, fueled by consumer demand and industry innovation. In 1990, for example, a new North American draft specification was issued for a dual-mode (analog and digital) cellular standard called *Code Division Multiple Access* (CDMA), offering the potential for up to twenty-fold increase in capacity. This would enable carriers to meet increasing consumer demand while lowering per-minute call costs. Unfortunately, the proposed standard clashes with the *Time Division Multiple Access* (TDMA) standard adopted by the U.S. in 1989.

Motorola has also presented a proposal for its Iridium system, a constellation of 77 low-flying satellites to support global mobile telephone service. The service will use 4800 bps narrowband digital voice encoding (see Chapter 2) and reportedly will require a $2.4 billion investment. In 1992, the World Administrative Radio Consortium (WARC) approved a spectrum allocation request for the service. Single integrated mobile subscriber sets would automatically select terrestrial or satellite system access in accordance with called party location.

In a move to provide uniform service to mobile subscribers traveling across the country, Bell Atlantic, Ameritech Corp., Nynex, and GTE Corp., an independent telephone company owning two cellular firms, have plans to weave together cellular telephone networks in 130 U.S. cities. This action represents continuing efforts toward an industry objective of providing seamless nationwide incoming and outgoing call service. McCaw Cellular Communications Inc., the nation's largest cellular company, is already installing equipment to notify and serve traveling subscribers.

Personal & wireless communications systems

Many industry experts believe that wireless communications for voice, data and other services is emerging as the most significant telecommunications development since fiber optics. The level of industry activity in research and development and the number of different technology and standards avenues being pursued are so great that it is difficult to provide a structured overview. There is no generally-agreed-upon definition of a personal communications system (PCS). To some it is a cellular system with smaller cells. For others it is an outgrowth of cordless telephone technology. Still others emphasize the potential for wireless LANs and PBXs.

Clearly, however, an extrapolation from existing cellular approaches to smaller and smaller microcells is an important aspect of future designs. As technology continues to evolve, the notion of having a cell site antenna on every street corner is not farfetched. As cell size diminishes, so does associated size and cost. Microcell transceivers, easily installed inside and outside of buildings will be linked by fiber optic cable to compact base stations, which during a transition period will be capable of supporting both analog and digital radio standards chosen by the Cellular Telecommunications Industry Association. For example, AT&T has announced that its AUTOPLEX Microcell, capable of supporting both analog and digital technology, will be available in 1992.

Elsewhere, initial experiments have begun in PCS technologies using various frequency bands and with numerous related techniques. The FCC recently granted a sizable number of experimental licenses for these activities. Research indicates that factors driving the PCS market include the demand for:

1. Immediate access to people and information

2. Ubiquitous, low-cost, user-friendly communications

3. A personal communications ID that can track a caller's location and the location of the called party without geographical constraints

4. A higher degree of security than what has been available in older systems

5. Release from the constraints of wired networks, even at the expense of personal privacy

PCN America, Inc., a subsidiary of Millicom, has petitioned the FCC to set aside the 1.7 to 2.3 GHz bandwidth for their personal communications network. The new service would offer features such as "smart cards" to be used with portable handsets, thus identifying the subscriber rather than a terminal. Motorola, the largest player in cellular and radio communications, has announced that it is developing a *Wireless In-Building Network* (WIN) using 18 GHz microwave technology that will deliver 15 Mbps to the desktop.

Anticipating an upward trend in the popularity of wireless communications, common carriers during the 1980s had already begun developing new marketing strategies, including:

1. The migration of cellular telephone technology from analog to digital
2. The evolution of the residential cordless telephone concept to a digital personal communicator
3. The expansion of traditional professional and commercial mobile data services to consumer products and services
4. The application of wireless communications to the office, including wireless LANs and PBXs

Apart from user demand, these developments are impelled by advances in technology. One such advance is the development of new, powerful voice processing algorithms that enable human speech to be transmitted at near toll quality, at data rates employing compression ratios of up to 16 times what was standard in the traditional telephone network. In addition to increased capacity and lower costs, this advance also permits the use of lower powered amplifiers, with reduced device size and weight.

These improvements have driven the development of new generation digital signal processors (DSPs) which can support the complex new algorithms. Employing submicron CMOS fabrication techniques, semiconductor producers such as Texas Instruments, AT&T, and Analog Devices have introduced processors capable of 20 to 40 MIPS (million instructions per second), while maintaining compact size, low cost, and low power consumption. Customized processors are supporting more integrated functions, to the point where internal antenna circuitry and battery design have become limiting technologies.

Although these technologies are important, most industry pundits believe that personal communications services will be driven—as is usual in the U.S.—by the marketplace. New systems will build upon elements of existing fixed and mobile networks. Continued downward trends in usage and equipment costs may well be the most important factor in accelerating user demand for personal communications products and services.

Video Conferencing & Imagery

Status and trends

Video conferencing is defined as the real-time, usually two-way transmission of voice and images between two or more locations. Today, both voice and video analog signals are digitized by video codecs before transmission which, as noted in Chapter 14, can involve wide bandwidths. To conserve bandwidth, some systems employ "freeze frame" where a television screen is only "repainted" every few seconds. Codecs for higher quality full-motion video attempt to minimize bandwidth requirements by taking advantage of intervals with relatively little motion (which require smaller bandwidths), and by trading off smooth motion tracking and picture resolution.

Videophones continue to draw skepticism, including AT&T's renewed efforts to capture the residential picture market, reflecting the slow start that this relatively new industry has experienced in the past. However, experts close to the scene predict rapid growth in the next several years. Figure 18.3 summarizes some of the factors behind these predictions. The market expansion from $500 million in 1991 to $1.5 billion in 1995 reflects not only an order of magnitude reduction in the cost of equipment, but growing awareness among business people that videoconferencing greatly enhances productivity. In the past, attempts were made to justify videoconferencing costs on the basis of merely offsetting travel expenses.

More recently, work at MIT and elsewhere has supported the conviction that in face-to-face or videoconferenced meetings, 40% of the information conveyed is visual. This establishes the value of video versus voice only audio conferencing. The productivity enhancement comes from the fact that videoconferencing permits high value meetings to occur on schedules that are simply not possible otherwise. With video conferencing, busy west coast executives can routinely schedule critical, time-sensitive meetings with European, east coast and Asian counterparts, within the span of just a few hours.

Other growth factors include the emergence of affordable digital telecommunications carrier services, described throughout this book, and the CCITT P × 64 standard that is leading to Group 3 fax-like universal connections not heretofore possible. In conjunction with switched 64 kbps digital services and inverse multiplexer arrangements, cited in Chapter 15, P × 64 standards permit users to select video conferencing services at selectable quality and cost increments. As shown in Fig. 18.3, within cor-

VIDEOCONFERENCING

- Rapidly expanding market - $500M in 1991 to $1.5B in 1995.

- Significant price drop for group systems:

  - $250K in 1985 for a videoconferencing center, and

  - $25K today for a roll-about, go anywhere cabinet.

- The motivation? Productivity enhancement, not just offset travel expenses.
  40% of information conveyed in a conversation is visual.

- Adjunct desktop video market emerging.

- Timely availability of affordable digital telecommunications carrier services.

- CCITT "P X 64" standards established in 1990.

- Progress towards ubiquitous, inexpensive easy to use service.

Fig. 18.3 Videoconferencing status and trends.

porations, the trend is away from expensive dedicated videoconferencing rooms towards roll-about, go anywhere cabinet mounted equipment, although the number of carrier/vendor provided public videoconferencing rooms is growing. Also, personal videoconferencing, employing pc-based desktop video equipment is seen by some to be the market segment with the greatest potential.

Technology

As described in Chapter 14, "lossless" codecs without bandwidth compression convert standard National Television System Committee (NTSC) analog TV signals into digital signals at rates of about 140 Mbps. Lossless means that no degradation in either motion tracking or resolution is perceived by most people viewing TV displays using digital versus NTSC analog input signals. Electronic storage of high-quality still images, e.g., high-resolution color photographs or medical X-rays, requires about 3 megabytes of storage capacity per image.

Codecs incorporating compression algorithms can reduce output bit rates for moving pictures and storage requirements for still images. The lossless class of compression algorithms reduces output bit rates by factors of 3 to 1. Lossy video and imagery algorithms result in reduction factors of 50–100 to 1, and 10–50 to 1 respectively. The capability of processors embedded in video compression equipment today ranges upwards to 60 MIPS (million instructions per second) using up to ten separate microprocessors. Future video encoders may employ processors with 1–4 BIPS (billion instructions per second) capabilities.

Lossless variable bit rate (VBR) video encoders under development at the Sarnoff Laboratory are described in Chapter 14. VBR encoders with appropriate VBR capable, ATM-based broadband integrated digital networks, will offer videoconferencing service users an option to reduce average transmission rates (and transmission costs) with no loss in quality.

Note that *videoconferencing peripherals* such as document cameras, VCRs, electronic still cameras, and facsimile machines not currently integrated in basic equipment, can add significantly to system acquisition costs.

Standards

CCITT videoconferencing recommendations, approved in 1990, are:

- H.261—specifying common compressed video formats
- H.230—outlining intercodec signaling
- H.221—Establishing standards for multiplexing voice, audio and graphics over a single link
- H.242—stipulating how to initiate and disconnect links.

Standards are necessary if ubiquitous public calling service, on a scale comparable to common telephone and facsimile services, is to be achieved. Cleverly, the above recommendations while standardizing output formats, do not constrain vendor-unique, quality-enhancing encoder developments.

Other groups contributing to standards include:

- *Joint Photographic Experts Group* (JPEG). Industry experts from companies worldwide, formulating still-frame digital image compression standards

- *Moving Pictures Expert Group* (MPEG). Industry experts from companies worldwide, formulating video and associated audio and digital storage compression and coding format standards.

JPEG and MPEG work in coordination with or under the auspices of the International Standards Organization (ISO), Electronic Industries Association (EIA) and IEEE's Consumer Electronics Society.

Market participants

Key equipment vendors include Compression Labs, with a 63% market share; PictureTEL Corp. with a 28% market share; NEC America, GPT Video Systems, and British Telecom. Carrier and service providers include AT&T, US Sprint, MCI, Hughes and PTTs in England, France, Germany, Australia, and Japan.

Statistical & Cable Multiplexers

Statistical multiplexers

Statistical multiplexers, also called *statmuxes* or *intelligent multiplexers*, implement an advanced form of time division multiplexing (TDM), which allocates bandwidth only to active attached devices. These devices, normally used in conjunction with point-to-point lines and modems, capitalize on the fact that on the average, asynchronous and synchronous terminals are active only 10% and 30% of the time respectively. Statmuxes are inserted between modems and devices normally attached one-for-one directly to modems.

Statmuxes can result in sizable savings relative to their cost. For example, four remote terminals located in New York City and served by four separate 4800 bps modems with dedicated (leased) telephone lines to Boston can be replaced by a statmux and a single 9600 bps modem and leased line, resulting in a monthly savings of about $1700, or 65% of the original cost. Micom offers its Micom Box 5, a 32-channel statistical multiplexer that can interface with modems of up to 19.2 kbps data rates, or equivalent digital channels, for $900.

Statmuxes create *virtual channels* and dynamically allocate them in accordance with real-time traffic requirements of attached devices. Some statmuxes incorporate fast packet switching to support statmux functions.

Statmuxes can be combined with data compression to achieve even greater savings. A simple example of how data compression works is as follows. The letter A occurs much more frequently than the letter Z, so an 8-bit representation for A is replaced by a 4-bit representation, and a 12-bit representation is substituted for the letter Z's normal 8-bit pattern. When combined with statmuxing, data compression saves another 20% to 30% in transmission bandwidth.

Cable multiplexers

RS-232-C connectors, long the standard for terminals and other DTEs, incorporate a 25-pin connector, and at 19.2 kbps, a 50 foot distance limitation. Cable multiplexing overcomes RS-232-C limitations and reduces the number of connecting wires to two twisted pairs (4 conductors). *Cable multiplexing* takes source signals at each interface connector pin, multiplexes them, and transmits them via twisted-pair cable to remote demultiplexers. Here the demultiplexers restore RS-232C levels at each of the 25 connector contacts. In the process the distance between devices can be extended to 1000 feet, and significant savings in cable size and cost can be realized.

Perhaps the greatest advantage, however, is that device input and output signals can be directed to centralized node monitoring and control positions which permits a level of local network and system management not otherwise possible. With matrix switching (switches that interconnect serial interfaces such as RS-232C under operator control, usually for network management purposes), cable multiplexing permits extensive remote test, diagnostic and equipment substitution capabilities that can significantly reduce downtime due to failures. Further, such designs permit the detection of deteriorating conditions so that corrective actions can be taken before serious failures occur.

Cable multiplexing products are available from Telenex, Equinox, American Photoetics, and Interoptics.

Outlooks for the Future

You are approaching the end of the *McGraw-Hill Telecommunications Factbook.* Looking back, you should now be able to discern clearly the "building-block" structure of the book and its focus on the relationship between advanced technology applications and modern business needs. Part 1 began by presenting telecommunications background material, then moved into fundamentals of transmission, circuit switching products, signaling concepts, and premises distribution systems, ending with tariff structures, voice network design principles, and their implications on cost effective business networks.

Part 2 laid the foundation for understanding public and private voice networks in the post-divestiture environment, followed by a tutorial on how to be an informed consumer of business telecommunications products and services.

Parts 3 and 4 treated data communications, examining the burgeoning field of local, metropolitan, and wide area networks. Part 5 discussed emerging applications for integrated voice, data, and image information exchange, and advanced technologies underlying integrated digital LEC and IXC networks.

Part 6 covered telecommunications sectors that do not readily fit into the preceding categories, but which are currently exhibiting explosive growth.

Virtually all important services and technologies have been treated in these chapters, and a special effort has been made not just to define and explain each topic, but to illustrate how individual technologies, products and services relate to one another, and how they reflect industry patterns and trends.

We saw, for example, how switched 64 kbps services complement the new CCITT P × 64 video conferencing standards. We observed how the DS"N" digital signal hierarchy and related equipment, evolving from DS1 rate channel bank for analog voice has become the dominant LEC/IXC mechanism for providing switched and dedicated data communications services.

We next covered one of the most important telecommunications developments yet to take place, i.e., the progression towards integrated digital networks. As noted earlier, the word *integrated,* when used in this context, takes on the broadest possible meaning encompassing both the ability to transparently accommodate voice, data, video, and other services, as well as the integration of switching and multiplexing within network facilities. Underscoring this development, we observed the recent confluence of switching and multiplexing techniques within asynchronous transfer mode (ATM) technologies, exemplifying the relationship between CCITT standards and future market offerings.

All of which brings us now to Part 7—containing summary remarks and a brief discussion of prospects for the future. At this point it is always tempting to wax omniscient, confidently predicting future developments. In fact however, if you have grasped the implications of material already presented, your prognostications are as reliable as ours, since we now possess a common view of historical events, and are faced with the same marketplace uncertainties. Nevertheless, we offer some thoughts based on lessons learned.

Current "Key" Applications

The book cites a variety of business applications with the potential to shape future telecommunications product and service offerings. Included are LAN-to-LAN traffic, medical imagery, and various applications best served by wide-bandwidth, variable-bit-rate (VBR) services (such as video conferencing employing new lossless VBR algorithms). If any of these initiatives attain what we term *key* application status, progress towards ATM/Sonet-based integrated digital networks (IDNs) will be greatly accelerated.

However, while there is evidence indicating that these and other applications will proliferate, recall that some years ago, industry pundit's predictions for the *office of the future,* the *paperless environment,* and *information integration* have yet to materialize. On the other hand, perhaps the dominant application of the late 80s initially went unrecognized by experts, taking decades to fully develop. Of course, we are referring to the facsimile (fax) explosion, an impressive example of telecommunications growth.

It is useful to consider the development of the fax market. Although the technological basis for fax existed years ago, its potential remained dormant. At first, when few organizations possessed fax equipment, new users had to justify investment based on in-house demand. Following adoption of Group III standards by fax equipment manufacturers, things changed rapidly. As ever-increasing numbers of Group III machines appeared, the application broke free of in-house limitations and soon, faxing between independent enterprises became commonplace, and expected.

Growth was fed by regenerative forces—as more units were sold, more buyers felt compelled to purchase. Where at first, most small businesses considered fax machines optional, today small enterprises and even individuals view fax equipment ownership as essential.

Looking back, it is easy to spot the critical ingredients in the fax success story. First, Group III standards permitted interoperation between fax machines made by different manufacturers, essential for the market to encompass intercompany applications through universal connectivity. Moreover, universal connectivity depended on the ability of Group III machines to operate reliably over the ubiquitous voice telephone network. Today's Group IV digital machines are faster and produce higher quality images, but will not match Group III popularity until end-to-end digital telephone networks become a reality.

A final factor contributing to fax's growth involves pricing. All potential telecommunications applications have threshold price levels. No matter how appealing or novel a technology may be, there is a market sustaining price level above which only a small number of consumers—either for status or special operational reasons—will participate. Prices which elevate applications to key status are usually significantly below market sustaining prices. We witnessed this in the consumer VCR market. Above $1000, sales languished and only relatively affluent, technology oriented users entered the market. At $200–300, a large percentage of television set owners purchased one or more units.

There is little argument that in breaking the $500 price barrier, fax machines overcame the last obstacle to achieving key application status. Note the interdependency among the pricing, standards and universal connectivity factors cited above.

Future Key Applications

It is interesting to ponder what key applications are on the horizon. Two promising candidates are: affordable LAN-to-LAN public connectivity in the business sector and video-on-demand in the consumer arena. Fueling growth in both areas is the potent trend towards personal computing and personal communications. As noted in Chapter 18, personal communications, the fastest growing telecommunications industry segment, refers to a broad range of products and services, e.g., facsimile machines, cellular telephones, emerging personal communication system (PCS) adjuncts, and a variety of radio systems including pagers, hand-held remote data entry terminals and citizens-band-like radio systems.

While precise definitions do not yet exist, personal communications provide at least one human operator with direct terminal access and real-time interactive communications with a remote human operator or other information resource. For example, the Department of Defense is transitioning from over-the-counter message services to equivalent services based upon direct PC-to-PC messaging.

LAN-to-LAN connectivity potential derives from the overwhelming business commitment to personal computers (PCs) (about 100 million units so far), and the subsequent decision to attach PCs to LANs (some predict 50% of PCs will be connected to LANs).

The motivation for video-on-demand is similarly traceable to growth in personal services. Here, the impact of appropriate pricing, and continued FCC and court approval in the video dial tone decisions are clearly factors. Given that LECs are permitted to provide on-demand value-added video services, that they are able to charge for programming material as well as telecommunications services, and that the service is priced right, they can in all likelihood compete with the retail video rental industry. For example, at five times current rental prices, few people would opt for LEC services. At two times the cost or less, video-on-demand could be another key application. Apart from the inconvenience of traveling to pick up and return video tapes, an additional LEC advantage would be near 100% availability of popular movie titles, a capability which few rental stores could meet. The advantage of video-on-demand versus existing pay-per-view services, delivered for example via cable, is that viewers can specify viewing times, perhaps in intervals as small as fifteen minutes.

Both LAN-to-LAN and video-on-demand applications require wide bandwidth transmission direct to premises, and both involve bursty or intermittent use of that bandwidth. When a user requests video service, he needs wide bandwidth for the duration of his viewing session, but may not make such bandwidth demands again for hours or even days. Similarly, some 85% of the data exchanged among LAN attached PCs, stays within the same LAN. On those occasions when LAN-to-LAN (inter-LAN) connectivity is required, as noted in Chapters 13 and 14, users expect the same rapid response-time performance experienced with multimegabit intra-LAN facilities. Thus, for the duration of LAN-to-LAN sessions, wide bandwidth MAN and WAN networks are required. However, again due to the relatively long intervals between LAN-to-LAN sessions, high-peak but low-average bit rate services are needed.

Wide Bandwidth Local Loops

To deliver LAN-to-LAN and video-on-demand services to users, LECs must upgrade today's pervasive twisted-pair local loops. For video-on-demand, this means fiber optic cable to the premises. For large telecommunications users, such installations are feasible and already beginning to occur. For small businesses and residences, cost recovery and regulatory approval will not occur rapidly. As noted above, LEC tariff levels will be critical.

What is certain is that linear extrapolation of current private line LEC and IXC bandwidth-dependent tariffs to the bandwidths needed, will not be affordable. Consequently, LECs and IXCs must establish usage-based tariff structures for tomorrow's intermittent bandwidth-on-demand services. Such structures would be based on kilobits-per-second × hours used, much the same as electric utility kilowatt-hour charges, with discounts for high volume usage.

ATM/SONET-Based Integrated Digital Networks

Beyond providing appropriate access service, to proliferate bandwidth-on-demand services, LECs and IXCs must incorporate ATM/SONET core network technologies into their integrated digital networks (see Chapter 17). Progress towards SONET (synchronous optical network) based transmission is well underway, supported by the ever-increasing deployment of fiber optic cable. What remains is 1) the replacement of today's digital signal hierarchy (and its link-by-link synchronous operation), with standards-based SONET network-wide synchronous digital hierarchy facilities, and 2) the migration to ATM-based switching and multiplexing. If manufacturers' research and development investments are any indication, there is now a clear trend towards standards-based ATM and SONET equipment. As in the past, their rate of deployment will be driven by carrier pricing, return on investment realities, and user acceptance.

Chapter 7 presented the network design axiom that traffic aggregation leads to more efficient use of network resources. Applying this axiom to public versus private network trade-offs, one can legitimately conclude that it is in everyone's (society's) best interest to ultimately replace all facilities-based private networks with virtual private networks (VPNs). Recall that VPNs use public facilities, but supply services that appear to users as being delivered over dedicated facilities.

In the voice network market, history has translated this hypothesis into reality. Keen competition among the IXCs for voice business revenue has led to aggressive pricing. As noted in Chapter 10, the evolution of public IXC infrastructures towards digital, software-driven architectures has made virtual private network service the IXC's most economical offering. This phenomenon has all but driven facilities-based private voice networks, prevalent in the 60s and 70s, from the large user telecommunications scene. However, the migration from private to public data networks has not been so dramatic since large data users can still obtain more flexible and lower priced service from dedicated facilities.

As previously noted, the reason for the persistence of private data networks has been based on the state of switching technology and the profile of data user demand. In other words, switch processor limitations and the complexity of the X.25 protocol have made it unfeasible to build large central office packet switches, capable of aggregating and handling the quantity of traffic that would make public facilities more economical than private facilities. Consequently, today's large X.25 users operate private networks with packet switches located on their premises, an arrangement with both technical and cost advantages.

ATM/SONET based LEC and IXC integrated digital networks are needed to enable the carriers to offer virtual private data network services more economically than they can offer private network facilities. We also noted that today's demand for nonvoice services has not yet reached levels that alone justify migration to ATM/SONET based IDNs. So the transition to networks supporting new applications, at prices users are willing to pay, will be based on LEC and IXC investment, availability of competitive, standards-based ATM and SONET equipment, incorporation and interworking of those facilities within existing IDNs, and to some extent, the longevity of interim solutions to data problems, such as frame relay and FDDI based MAN service offerings.

General Observations

The past decade has produced profound technological and social developments, which, upon reflection, many believe to be interrelated. The spirit of democracy, free enterprise, and markets driven by competition has received world-wide endorsement, not with words alone, but by history shattering actions. In the United States, milestone judicial and executive department actions have underwritten and reenforced deregulation and competition as the guiding principles governing industries ranging from transportation, to banking, to telecommunications.

In the telecommunications arena, the most consequential action has been Judge Greene's Modification of Final Judgment (MFJ), which inextricably links legal proclamations with the technical aspects of public service provision. Whatever your net assessment of divestiture may be, an undeniable outgrowth of the MFJ is that we now have competition among interexchange carriers, and if competitive access providers (CAPs) have their way, we will also have competition at the local exchange carrier level.

In the equipment arena, the Carterphone and other regulatory authority and judicial decisions have prompted intense competition among manufacturers and vendors. As a result, once unique telecommunications products and services have been relegated to a commodity status. Telephone sets, facsimile, cellular telephone and PBX switching equipment are well documented examples.

As a consequence, service providers, manufacturers and vendors continually seek new ways of differentiating their offerings in an increasingly crowded marketplace. Coupled with the dizzying pace of technological innovation, the array of telecommunications options is expanding and changing at an unprecedented rate. Because service providers, manufacturers and vendors are not only motivated to provide offerings that substantively improve telecommunications, but also to make a profit, it is often difficult to discriminate between truly effective offerings, and those calculated to merely increase market share.

The bottom line is that an informed telecommunications consumer is more likely to make the right choice than one who relies exclusively on other's guidance. Though this has always been the case, it has become increasingly important as new worldwide competitive environments and new technology applications continue to proliferate, and business dependence on information exchange continues to grow.

In a very real sense, the examples presented in part of the book illustrate the important role users play in determining which products and services will survive, and which will dominate the array of future telecommunications options. Simply put, the ultimate factor affecting telecommunications service availability and success is the vote users cast through service and product selection. Again, the ideal way to assure availability of the best possible future options, is to be an informed buyer today. It is the sincerest wish of the authors, that this *McGraw-Hill Telecommunications Factbook* will contribute mightily to that cause!

Standards Setting in the United States

Successful provisioning of facilities for telecommunication services, whether public or private, is largely dependent on the interconnectivity and interoperability of those facilities. Interconnectivity and interoperability are cardinal objectives of standards-setting processes. The subject of telecommunications standards is many-faceted, often surrounded by confusion and misunderstanding. This appendix examines the "what, how, when, and why" of standards. From a historical perspective, the telecommunications industry grew up around a "standards"-based approach.

In the past several decades the data processing industry and more particularly the discipline of data communications—itself an intersection of the data processing and telecommunications worlds—has greatly complicated the traditional standard-setting proclivity of the telecommunications industry. To gain an appreciation of the role of standards in telecommunications and data communications, some background is necessary. Data processing standards are addressed only insofar as they affect data communications.

Background

Telecommunications standards date back to the 1880s with the founding of the International Telegraph Union (ITU) based in Geneva, Switzerland, as the driving force to achieve interconnectability and interoperability among European telegraph systems—then mostly government appendages to railways or postal administrations. The ITU constituted a collegial process for establishing recommendations, which while not binding on member telegraph administrations, were considered as treaties by foreign ministries. The Transatlantic Telegraph Cable linking North America (essentially the United States) with European international networks greatly accelerated the significance of ITU "standards" (actually recommendations). Thus, the U.S. through the State Department became a signatory to ITU recommendations, an arrangement which persists to this day.

With the advent of the telephone, which used wire media much as telegraphy did, and with the appearance of wireless radio communication, the ITU split into two principal bodies and a third offshoot body. The first two bodies divided the telecommunications domain into Radio and Telephone and Telegraph with acronymic "names" derived from the initials of their formal names in French: the CCIR—the International Consultative Committee for Radio, and the CCITT—the International Consultative Committee for Telephone and Telegraph. The radio domain also merited a third entity—the IFRB—the International Frequency Registration Board—one of the first attempts at fostering international cooperation between often hostile, and sometimes warring states.

The U.S. and most other nations are members of the ITU out of necessity, since electromagnetic waves respect neither geographic nor political boundaries. Today the ITU—the International Telecommunications Union is a part of the UN and remains based in Geneva.

Early on, the U.S. domestic scene, being essentially private enterprise driven, did not see the need for direct government involvement in standards setting, other than the State Department's continuing involvement in treaty-making through the ITU. The rise of the Bell System colossus under the leadership of Theodore Vail gave the U.S. a standards-making engine of awesome speed and power. Until its demise by divestiture in 1984, the old Bell System drove standards for telephone and also to some extent for telegraphy—despite the fact that telegraphy was controlled by Western Union and individual private rail companies.

The Bell System established the world's best and lowest-cost universal telephone service—albeit in the guise of a monopoly, courtesy of the 1916 Kingsburg agreement between the U.S. Government and AT&T (the proprietor of the Bell System). Prior to divestiture, government impact on standards-making took the form of FCC initiatives loosely termed *deregulation*. There were also some true standards-making bodies, such as the IEEE (Institute of Electrical and Electronic Engineers, a U.S. based professional society), the EIA (Electronic Industry Association, trade group), and the ANSI (American National Standards Institute). The FCC codified existing standards developed by the Bell System into Rules and Regulations. Two principal standard areas were Customer Premise Equipment (CPE)—the purpose of Part 68 Rules—and interconnection of other common carriers as a consequence of deregulation.

Impact of U.S. Developments

From the late 1960s, telecommunications standards have been greatly affected by federal judicial and regulatory initiatives directed toward the Bell System and its workings. These activities can be grouped under two categories: *Deregulation*, the regulatory domain; and, *divestiture*, the consequence of the U.S. Department of Justice anti-trust suit against the old AT&T. The combined effect of these initiatives drastically altered the way standards are established and administered in the U.S., and to some extent how the CCITT/CCIR operates, reflecting U.S. influence on its activities.

Pre-divestiture, U.S. telecommunications was run under the rubric of the Bell Systems—a benign monolith, but a monolith nevertheless. On the plus side, this made for controlled introduction of new technology in a manner that made standardization a built-in objective of the process. Undeniably, this approach led to the establishment of the world's finest, large-scale domestic telecommunications system.

Also, undeniable was the cost-effectiveness of the service provided by the Bell System. This was attributable to the lack of competitive pressure on the financial structure of the Bell System, which conformed only to a regulatory rate-of-return philosophy. The negative side of this arrangement, however, was that standardization was also used to delay introduction of new technology having cost and/or performance advantages.

Early federal initiatives were designed to mitigate the negative aspects of standardization, increasingly viewed as skewed to AT&T's financial advantage. Furthermore, Bell System standards were not visible to the community at large except for those affecting interfaces with CPE, and those visible by virtue of the Bell System's relationship with independent telephone (i.e., non-Bell affiliated) companies, represented then by the United States Independent Telephone Association (USITA).

Joint Bell System (through Bell Labs) and USITA issuance of *compatibility specifications* was a major standards accomplishment. Their influence on the engineering and operation of the physical plant of the Bell System, were of inestimable value. They made possible the levels of both customer and competitors interconnection that ultimately led to today's competitive telecommunications environment.

Divestiture was the instrument causing the dissolution of the Bell System and its unilateral standard-setting activities. The loss of the concept of "system" in the U.S. telecommunications infrastructure has turned what was heretofore a militaristic approach to standards, into a more democratic, collegial system, much like the ITU in Europe before its replacement by CEPT (the Conference of European Posts and Telecommunications—a grouping of administrations usually associated with governments, striving for consensus).

As might be expected, today's standards-setting environment leaves something to be desired in terms of timeliness in adopting standards agreements among affected parties. In the early days, as we noted, the charge was often made that technology advances were being delayed in the interest of one company's financial benefit. Now, the length of time involved with establishing consensus for standards adoption, often results in technology racing past the standards process, frequently rendering it moot.

Current Status

U.S. telecommunications (including data communications) standards today are no longer dominated by any single entity, including the U.S. government. To the extent the Department of State participates in ITU standards recommendations and that the resulting *treaties* are ratified by Congress, the international standards-setting process has changed little from two decades ago, except for a quickening of pace. The CCITT and CCIR carry out their work in Study Groups that meet over a four-year period, culminating in a plenary session that adopts recommendations in categories identified by letters, as shown in Figs. A.1 and A.2. Major representation changes, however, have resulted from the replacement of the Bell System by various

SERIES

A. Organization and Working Procedures

B. Terms and Definition

C. General Telecommunications Statistics

D. General Tariff Principles

E. Telephone Operation and Tariffs

F. Telegraph, Mobile Services, Telematic Services

G. Transmission - Analog and Digital

H. Line Transmission of non-Telephone Signals

I. ISDN - General Structure and Service Capabilities

J. Program and Television Transmission

K. Protection against Interference

L. Protection of Facilities

M. Maintenance: Telephone, Telegraph, Data Transmission

N. Maintenance: Sound Program and Television

O. Measurements

P. Telephone Transmission Quality

Q. Signaling and Switching

R. Terminal Equipment for Telegraph Series

S. Printing Telegraph Equipment

T. Terminal Equipment for Telegraph Services

U. Telegraph Switching

V. Data Communication over Telephone Network

X. Data Communications Networks

Z. Functional Specification and Description Language

Note: These recommendations are published in different color books every four years.

Fig. A.1 CCITT recommendation categories.

federal agencies and a variety of quasi-nongovernmental organizations (QUANGOs) such as ANSI, IEEE, and EIA/TIA.

The international arena is bifurcated between the ITU and its constituents with its *standards* cast as *recommendations*, together with the International Organization

VOLUMES

I.   Spectrum Utilization and Monitoring

II.   Space Research and Radioastronomy

III.   Fixed Services below 30MHz

V.   Fixed Satellite Services (and Frequency Coordination between Terrestrial Radio Relay and Satellite Services

VI.   Propagation in Ionized Media

VII.   Standard Frequencies and Time Signals

VIII.   Land Mobile Services, Maritime Mobile Services, Radio Determination Satellite Services, Amateur and Mobile Satellite Services

IX.   Fixed Service using Radio Relay System

X.   Sound Broadcasting Service, and Sound and Broadcasting Satellite Services and Recording

XI.   Television Broadcasting Service

XII.   Television and Sound Transmission

XIII.   Vocabulary

XIV.   Administration of CCIR

XV.   Questions for coming Plenary

Note: CCIR recommendations publications are referred to as "green" books.

Fig. A.2 CCIR recommendation categories.

for Standardization (ISO). The ISO is much more active in data communications than telecommunications. A third entity, the International Electrotechnical Commission (IEC), works largely through the ISO. The designated U.S. representative to ISO is ANSI, which, through its accreditation of the Exchange Carriers Standards Association (ECSA), has made it the primary vehicle for telecommunications—and to a limited extent—data communications standardization in the U.S. today. The formal entity of ECSA is known as the T1 (for telecommunications—not T Carrier!) Committee composed of interested parties from LEC and IXC service providers, equipment manufacturers, the government, and end users when they elect to participate. While under the ISO aegis, through its loose coupling with ITU and the CCITT, many ANSI (i.e., ECSA) initiatives are also considered in ITU proceedings.

U.S. data communications standards are influenced by CBEMA (Computer and Business Equipment Manufacturer Association) working on the ISO X3 Committee in the areas of data communications and open system interconnection (OSI)—a major initiative of the ISO. Other entities such as the IEEE and the EIA/TIA (Electronics In-

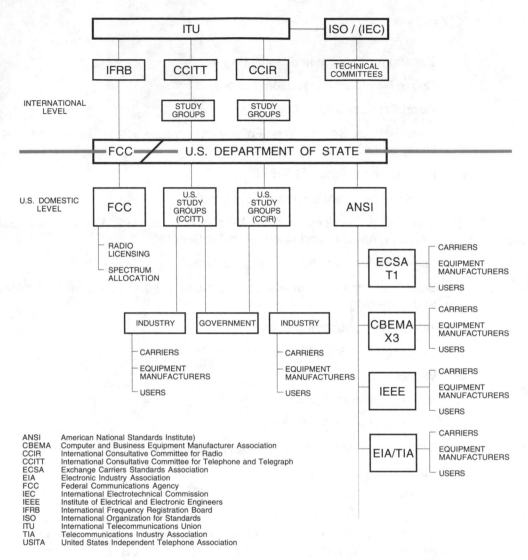

ANSI   American National Standards Institute)
CBEMA   Computer and Business Equipment Manufacturer Association
CCIR   International Consultative Committee for Radio
CCITT   International Consultative Committee for Telephone and Telegraph
ECSA   Exchange Carriers Standards Association
EIA   Electronic Industry Association
FCC   Federal Communications Agency
IEC   International Electrotechnical Commission
IEEE   Institute of Electrical and Electronic Engineers
IFRB   International Frequency Registration Board
ISO   International Organization for Standards
ITU   International Telecommunications Union
TIA   Telecommunications Industry Association
USITA   United States Independent Telephone Association

Fig. A.3 Organization for telecommunications/data communications standards in the U.S.

dustry Association/Telecommunications Industry Association) also produce standards which are often adopted by ANSI. Figure A.3 depicts these relationships and indicates the presence or absence of the major U.S. federal government players—the FCC and the Department of State. Most federal inputs to the standards bodies come from the Department of Commerce's National Institute of Science and Technology (NIST—the old National Bureau of Standards) and the National Telecommunications and Information Agency (NTIA). Other government participants include components of the Department of Defense, and other interested agencies via the National Communications System (NCS)—an executive-branch agency operating at White House level.

Inspection of Fig. A.3 leads to the conclusion that the speed and decisiveness of the old Bell System standards-making process can hardly be expected from today's cumbersome, consensus-based arrangements. This accounts for the rise of de facto standards, growing out of commercial products and services. Examples of these proprietary standards precursors include IBM's SNA and token ring, Xerox's Ethernet, Datapoint's Arcnet, and AT&T's ISDN. De facto standards serve to furnish input to the formal standards-setting process, which attempts to generalize them into universally adaptable rules. Sometimes this occurs quickly, as in the case of the 802.X series of recommendations. More often, delays are encountered, which subject the technology supporting the standard to potential obsolescence, e.g., ISDN.

ISDN Standards Activities in the U.S.

The foregoing has described telecommunications standards setting both at international levels and within the United States. The purpose of the present section is to illustrate how the cumbersome standards setting process, rulings emanating from deregulation and divestiture, and market forces have combined to delay U.S. migration towards Integrated Services Digital Network (ISDN), placing us well behind Europe and other industrial countries. Although this overview focuses on ISDN standards as they have developed to date, it also depicts the complex way in which international standards-setting interacts with U.S. market realities.

The concept of an ISDN was first formulated in the councils of the CCITT before the 1976 plenary session. While the concept incorporated significant contributions from the U.S., it can fairly be said to have reflected a European view of telecommunications. In the early 1970s European telecommunication administrations, and to some extent Japan, viewed evolving digitization of their telecommunication infrastructure, initiated by U.S.-based activities, as needing a standards framework to ensure future interoperability. In those days of analog technology, switching and transmission systems were viewed as independent network segments. Digitization of the switching and transmission plants, made possible the integration of many functions, particularly multiplexing, leading to the *integrated digital network* (IDN) concept. But this technological improvement could not reach customers until a systematic provision for access was established. The concept of user access to the IDN underlies the Integrated Service Digital Network—ISDN.

By the 1976 CCITT Plenary Session, the concept had gelled sufficiently to warrant creation of recommendations within the I Series heading (coinciding with ISDN) in the Green Book issued by the Plenary. The pattern for evolution was set, and has persisted through four successive Plenaries—1980 (Yellow), 1984 (Red), 1988 (Blue), and the 1992 Plenary. Extensive coverage of ISDN appears in not only the I Series but also the G (Transmission), Q (Switching and Signaling) Series, and to a lesser extent in the X (Data Communication Networks) and E (Tariffs) Series. At present, the topic has been well developed from a *recommendations* (the polite term for *standards*) perspective. ISDN is proceeding rapidly in Europe and Japan—but not in the USA. Why?

To answer this question, we must look to what makes telecommunications and data communications in the U.S. different from other industrialized nations. The straightforward answer is that an integrated, monolithic telecommunications administration corresponding to foreign PTTs no longer exists in the U.S. to rapidly define and enforce standards. Aside from the loss of this standards-setting capability, specific events arising from deregulation impeded U.S. developments to support access to an IDN.

Underlying the concept of ISDN is a notion of the NT—the *network termination.* As originally conceived, NT is service provider owned equipment on customer premises which establishes the point of telecommunications service delivery by a carrier (USA) or an administration (most of the rest of the world). This would have left the details of how the NT and central office equipment would be designed, permitting the classic 2-wire loop to a customer premises to support basic and primary rate signals, entirely in the hands of the service provider. Thus, basic and primary rate access to CPE is standardized at the customer side of the NT—identified as the S or T reference points in I Series parlance. This necessitates carrier (or administration) owned equipment on customer premises, a factor considered integral to service provision by the framers of CCITT ISDN recommendations.

The ISDN primary rate interface for the U.S. reflects the 4-wire DS1 level of the digital signal hierarchy. Provisioning for the basic rate interface reflects the great amount of 2-wire plastic insulated copper (PIC) twisted pair facilities composing local telephone distribution in this country. These decisions represented the status of CCITT recommendations at the time of the 1980 Yellow Book. During the four-year period leading to the 1984 Red Book, the momentous deregulation and divestiture changes took place in the U.S. These changes are not reflected in the Red Book, which largely followed and expanded upon the Yellow Book.

The FCC's computer Inquiry II decision in 1979, the primary vehicle for deregulation, mandated the separation of all and any forms of terminal equipment at a customer premise from the service provided by a carrier. By the early 1980s, the FCC was confronted with a highly disputed proceeding with a deceptive formal title under Docket 81-216 variously styled as the *NCTE Order* or *NCTE Deregulation.* The Commission's decision, rendered in 1983, took the position that all NCTE was to be treated as CPE (the Computer Inquiry II term for terminal equipment subject to Part 68 of the FCC Rules and Regulations). This decision reflected vociferous arguments, primarily from the data processing industry that "separation of NCTE from service provision will not retard network innovation."

Note that the NCTE (network channel termination equipment) is exactly the same NT discussed earlier in the context of the ISDN concept. As a consequence of this decision, the introduction of ISDN has been severely impeded because the NT and its network side now had to be standardized as CPE. Considering the evolution of the U.S. telecommunications infrastructure toward ISDN as "network innovation," one could readily say that treating NCTE (the NT) as CPE has seriously suppressed network innovation.

Following the NCTE Order decision and the January 1, 1984, Bell System divestiture date, the U.S. embarked on serious, complex standards-setting activity to define what is now termed the U reference point involving an army of interested parties under the auspices of the Exchange Carriers Standards Association (ECSA) T1 Committee. This activity was to develop a standard for U.S. applications that bore some

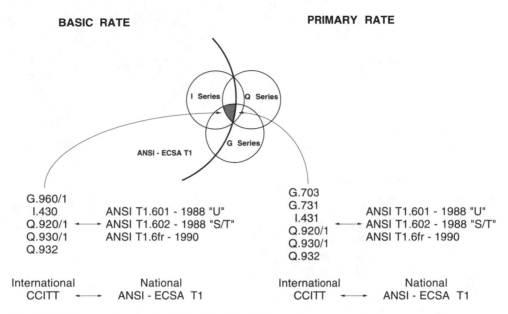

BASIC RATE   **PRIMARY RATE**

G.960/1
 I.430
Q.920/1 ←——→   ANSI T1.601 - 1988 "U"
Q.930/1   ANSI T1.602 - 1988 "S/T"
Q.932   ANSI T1.6fr - 1990

G.703
G.731
 I.431
Q.920/1 ←——→   ANSI T1.601 - 1988 "U"
Q.930/1   ANSI T1.602 - 1988 "S/T"
Q.932   ANSI T1.6fr - 1990

International   National   International   National
  CCITT   ←——→   ANSI - ECSA T1   CCITT   ←——→   ANSI - ECSA T1

Fig. A.4 ISDN access standards relationships in the U.S.

resemblance to the CCITT ISDN definition in the 1984 Red Book. The result, which reflects significant communication and semiconductor technology advancement arrived in 1988 as ANSI TI.601-1988, entitled *ISDN-Basic Access Interface*. This document in an earlier draft form was submitted in time for inclusion in the 1988 Plenary (the Blue Book) and now appears in a similar form in G.960 and G.961 of the G Series of Recommendations.

Developers of CPE containing an NT could finally incorporate standards into product developments with a reasonable level of confidence. However, five to six years of ISDN service evolution had been lost. Europe, meanwhile, unencumbered by the above problems had expanded ISDN service and the supporting facilities infrastructure to a far greater extent than the U.S. The basis for the complexity of standards-setting activities for ISDN access is graphically portrayed in Fig. A.4. The figure illustrates the tremendous amount of coordination that must take place among activities developing related standards and the participating international organizations.

B

Premises Distribution Systems

A *premises distribution system* (PDS) is composed of wire, cable and other equipment necessary to link desktop terminals (workstation locations) with telephone switch, computer, image, video and other information hosts, and with external telecommunications networks. These hosts may all be located in a single building, or distributed among multiple buildings in campus environments. The cable system extends from the workstation to the host port, enabling the telecommunications user to literally plug into the desired information channel. For many years, cable system architecture was relatively static. Technological advances, however, have created an environment in which cable system design and implementation is as critical as that of the voice, data or video systems themselves.

Chapter 6 introduced PDS systems and described PDS components as defined by the Electronic Industries Association (EIA) and the Telecommunications Industry Association (TIA) Commercial Building Wiring Standard 568. Chapter 6 established the concept of *uniform wiring,* capable of supporting voice and data in multiple product/vendor environments, and minimizing the need for future modifications to accommodate user requirement changes or technology upgrades. These are the primary PDS design objectives. This appendix adds practical considerations gained from experience with previous approaches and focuses on PDS planning and implementation factors and techniques needed to achieve PDS objectives.

Background

Wire pairs are the predominant means by which information is carried through the premises distribution system. A pair is the basic increment of a telephone circuit. Communications wire and cable made of copper is normally specified, purchased, and installed according to pair count, e.g., 4 pair, 100 pair, etc.

Early cable systems were based upon one pair assigned to each telephone number, running between the switching equipment and the telephone. Where key (multibutton) telephones were required, cables were installed in 25-pair increments between the key service unit (KSU) and the individual telephones: 6-, 10-, and 12-button sets used a 25-pair cable; 18- and 20-button sets used a 50-pair cable; and 30-button sets used 75-pair cable. As you might expect, building conduits and raceways filled up quickly.

With the development of electronic key telephones, which connected directly to a PBX station card, the KSU was eliminated, and pair counts were drastically reduced. The Rolmphone ETS series, one of the first such electronic key telephones, used three pair—one for voice, one for signaling, and one for power.

Similar products emerged from other manufacturers as the skinny-wire key telephone race heated up. AT&T introduced the Dimension EKTS telephone, which used 4 pair, NEC announced its Electra series, which used 3 pair, and NTI offered the Vantage series, which used 3 pair.

These electronic key telephones were not digital but did represent a substantial advance in technology, bringing about cost savings in cable material and labor, and taking pressure off the building's cable access areas. Of course, a corresponding price was exacted in terms of having to furnish more complex and expensive PBX or Centrex printed circuit boards (PCBs) necessary to drive these new key telephones.

The idea of a "universal" cable system grew out of the reduced pair count needed by switch-driven electronic key telephones. Since a single line telephone could run on one pair, the cable system universal wiring plan was based upon the key telephone requirements. Every workstation would be wired for a key telephone, which greatly simplified moves, additions, and changes. Thus, in the early 1980s, AT&T standardized on a 4-pair wiring plan, with Rolm, NTI, and NEC adopting a 3-pair scheme. By that time, the FCC had already implemented the Part 68 Standard Plug and Jack program, which specified for the industry a limited number of modular connectors for linking voice and data terminals to a premises distribution system.

It was not long before problems caused by this new architecture surfaced, which persist to this day. One arose from the *66-type* cross-connect terminal block used by the industry. This item (still widely installed in cross-connect fields today), is a passive device which physically and electrically joins segments of wire and cable. It is a *live front* device, meaning that its current-carrying parts are exposed on the front of the block.

The 66 blocks are installed in a free-standing metal frame or on a wall-mounted backboard which is often located in an electrical closet or janitorial space. Trades and building maintenance people tend to lean pipes, tools, mops, and brooms up against these frames and backboards.

Early KSUs and electromechanical key telephones had been quite robust when subjected to the resulting electrical shorts. Not so the new equipment. PBX station cards and even telephones can fail with alarming regularity if short circuits "zap" their semiconductor components. Sheets of acrylic plastic placed over the frames and backboards eliminate this problem.

Also, plugging a single-line K2500 telephone in an electronic multiline telephone jack could destroy the PBX station card because of differences in pin connections for power leads within the telephone plug and jack assembly.

Yet another problem was based upon the fact that the new electronic telephones used a data link to the PBX station card for signaling and control. This data link could only be extended over a limited distance, before low-level signals became distorted, and telephones no longer functioned properly. This *loop limit* was originally on the order of 500 to 1000 feet, which is not a long way in a large low-wide building, or in a multibuilding campus environment.

Only within the past five years, as solid state components capable of working at very low current levels were developed, has the loop limit problem been mitigated. Most of today's electronic telephones are capable of satisfactory operation at distances of at least 2500 feet. Terminal loop limits should nonetheless be verified against actual distances to be encountered during installation.

With the emergence of specifications and standards beginning with the IBM Type 3 Media Specification, and the IEEE 10BaseT standard, allowing the twisted pair cable system to become a local area network (LAN) component, the loop limit issue has become even more critical. Despite the fact that the electronic telephones now work at extended distances, the recommended loop limits for LAN connections are only about 330 feet between the terminal and the serving telecommunications closet.

This does not create a compatibility problem in integrating voice and LAN equipment within the same premises distribution system, but it does mean that more design engineering and installation oversight time (equating to increased cost) must be expended to ensure compliance with the new standards.

Cable System Planning & Implementation

The principal considerations in planning a communications cable system involve the physical aspects of building construction or renovation, operational requirements arising from user applications, and the financial implications of design alternatives.

Objectives of the process are to correlate cable system design and implementation efforts with the building occupancy schedule, to ensure accommodation of growth in present user arrangements, to integrate new applications with minimal disruption, and to reduce life cycle costs. In order to achieve these objectives, cable system designers must consider the factors described below.

Physical factors

Physical factors are those driven by architectural, mechanical, electrical, and interior design decisions that affect the building. Since these factors are interrelated, a change to one will tend to affect the others, e.g., further subdividing a floor, creating heavy workstation density, or selecting carpet tile for certain workstation areas. Physical factors include:

- Number of occupied floors
- Slab-to-slab floor height
- Height of building

- Square footage per floor
- Workstations locations per floor
- Number and location of communications closets per floor
- Number and location of equipment rooms
- Location of communications service entrance rooms
- Cable access arrangements
- Building structures, fixtures, and systems

The following building design characteristics frequently complicate design and implementation of wire and cable systems:

- Individual conduits for workstation cables
- Nonenterable spaces above ceiling
- Inadequate telecommunications closets
- Placement of sprinkler systems, HVAC ducts, and other intrabuilding utilities.

Operational factors

Operational factors are those largely driven by computer and communications decisions made by or for the building occupant. Operational and physical factors are interrelated, so that a change to one will tend to affect the other, e.g., changing density of LAN terminals necessitating expansion of communications closets on a floor, or the need to provide broadband services to certain workstation locations. Operational factors include:

- Type of telephone system
- Type of computer system(s)
- Type of LAN system(s)
- User access to systems
- Terminals per workstation
- Equipment compatibility
- Networking of dissimilar systems
- Current and future transmission speeds
- Rate of move, addition, and change activity
- Cable capacity vs. expense vs. availability
- Vertical riser vs. horizontal feeder for high-rise building vs. low-wide building
- Expansion, e.g., to new floors, without changeout of core system workstation wiring or distribution cable

- Prewiring for all potential services at every potential workstation location
- Avoidance of single-thread failure points
- Reconfiguration capability for fallback and recovery
- Universal cable system concept, based upon industry standards and off-the-shelf availability (i.e., one cable type linking all terminal and host equipment; one outlet type for all services at every workstation location)
- Conformity with national standards, e.g., EIA/TIA, ANSI, ICEA
- Evolution to emerging technologies by adding components to core system, e.g., ISDN interfaces
- Intelligent terminal interfaces and cross-connect/patch fields

Administrative & financial factors

Administrative and financial factors are also affected by computer and communications decisions made by or for the building occupant. Administrative and financial factors are tied back to physical and operational factors, e.g., more communications closets on a floor will require a more powerful cable management system, or prewiring workstation locations to lower system life cycle costs. Administrative and financial factors include:

- Leased vs. owned building
- Building occupancy schedule and stacking plan
- Cable system life cycle period
- Terminal support per cable type
- Uniform wiring
- Standard plug and jack connections
- Ease of operation, administration, and management
- Building life cycle

System management

Paper records and manual procedures may not be adequate for planning and control of cable systems serving buildings of even moderate size and population, where these conditions exist:

- LAN terminals in addition to telephones
- High per-floor workstation density
- Low square footage per workstation area
- Annual rate of moves, additions, and changes greater than 25% of workstation population.

Steps in cable system management, which may require the use of computer based cable management systems as described in Chapter 6, are:

1. Identification of baseline configuration, from as-built cable records:

   - Bill of Materials
   - Cable route map/running list
   - Termination scheme
   - Cross-connect scheme
   - Designation and labeling scheme

   These records must reflect any changes from original engineering (bid) specifications that resulted from:

   - Inability to use planned cable routes
   - Last-minute changes to interior layout
   - Final system design revisions

2. Development of a management system, with the following minimum baseline data elements:

   - Horizontal workstation wiring, with each run uniquely numbered
   - Backbone distribution cable, with each run uniquely numbered
   - Cross-connect blocks and patch panels within wiring closets and equipment rooms, organized into functionally related cross-connect and patch fields with blocks and panels designated by field
   - Individual workstation and distribution cable pairs, identified by cable number, color code, and/or termination sequence, and designated on cross-connect blocks and panels
   - Host equipment ports (voice, data, etc.), identified by cable number, color code, and/or termination sequence, and designated on main cross-connect blocks and panels.

3. Specification of management system functions, which at a minimum should:

   - Map all terminal connection points
   - Assign cable and pair numbers
   - Monitor connection and disconnection of circuits between terminals and host equipment
   - Perform end-to-end circuit tracing
   - Prepare trouble tickets and work orders
   - Produce directories, inventory lists, installation schedules, and reports.

4. Development of system enhancements:

   - Design models
   - Routing assignments
   - Workstation profiles
   - On-line directory and inventory services
   - Cable fill analyses
   - Financial models

C

Modern PBX/Centrex & Key/Hybrid Telephone System Features

This appendix contains feature lists for PBX, Centrex and KTS/Hybrid premises services. The lists are generally organized in the fashion of a vendor proposal. Certain differences exist between PBX and Centrex features, because of switch location, ownership and influence of residential service on Centrex, e.g., call waiting. At present, the Aries/MPSG *Master Glossary of Terminology* does not cover Centrex features.

PBX features tend to be described much the same across the competing vendor product lines. Exceptions do exist; for example, AT&T's Automatic Callback is Northern Telecom's Ring Again feature. By comparing a vendor feature with the MPSG *Master Glossary of Terminology,* however, comparability will be achieved. Note that each vendor's product exhibits subtle differences in the purpose, programming, capacity, and interaction of individual features. These differences may change as new software releases or "generics" are introduced, so verification of precise feature details should be undertaken in the proposal evaluation stage.

Centrex features are identified against specific central office switches and software generics because less information is generally made available by the LECs. Within a LEC franchised territory, the same switch models may not offer identical generics. The LEC should be queried about the feature characteristics described above, to ensure that user needs are adequately met. Centrex features may undergo more enhancements and generic releases than those delivered by PBXs, which may have shorter operational lifetimes.

KTS/Hybrid features are listed similarly to those of the PBX. Although feature counts are increasing as more hybrid systems enter the market, there are still fewer KTS features than PBX features. Feature implementation in KTS/Hybrid systems is somewhat different because of smaller size, largely nondigital stored program control technology, and trunking arrangement. Of the three market segments, KTS/Hybrid features are evolving most rapidly, as additional value-added functions are integrated within the system. Even though the distinction between KTS/Hybrid features and PBX features is blurring, subtle differences exist—most involving intercom operations and trunk-related functions—which make it important to verify precise characteristics prior to implementing a telecommunications project.

NOTE: In the lists below, the features preceded by (•) are standard, while those preceded by (*) are optional.

PBX Feature List

System features

* * Automatic Call Distribution
* • Automatic Program Reloading
* • Automatic Route Selection
* • Automatic Station Release with Howler
* * Autovon Access
* • Call Forwarding: Busy Line
* • Call Forwarding: No Answer
* • Call Forwarding: Remote
* • Call Forwarding: Private Facilities
* * Call Number Buffering
* * Circuit Compatibility
* * DC Power
* * Direct Inward Dialing
* * Direct Inward System Access
* * Direct Outward Dialing
* * Distinctive Ringing
* * End-to-End DTMF Signaling
* * Flexible Hunting Arrangements
* * Flexible Intercept
* * Flexible Release
* * Flexible Station Restrictions
* * Hot Standby
* * Off-Premises Stations
* * Power Failure Cut-Through
* * Queuing on Facilities
* * Reserve Power
* * Special Circuit Interface
* * Speed Calling: System
* * Ten Digit Telephone Number Restriction
* * Tenant Service
* * Time-of-Day Routing
* * Uniform Call Distribution

Attendant features

* Alphanumeric Display
* Attendant Abbreviated Dialing
* Attendant-Assisted Dialing
* Attendant-Controlled Conference
* Automatic Recall
* Busy Override
* Calling Number Identification
* Call Processing Indications
* Call Waiting Identification
* Camp-On Busy
* Centralized Attendant Service
* Console Headsets
* Console Signal Volume Control
* Dial "0" Access to Attendant
* Direct Trunk Group Selection
* Flexible Night Service
* Individual Trunk Access
* Locked Loop Operation
* Loop Transfer
* Message Center Service
* Multiple Console Operation
* Position Busy
* Station Busy Verification
* Switched Loop Operation
* Trunk Group Busy Lamps
* Two-Way Splitting

Station features

* Add-On Conference
* Automatic Callback
* Automatic Re-Ring on Held Call
* Call Forwarding, Variable
* Call Hold, Hard
* Call Park
* Call Transfer
* Call Transfer via Camp-On
* Call Waiting
* Cancel Call Waiting
* Classes of Service
* Consultation Hold
* Data Privacy
* Data Terminal Connection
* Dial Call Pickup
* Dial Trunk Transfer

* Dial Intercom
* Directed Call Pickup
* Double Camp-On Indication
* DTMF Dialing
* Executive Override
* Feature Acceptance Indication
* Flexible Station-Controlled Conference
* Individual Speed Calling
* Last Number Redial
* Optional Feature
* Malicious Call Trace From Station
* Music on Hold
* Portable Stations
* Saved Number Redial
* Secretarial Intercept
* Station Call Waiting
* Station Forced Busy
* Station-to-Station Calling

Management features

* Account Codes
* Authorization Codes
* Automatic Circuit Assurance
* Call Accounting
* Diagnostic and Maintenance Programs
* Feature Usage Monitoring
* Line Load Control
* Line, Trunk, and Network Testing
* Machine Readable Output
* Maintenance Busy
* Malicious Call Trace
* PCB Status Indications
* System Administration and Control
* System Alarm Indications
* Remote Diagnostics and Maintenance
* Traffic Measurement

Centrex Feature List

The following is the key for the lists of Centrex features:

x = Available: 1AESS, 1AE8 or earlier; 5ESS, 5ESS, 5E2.2 or earlier; DMS-100, BCS20 or earlier
@ = Requires CPE
• = Standard
* = Optional
A = Accept digital

Attendant features	Software generic		
	1AESS	5ESS	DMS100
• Dial "0" Access to Attendant	x	x	x
• Direct Trunk Group Selection	x	x	x
• Flexible Night Service	x	x	x
• Individual Trunk Access	x	x	x
• Locked Loop Operation	x	x	x
• Loop Transfer	x	x	x
* Message Center Service	1AE.6	5E4.2	BCS15
* Multiple Console Operation	x	x	x
• Position Busy	x	x	x
• Station Busy Verification	x	x	x
• Switched Loop Operation	x	x	x
• Trunk Group Busy Lamps	x	x	x
• Two-Way Splitting	x	x	x

Station features	Software generic		
• Add-On Conference	x	x	x
• Automatic Callback	1AE.6	5E2.2	BCS14
• Auto Re-Ring on Held Call	x	x	x
• Call Forwarding, Variable	1AE.6	5E2.2	BCS8
• Call Hold, Hard	N/A	5E3	BCS23
• Call Park	x	x	x
• Call Transfer	x	x	x
• Call Waiting	1AE.4	5E2.2	BCS8
• Cancel Call Waiting	1AE.4	5E2.2	BCS8
• Classes of Service	x	x	x
• Consultation Hold	x	x	x
• Data Privacy	1AE.9	5E5	BCS9
• Data Terminal Connection	1AE.9	5E5	BCS9
• Dial Call Pickup	x	x	x
• DTMF Dialing	x	x	x
• Executive Override	x	x	x
• Feature Acceptance Indication	1AE.4	5E2.2	BCS14
• Flex Station-Ctrld Conf	x	x	x
• Individual Sppd Calling	x	x	x
• Last Number Redial	x	x	x
• Malicious Call Trace Fr Sta	x	x	x
*@Music on Hold	x	x	x
• Portable Stations	x	x	x
• Saved Number Redial	x	x	x
•@Secretarial Intercept	x	x	x
• Station Forced Busy	x	x	x
• Station-to-Station Calling	x	x	x

		1AESS	5ESS	DMS100
Management features		**Software generic**		
•	Account Code	1AE.6	5E3	BCS10
•	Authorization Codes	1AE.6	5E3	BCS13
•	Automatic Circuit Assurance	x	x	x
*@	Call Accounting	1AE.4	5E3	BCS13
•	Diagnostic and Maint Progs	x	x	x
•	Feature Usage Monitoring	N/A	x	x
•	Line Load Control	x	x	x
*@	Line, Trunk, and Network Test	x	x	x
•	Machine Readable Output	x	x	x
•	Maintenance Busy	x	x	x
•	Malicious Call Trace	x	x	x
•	PCB Status Indications	x	x	x
•@	System Admin and Control	x	x	x
•	System Alarm Indicators	x	x	x
•@	Traffic Measurement	x	x	x

KTS & Hybrid Feature List

System features: Traditional KTS

- • Automatic Line Selection
- • Exclusive Hold
- • Flexible Line Appearances
- • Flexible Ringing Arrangements
- * Line Extension
- * Music on Hold
- * Reserve Power
- * Ringdown Circuits
- • Special Circuit Interface
- • Toll Restriction: "1" and "0"

System features: Hybrid & digital KTS

- • Automatic Program Reloading
- • Automatic Route Selection
- • Automatic System Redial
- • Call Forwarding: Busy Line
- • Call Forwarding: Intercom
- • Call Forwarding: Remote
- • Direct Inward Dialing
- • Direct Inward System Access
- • Direct Outward Dialing
- • Distinctive Ringing

- End-to-End DTMF Signaling
- Flexible Hunting Arrangements
- Flexible Intercept
- Flexible Night Service
- Flexible Release
- Flexible Station Restrictions
- Multiparty Conference
- Off-Premises Stations
- On-Hold Timing
- Power Failure Cut-Through
- Queuing on Facilities
- Remote Access to System Lines
* Reserve Power
- Special Circuit Interface
* Option
- Speed Calling: System
- System Paging
- Toll Restriction: 6-Digit
- Toll Restriction: 10-Digit
- Tenant Service
- Uniform Call Distribution

Attendant features: Traditional KTS

Not Applicable

Attendant features: Hybrid & digital KTS

- Alphanumeric Display
- Attendant Abbreviated Dialing
- Attendant-Assisted Dialing
- Attendant-Controlled Conference
- Automatic Recall
- Busy Override
- Calling Number Identification
- Call Processing Indications
- Call Waiting Identification
- Camp-On Busy
* Centralized Attendant Service
* Console Headsets
- Console Signal Volume Control
- Dial "0" Access to Attendant
- Direct Trunk Group Selection
- Flexible Night Service
- Individual Trunk Access
- Locked Loop Operation

- • Loop Transfer
- * Message Center Service
- * Multiple Console Operation
- • Position Busy
- • Station Busy Verification
- • Switched Loop Operation
- • Trunk Group Busy Lamps
- • Two-Way Splitting

Station features:Traditional KTS

- * Automatic Dialing
- • Automatic Exclusion with Release
- * Option
- * Buzzer Signaling
- * BLF/DSS
- * Call Announcing
- * Call Progress Monitoring
- * Conferencing
- * Dial Select Intercom
- * Group Listening
- • Line Preference
- * Manual Intercom
- * On-Hook Dialing
- • Single Button Intercom Access
- * Speakerphone Operation

Station features: Hybrid and digital KTS

- • Automatic Callback
- • Automatic Exclusion with Release
- • Call Park
- • Call Reminder
- • Do-Not Disturb
- • Executive Override on Intercom
- • Executive Passwords
- • Facsimile Interface
- • Flexible Classes of Service
- • Hotline Service
- * Integral Modem
- • Last Number Redial
- * Listen on Hold
- • Manager/Secretary Hotline
- • Message Waiting Indication
- • Music-on-Hold Disable
- • Number Copy/Autodial

- Saved Number Redial
* Secure Telephone Interface (STU III)
- Station Lockout with Howler

Management features: Traditional KTS

- PCB Status Indications
- System Alarm Indications

Management features: Hybrid & digital KTS

* Account Codes
- Administrative Station
* Authorization Codes
- Automatic Circuit Assurance
* Call Accounting
- Diagnostic and Maintenance Programs
- Line Load Control
- Maintenance Busy
- Malicious Call Trace
- PCB Status Indications
- Programming at User Stations
- Remote Diagnostics and Maintenance
- Station Security Lock
- System Administration and Control
- System Alarm Indications
- Traffic Measurement

Glossary

access services specified sets of information transfer capabilities furnished to users at telecommunications network points-of-termination (POTs) to provide access to network transport services. Two examples are: 1. subscriber access lines, the connection between a network POT (in this case more commonly known as a Network Interface-NI) and a local exchange carrier switching system, and 2. trunks between interexchange carrier points-of-presence and local exchange carrier switching systems (The POT at the POP is identified as the point of interface—POI). End-to-end connections require originating and terminating access services.

access tandem (AT) a LEC switching system that performs concentration and distribution functions for inter-LATA traffic originating or terminating within a LATA.

address signals convey destination information such as a called 4-digit extension number, central office code, and when required, area code and serving IXC carrier code. These signals may be generated by station equipment, or by a switching system.

analog carrier system a transmission system that uses repeaters that compensate for analog medium impairments, and produce output signals that are linear-scaled versions of input signals. Analog carrier systems can carry speech, data, video and supervisory signals, although they are best suited for speech signals.

analog signal a continuous electrical signal that varies in direct correlation with an impressed phenomenon, stimulus, or event that bears intelligence. Sound waves and their electrical analogs, are characterized by loudness (amplitude) and pitch. Analog signals can assume any of an infinite number of amplitude values or states within a specified range, in accordance with, or analogous to, an impressed stimulus. Pitch refers to how many times per second the signal swings between high and low amplitudes, i.e., its frequency.

asynchronous transfer mode (ATM) a broad-bandwidth, low delay, packet-like (cell relay) switching and multiplexing technique. It is essentially connection oriented, although it is envisioned to support all services. ATM networks will accept or reject connections based on user's average and peak bandwidth requirements providing flexible and efficient service for LAN-to-LAN, compressed video and other applications that involve variable bit rate (VBR) traffic.

automatic call distribution (ACD) a means for efficiently directing and managing large numbers of incoming calls to specific departments/terminals within an organization.

backbone network a transmission facility designed to interconnect, often lower-speed distribution networks, channels, or clusters of dispersed terminals or devices.

backbone wiring in a premises distribution system, the cable connecting telecommunications closets and equipment rooms within a building, and/or between buildings in a campus. Backbone wiring is sometimes referred to as the riser subsystem.

bandwidth a frequency range, usually specified by the number of hertz in a band or between upper and lower limiting frequencies. Alternatively, the frequency range that a device is capable of generating, handling, passing or allowing.

binary digit see *bit*.

bipolar signals Signals in which positive and negative pulses, always alternating, represent one binary signal state only. The absence of pulses represents the other binary state. Bipolar has two forms, AMI and ASI. In alternate mark inversion (AMI) the pulses correspond to 1s; in alternate space inversion (ASI) the pulses correspond to 0s.

bit the most fundamental and widely used form of digital signals are binary signals, in which one amplitude condition represents a binary digit 1, and another amplitude condition represents a binary digit 0. Thus a binary digit, or bit is one of the members of a set of two in a numeration system that is based on two and only two possible different values or states.

bit error rate (BER) the ratio of the number of bits received with errors to the total number of bits transmitted. BER and the average number of error-free seconds are the principal impairment measurements for digital channels.

bit rate the capacity characteristic of digital signals as defined by the number of bits (or bytes) per second that a channel will support. For example, a transmission facility that can support information exchange at the rate of 1 megabit per second (1 Mbps or 1,000,000 bits per second) delivers the same quantity of information, i.e., throughput, as a 1 kilobit per second (kbps or 1,000 bits per second) facility, but, in only 1/1000 of the time.

BOC (Bell-operating company) the common term for one of 24 local exchange carrier telephone companies that were part of the Bell System prior to divestiture. All but two of the BOCs (Southern New England Telephone in Connecticut and Cincinnati Bell in Ohio) are owned and managed by one of 7 regional Bell holding companies (RBHCs). Approximately 80% of America's local exchange users are served by the BOCs.

bridges in IEEE 802 local area network (LAN) standards, devices that connect LANs, or LAN segments, at the data link layer. Bridges provide the means to extend the LAN environment in physical extent, number of stations, performance and reliability. Bridges perform three basic functions: frame (as opposed to packet) forwarding; learning of station addresses; and resolving of possible loops in the topology by participating in the spanning tree algorithm. Self-learning bridges construct tables of network addresses by "listening" to source address information contained in data signal frames. Other functions include: 1. the ability to filter traffic to keep traffic originating and terminating in one network segment from leaving that segment, 2. restricting specified traffic to one segment that might otherwise be routed to other segments, and 3) collecting and storing network management and control information obtained via traffic monitoring.

Broadband Integrated Services Digital Network (BISDN) CCITT is developing a BISDN umbrella standard, incorporating underlying standards for integrated digital network switching, multiplexing and transmission facilities, that will be able to meet expanding voice, data, video and other requirements well into the future. In one of the first draft CCITT documents, BISDN is simply defined as "a service requiring transmission channels capable of supporting rates greater than the primary rate". In the U.S. the primary rate for "narrowband" ISDN (as the current standard is sometimes referred to) is 1.544 Mbps.

brouters   devices that combine the functions of bridges and routers. See *bridges*, *routers*.

bus   in digital systems, e.g., time division multiplexing equipment, a bus (also referred to as a highway) is defined as one or more conductors (or some medium) that connect a related group of devices.

business applications   unique aggregations of telecommunications services that satisfy particular enterprise needs.

byte   generally, an 8-bit quantity of information, used mainly in referring to parallel data transfer, and data storage; also generally referred to in data communications as an octet or character.

cable   a group of metallic conductors or optical fibers that are bound together, usually with a protective sheath, a strength member, and insulation between individual conductors/fibers and for the entire group.

carrier   1. a local (intra-LATA) or long distance (inter-LATA) telecommunications service-providing organization. 2. A waveform, pulsed or continuous which is modulated by another information bearing waveform.

carrier system   a transmission system for transmitting one or more channels of information by processing and converting to a form suitable for the transmission medium used. Carrier systems are classified as either analog carrier systems or digital carrier systems.

CCITT   the International Consultative Committee for Telephone and Telegraph, a consultative committee to the International Telecommunications Union (ITU) which recommend international standards for telephone and telegraph services and facilities to aid international connectivity and interoperability.

cell relay   the process of transferring data in the form of fixed length packets called cells. Cell relay is used in high-bandwidth, low-delay, packet-like switching and multiplexing techniques. The objective is to develop a single multiplexing/switching mechanism for dividing up usable capacity (bandwidth) in a manner that supports its allocation to both isochronous (e.g., voice and video traffic) as well as packet data communications services. Standards groups have debated the optimum cell size. Small cells favor low-delay for isochronous applications but involve a higher header-to-user information overhead penalty than would be needed for most data applications. The current CCITT specification for BISDN is for a 53- byte cell which includes a 5-byte header and a 48-byte payload.

centi call seconds (CCS)   a unit of the average traffic intensity of a facility during a period of time, a CCS is 100 call seconds of traffic during one hour. Therefore a single traffic source, e.g., one call that generates traffic 100% of the time produces 36 CCS of traffic per hour, i.e., 3600 seconds of traffic every 3600 seconds. An equivalent amount of traffic could also be generated by 10 sources that only generate traffic 10% of the time. That is, 10 sources of traffic generating 3.6 CCS each, contributes the same total traffic as a single 36 CCS traffic source. An alternative measure for traffic is erlangs where 1 erlang equals 36 CCS.

CENTRal EXchange (Centrex)   a LEC provided switching service for business customers that permits station-to-station dialing, listed directory number service, direct inward dialing and station number identification on outgoing calls. The switching functions are usually performed in a central office. Digital Centrex offers the advanced features of fourth generation PBXs, without the need to purchase or lease equipment, and, in most cases, eliminates the need for floor space, electrical prime power and heating, ventilation and air conditioning.

central offices (CO)   a telephone company building in which network equipment such as switches are installed.

channel a single communications path in a transmission medium connecting two or more points in a network, each path being separated by some means; e.g., spatial or multiplex separation, such as frequency or time division multiplexing. "Channel" and "circuit" are often used interchangeably, however circuit can also describe a physical configuration of equipment that provides a network transmission capability for multiple channels. The characteristics of channels and circuits are determined by the network equipment and media used to support them.

channel service unit (CSU) Channel Service Units (CSUs) and Data Service Units (DSUs) are required to connect digital customer premises equipment (CPE) to carrier networks. A CSU is network channel terminating equipment (NCTE) attaching as CPE to telephone company's digital circuits, and protecting the network from harm. Other CSU functions include line conditioning and equalization, error control (e.g., bipolar signal violations), and the logical ability to respond to local and network loop-back circuit testing commands. See *Data Service Units (DSU)*.

circuit switching a process that establishes connections on demand and permits the exclusive use of those connections until released. Packet and message switching, primarily used in data communications networks are alternative switching techniques.

circuit-associated signaling a technique that uses the same facility path for voice and signaling traffic. Historically this approach was selected to avoid the costs of separate channels for signaling and because the amount of traffic generated by signaling is small compared to voice, minimizing the chance for mutual interference. Circuit associated signaling can be contrasted with some common-channel signaling systems which use completely separate packet switched networks for signaling traffic.

city-wide digital Centrex a capability to serve multiple business locations within a single NXX (exchange code), using multiple LEC central office Centrex switches. Outside callers are unaware that multiple business locations are involved.

coaxial cable consists of an insulated central conductor surrounded by a second cylindrical conductor which is clad with an insulating sheath. The outer conductor usually consists of copper tubing or copper braid.

codec a contraction of *co*der and *dec*oder; a device that encodes analog signals into digital signals, for transmission through a network in digital format, and decodes received digital signals back into analog signals.

common management information protocol (CMIP) the OSI protocol for network management. A structure for formatting messages and transmitting information between reporting devices (agents) and data collection programs, developed by the International Standards Organization and designated ISO/IEC 9596.

common-channel signaling (CCS) a signaling system developed for use between stored program control digital switching systems, in which all of the signaling information for one or more trunk groups is transmitted over a dedicated signaling channel, usually, but not always completely separate from the user traffic bearing facilities.

communications the process of representing, transferring, interpreting or processing information (data) among persons, places, or machines. Communications implies a sender, a receiver and a transmission medium over which the information travels. The meaning assigned to the data must be recoverable without degradation.

conductor in electrical circuits, any material that readily permits a flow of electrons (electrical current) through itself. Analogously, optical fibers are sometimes said to conduct light-waves and are also referred to as conductors.

Corporation for Open Systems (COS) a nonprofit organization composed of manufacturing, service, and user organizations in the computer-communications area. COS seeks to facilitate the development of the international, multivendor marketplace through the development, introduction, and verification of OSI and ISDN standards and by ensuring vendor equipment interoperability.

cross-connect 1. in a premises distribution system, equipment used to terminate and administer communications circuits. In a wire cross-connect, jumper wires or patch cords are used to make circuit connections, between horizontal and backbone wiring segments. 2. in transmission systems a patch panel for connecting circuits.

CSMA/CD carrier sense multiple access with collision detection; a local area network contention-based access-control protocol technique by which all devices attached to the network "listen" for transmissions in progress before attempting to transmit themselves and, if two or more begin transmission simultaneously, are able to detect the "collision". In that case each backs off (defers) for a variable period of time (determined by a preset algorithm) before again attempting to transmit. (Defined by the IEEE 802.3 standard).

customer premises equipment all telecommunications terminal equipment located on the customer premises, except coin operated telephones.

D-type channel bank channel termination equipment used for combining (multiplexing) individual analog channel signals on a time division basis. D-type channel banks provide interfaces for "n" analog signal inputs. Each analog input signal is directed to a codec for encoding to PCM samples. A part of a T1 carrier system.

data compression techniques which remove redundancy in transmitted bit patterns to reduce transmission rates by 20% to 200%. For example, a modem designed to send and receive data at 1200 bps without data compression may be capable of supporting 2400 bps with data compression, using the same network analog voice-grade channel.

data service units (DSU) channel service units (CSUs) and data service units (DSUs) are required to connect digital customer premises equipment (CPE) to carrier networks. A hardware device providing an interface between a digital line and a unit of data terminal equipment. DSUs provide transmit and receive control logic, synchronization and timing recovery across data circuits. DSUs may also convert ordinary binary signals generated by CPE to special bipolar signals. Bipolar signals are designed specifically to facilitate transmission at up to 1.544 Mbps rates over UTP cable, a media originally intended for 3 kHz, voice bandwidth signals. See *Channel Service Units (CSU)*.

data terminal equipment (DTE) any device that can send data, receive data or perform both functions. (Note: sometimes DTE implies digital terminal equipment, a type of CPE used with digital service—see *CSU* and *DSU*.

digital carrier systems a carrier system for digital signals that uses regenerative versus linear repeaters and time division multiplexing.

✳ **digital cross-connect system (DCS)** a new generation of switching/multiplex equipment that permits per-channel DS0 (64 kbps) electronic cross-connection from one T1 transmission facility to another, directly from the constituent DS1 signals. Commonly referred to as "DACS," (digital access and cross-connect system), although this is a trademark of AT&T.

digital signal a signal (electrical or otherwise) in which information is carried in a limited number of different (two or more) discrete states. The most fundamental and widely used form of digital signals are binary signals, in which one amplitude condition represents a binary digit 1, and another amplitude condition represents a binary digit 0. See *binary digit* or *bit*.

digital termination service (DTS)   a service provided by some carriers permitting operators of private networks to use digital microwave equipment to gain access to carrier networks. The FCC has allocated a special microwave band for DTS.

direct inward dialing (DID)   PBX-to-central office trunks that allow incoming calls to a PBX to ring specific stations without attendant assistance. DID greatly reduces the number of required console attendants, compared with systems in which all calls must be extended by console attendants.

direct outward dialing (DOD)   PBX-to-central office trunks that allow outgoing calls to be placed directly by PBX stations.

dispersion   in dispersive media complex signals are distorted because the various frequency components which make up the signal have different propagation characteristics and paths. Due to the finite conductivity of copper, wire or cable media for guided wave transmission is fundamentally dispersive. Dispersion limits the upper bit rate that a medium can support by distorting the signal waveforms to the extent that transitions from one information state to another cannot be reliably detected by receiving equipment, (e.g., logical 1 logical 0 value changes).

DS"N" digital signal hierarchy   a time division multiplexed hierarchy of standard digital signals used in telecommunications systems. DS1 level in the hierarchy corresponds to a 1.544 Mbps TDM signal which comprises 24 DS0 signals. DS0 refers to individual digital signals at channel rates of 64 kbps. Four DS1 signals digitally multiplexed produce a DS2 level signal, containing 96 DS0 channels, and requires a transmission medium that supports 6.312 Mbps. A DS3 level signal results from the digital multiplexing of 7 DS2 signals, supports 672 DS0 signals and requires a 44.736 Mbps transmission medium. Finally a DS4 level signal supports 6 DS3 level signals, 4032 DS0 signals and requires a 274.176 Mbps transmission medium. The DS hierarchy accounts for non-synchronism in the multiplexing plan, hence the term "asynchronous digital hierarchy" and the use of overhead bits—note that bit rates at higher levels are not integer multiples of 64 kbps.

dual-tone multiple frequency (DTMF)   the generic name for the tone signaling scheme used to signal from telephones to switching equipment, in which 10-decimal digits and two auxiliary characters are represented by selecting two frequencies of the following group: 697, 770, 852, 941, 1209, 1336, 1447 Hz.

E&M leads signaling   an interface, used for connections between switches and transmission systems and between transmission systems themselves. Signaling information is transferred across the interface via 2-state voltage conditions on two leads, each with a ground return, separate from the leads used for message information. The message and signaling information are combined and separated by means appropriate to the transmission facility.

electrical signal   a signal consisting of an electrical current (i.e., a flow of electrons) that varies with time or space in accordance with specified parameters.

electronic mail   a generic term for non-interactive communication of text, data, image or voice messages between a sender and designated recipients using telecommunications.

electronic switched network (ESN) service   a private network service that provides user organizations with a uniform numbering plan and numerous call-routing features. The electronic tandem switching functions are furnished by either PBX or Centrex switching equipment.

end office (EO)   a LEC (BOC or an ITC) switching system within a LATA where local loops to customer stations are terminated for purposes of interconnection with each other and with trunks. CO (central office) and EO are often used interchangeably.

entrance facilities 1. in a premises distribution system, the point of interconnection between the building wiring system and external telecommunications facilities (LEC networks, other buildings, etc.). Bellcore defines the interface with LEC networks as end-user points of termination (POT). 2. has a further specific meaning in interstate access, entrance facilities for interstate access (ENFIA).

equipment room in a premises distribution system, a special purpose room(s), with access to the backbone wiring, for housing telecommunications, data processing, security, and alarm equipment.

erlang an international dimensionless unit of the average traffic intensity of a facility during a period of time; one erlang of traffic is equivalent to a single user who uses a single resource 100% of the time. See *centi call seconds*.

exchange carrier (or local exchange carrier-LEC) any company, BOC or independent, which provides intra-LATA telecommunications within its franchised area.

extended superframe format (ESF) an extension of the superframe format of T1 carrier systems from 12 to 24 frames and the use of framing bits for error checking, a facilities data link (FDL) as well as frame synchronization. See *superframe format*.

facilities-based private switched network services a private network for which LECs and IXCs dedicate physical switching and transmission facilities for the exclusive use of a particular customer.

fast packet a term referring to a number of broadband switching and networking paradigms. Implicit is the assumption of an operating environment that includes reliable, digital, broadband, nearly error free transmission systems.

FCC-Federal Communications Commission a board of commissioners empowered by the U.S. Congress to regulate all interstate and international communications, as well as use of the radio frequency media.

foreign exchange (FX) a service that provides a circuit(s) between a user station, a PBX, or a Centrex switch, and a central office other than the one that normally serves the caller.

frame in time division multiplexing systems, a sequence of time slots each containing a sample from one of the channels carried by the system. The frame is repeated at regular intervals, (normally the sampling rate used in analog-to-digital conversion processes for signals being multiplexed) and each channel usually occupies the same sequence position in successive frames.

frame relay a network interface protocol defined in CCITT Recommendation I. 122 "Framework for additional packet mode bearer services," as a packet mode service. In effect it combines the statistical multiplexing and port sharing of X.25 packet switching with the high speed and low delay of time division multiplexing and circuit switching. Unlike X.25, frame relay implements no layer 3 protocols and only the so-called core layer 2 functions. It is a high-speed switching technology that achieves ten times the packet throughput of existing X.25 networks by eliminating two-thirds of the X.25 protocol complexity. The basic units of information transferred are variable length frames, using only two bytes for header information. Delay for frame relay is lower than for X.25, but it is variable and larger than that experienced in circuit switched networks. This means that currently frame relay is not suitable for voice and video applications where excessive and variable delays are unacceptable.

frequency acoustic waves and electrical signals might be made up of only a single tone, like a single note on a piano. In this case the signal waveform is made up of repeating identical "cycles" and is said to be of a single frequency, equal to the number of cycles that occur in one second of time. In communications, frequency was traditionally expressed in cycles per second, but is now expressed in hertz (Hz), still equal to one cycle per second. Thus, one thousand cycles per second is equal to one thousand hertz, or a kilohertz (kHz).

frequency division multiplexing (FDM)   divides the frequency bandwidth (spectrum) of a broadband transmission circuit into many subbands, each capable of supporting a single, full time communications channel on a non-interfering basis with other multiplexed channels. FDM multiplexing is generally suitable for use with analog carrier transmission systems.

full duplex   a transmission path capable of transmitting signals in both directions simultaneously.

gateway   devices that interconnect otherwise incompatible nodes or networks, by performing protocol conversion.

grade of service (GOS)   an estimate of customer satisfaction with a particular aspect of service such as noise, echo or blocking. For example the noise grade of service is said to be 95% if, for a specified distribution of noise, 95% of the people judge the service to be good or better. In traffic networks, GOS defines the percentage of calls that receive no service (blocking) or poor service (long delays). GOS measures apply to all aspects of telecommunications networks. In many cases the literature equates GOS only with the probability of a blocked call. When used without further explanation, GOS generally refers to blocking probability.

ground start   a supervisory signal given at certain coin telephones and PBXs by connecting one side of the line to ground.

guided media   media that constrain electromagnetic or acoustic waves within boundaries established by their physical construction. Examples include paired metallic wire cable, coaxial cable, and fiber optic cable.

half duplex   a transmission path capable of transmitting signals in both directions, but only in one direction at a time.

header   control information appended to a segment of user data for control, synchronization, routing and sequencing of a transmitted data packet or frame.

hertz (Hz)   Measurement that distinguishes electromagnetic waveform energy; number of cycles, or complete waves, that pass a reference point per second; measurement of frequency, by which one hertz equals one cycle per second.

horizontal wiring   in a premises distribution system, the connection between the telecommunications outlet in work areas and the telecommunications closet.

hub   in local area networks (LANs), wiring concentrator equipment used in hierarchical star physical wiring topologies. Those directly connected to terminals or other user devices are often referred to as local hubs or concentrators. Central hubs are those at the highest hierarchical level. Hubs often provide the means for interconnecting 10BaseT, coaxial or fiber optic cable LAN segments. Intelligent hubs may implement multiport bridging and network management functions.

impairments   (e.g., transmission channel and signal impairments) degradation caused by practical limitations of channels, (e.g., signal level loss or attenuation, echo, various types of signal distortion, etc.) or interference induced from outside the channel (such as power-line hum or interference from heavy electrical machinery). The measurement of transmission impairments is an important aspect of predicting whether or not telecommunications systems will sustain the business applications they are intended to support. Signal-to-noise ratio, percent distortion, frequency response, and echo are measurements that define impairments most noticeable by users in analog voice systems.

inband signaling uses not only the same channel path as the voice traffic, but the same frequency range (band) used for the voice traffic.

Independent Telephone Company (ITC) a local exchange carrier that is not one of the 22 divested Bell-operating companies. ITCs are not generally subject to the restrictions of the MFJ, although some of the larger ones are bound by separate consent decrees. Southern New England Telephone and Cincinnati Bell are generally considered ITCs from a regulatory point of view.

Integrated Services Digital Network (ISDN) consists of a set of standards being developed by the CCITT and various U.S. standards-setting organizations. The CCITT formal recommendations, adopted in October, 1984, first defined ISDN as ". . . a network, in general evolving from a telephony integrated digital network, that provides end-to-end digital connectivity to support a wide range of services, including voice and non-voice, to which users will have access by a limited set of standard multipurpose user-network interfaces." The concept of user access to an existing integrated digital network (IDN) underlies the ISDN.

inter-LATA services, revenues, functions, etc., that relate to telecommunications originating in one LATA and terminating outside that LATA. An interexchange carrier (IXC) is a company which provides telecommunications services between LATAs. (The domain of IXCs.)

Intermediate cross-connects in a premises distribution system, cross-connects located telecommunications closets.

International Organization for Standardization (ISO) a worldwide federation of national standards bodies (ISO National Bodies). The work of preparing international standards is normally carried out through ISO technical committees. Draft proposals for international standards adopted by the technical committees are circulated to the National Bodies for approval before their acceptance as Draft International Standards by the committee.

Internet a large collection of connected networks, primarily in the United States, running the Internet suite of protocols. Sometimes referred to as the DARPA Internet, NSF/DARPA Internet, or the Federal Research Network.

Internet suite of protocols a collection of computer-communication protocols originally developed under DARPA sponsorship, including the transmission control protocol/internet protocol (TCP/IP).

interoffice channel in LEC tariffs, the channel connecting two serving COs (more accurately serving wire centers). In IXC tariffs, the channel connecting two serving IXC POPs.

interoffice transmission facilities used to connect LEC switching systems.

intra-LATA services, revenues, functions, etc., that relate to telecommunications originating and terminating within a single LATA. (The domain of LECs.)

isochronous signals periodic signals in which the time interval that separates any two corresponding significant occurrences or level transitions, is always equal to some unit interval or a multiple of that unit interval. For example, in digitized voice signals, ideally voice samples occur isochronously at precisely the sampling interval or frame rate. Packet data signals are not isochronous.

key telephone system (KTS) an arrangement of multiline telephones and associated equipment that permits the station user to depress buttons (keys) to access different central office or PBX lines, as well as to perform other functions. Typical functions include answering or placing a call on a selected line, putting a call on hold, using the intercom feature between phones at the same location, or activating a signal buzzer.

LAN (IEEE 802.3 10BaseT)   an Institute of Electrical and Electronic Engineers specification for a class of LANs using four-pair unshielded twisted pair (UTP) cable. See *local area network*.

LDN   listed directory number; generally an organization's main telephone number that appears in the telephone book.

line   see *loop transmission facilities*.

local access and transport area (LATA)   a geographic area (called an "exchange" or "exchange area" in the MFJ) within each BOC's franchised area that has been established by a BOC in accordance with the provisions of the MFJ for the purpose of defining the territory within which a BOC may offer its telecommunications services. In 1989, there were 198 LATAs, also referred to as market service areas (MSAs), in the United States.

local area network (LAN)   a premises high-speed (typically in the range of 10 Mbps) data communications system wherein all segments of the transmission medium (typically coaxial cable, twisted-pair or optical fiber) are contained within an office or campus environment.

local channel   in LEC tariffs, the local loop that connects customer premises to serving LEC wire centers. In IXC tariffs, the network components (transmission, switching, other) used to connect customer premises to serving IXC POPs.

loop signaling   a method of signaling over dc circuit paths that utilizes the metallic loop formed by the line or trunk conductors and terminating circuits.

loop start   A supervisory signal given at a telephone or PBX in response to closing the loop's DC current path.

loop transmission facilities   connect switching systems to customer premises equipment throughout the serving area. A loop is a transmission path between a customer's premises and a LEC central office. The most common form of loop, a pair of wires, is also called a line. A "loop" can be derived from digital loop carrier (DLC) systems also referred to as subscriber loop carrier (SLC) systems.

main cross-connects   in a premises distribution system, cross-connects located in an equipment room.

media   see *transmission medium; guided media; unguided media*.

message telecommunications service (MTS)   non-private-line intrastate and interstate long-distance that uses in whole or in part the public switched telephone network (PSTN).

microwave   in telecommunications, frequencies above 1 GHz.

modems (MOdulator/DEModulators)   devices that transform digital signals generated by data terminal equipments (DTEs) to analog signal formats, suitable for transmission through the extensive, world-wide connectivity of public and private, switched (dial-up) and non-switched telephone voice networks.

modification of final judgment (MFJ)   a ruling issued by U.S. District Court Judge Harold Greene which concluded the U. S. Justice Department's antitrust suit against AT&T by modification of an earlier (1956) consent decree's final judgment.

modulation   is the process of varying certain parameters of a carrier signal—i.e., a signal suitable for modulation by an information signal—by means of another signal (the modulating or information bearing signal).

multiline telephone   A telephone that incorporates visual displays and switches (keys) that permit the station user to access more than one central office or other line and to perform other desired functions. Typical functions include answering or originating a call on a selected line, putting a call on hold, operating an intercom feature, a buzzer, etc. Displays can indicate busy, ringing and message waiting status.

multimode optical fiber(s) multimode fibers, with much wider cores than single mode fibers, allow light to enter at various angles, and reflect (bounce off of) core-clad boundaries as electromagnetic (light) wave propagates from transmitter to receiver. From a technical performance trade-off point of view, single mode fiber exhibits bandwidths of up to 100,000 MHz (MHz = 1,000,000 hertz or cycles per second = one megahertz) while multimode band-width is in the range of 1,000 to 2,000 MHz (1,000 MHz = one billion hertz = one gigahertz = 1 GHz). See *optical fiber(s)*; *single mode fiber(s)*.

multiplexing a technique that enables a number of communications channels to be combined into a single broadband signal and transmitted over a single circuit. At the receiving terminal, demultiplexing of the broadband signal separates and recovers the original channels. Multiplexing makes more efficient use of transmission capacity to achieve a low per channel cost. Two basic multiplexing methods used in telecommunications systems, are frequency division multiplexing (FDM) and time division multiplexing (TDM).

network control point (NCP) in virtual private networks a centralized data base that stores a subscriber's unique VPN definition. Highly sophisticated, this database screens every call and applies call processing control in accordance with customer defined requirements.

network operating system (NOS) software that controls the execution of network programs and modules. Structurally, networking software comprises multiple modules, most residing in network servers, but some must be installed in each terminal/station that can access network resources. Peer-to-peer NOSs permit any terminal/station to act as a resource server or a client, and can be based on Microsoft's Disk operating system (MS-DOS) designed for IBM and compatible PCs. Since MS-DOS is not designed to run multiple programs and respond to many simultaneous users, most NOSs designed for large networks with dedicated servers/superservers, have a multitasking and multiuser architecture. Advanced NOS products support network management, diagnostics, and administration, as well as, primary server, client, device and external network driver functions.

network services specified sets of information transfer capabilities furnished to users between telecommunications network points-of-termination. Network services categories include access and transport, public and private, and switched and non- switched.

Open Systems Interconnection (OSI) standards for the exchange of information among systems that are "open" to one another by virtue of incorporating ISO standards. The OSI reference model segments communications functions into seven layers. Each layer relies on the next lower layer to provide more primitive functions and, in turn, provides services to support the next higher layer.

operating telephone company any Bell-operating company or independent telephone company (termed exchange carrier in the MFJ) operating in North America.

optical fiber(s) lightguides for electromagnetic waves in the infrared and visible light spectrum composed of concentric cylinders made of dielectric materials with different indices of refraction (i.e., velocity of propagation normalized to the velocity of light in free space). At the center is a core comprising the glass or plastic strand or fiber in which a lightwave travels. A low index of refraction clad surrounds the core and is itself enclosed in a light-absorbing jacket that prevents interference among multi-fiber cables. Multi-fiber cable can be purchased with between 2 and 136 fibers.

out-of-band signaling   uses the same channel path as the voice traffic but signaling is in a frequency band outside that used for the voice traffic. In digital systems, out-of-band signaling may take the appearance of an allocated bit position or a dedicated channel or time slot.

personal communications   provides at least one human operator with direct terminal access and real-time or near-real time interactive communications with a remote human operator or an information system resource. Personal communications can refer to a broad range of services, systems and equipment, e.g., facsimile machines, landline telephones, cellular telephone systems and emerging personal communication system (PCS) adjuncts, and a variety of radio systems including pagers, hand-held remote data entry terminals, and autonomous citizen-band-like radio systems.

point-of-presence (POP)   a physical location within a LATA which an IXC establishes for the purpose of gaining access to BOC/LEC networks within the LATA using LEC provided access services. An IXC may have more than one POP within a LATA and the POP may support public and private, switched and non-switched services.

premises distribution system (PDS)   the transmission network inside a building or among a group of buildings, for example an office park or a campus. PDS is used in this book as a generic term although AT&T used it to describe a specific product offering. The PDS connects desktop and other station equipment with common host equipment, (e.g., switches, computers and building automation systems), and to external telecommunications networks.

private branch exchange (PBX)   a premises switching system, serving a commercial or government organization, and usually located on that organization's premises. PBXs provide telecommunications services on the premises or campus, (e.g., internal calling and other services), and access to public and private telecommunications network services.

private network   a network made up of circuits and, sometimes, switching equipment, for the exclusive use of one organization.

protocols   strict procedures for the initiation, maintenance and termination of data communications. Protocols define the syntax (arrangements, formats and patterns of bits and bytes) and the semantics (system control, information context or meaning of patterns of bits or bytes) of exchanged data, as well as numerous other characteristics (data rates, timing, etc.).

public switched telephone network (PSTN)   denotes those portions of the LEC and IXC networks that provide public switched telephone network services.

pulse code modulation (PCM)   a modulation scheme involving conversion of a signal from analog to digital form by means of coding. See *modulation*.

quantizing noise   in any analog-to-digital conversion process, e.g., PCM, the difference between the converted binary value and the actual analog signal's amplitude.

regional Bell holding company (RBHC)   one of 7 regional companies created by the AT&T divestiture to assume ownership of the Bell operating companies. They are Ameritech, Bell Atlantic, Bell South, NYNEX, Pacific Telesis, Southwestern Bell and US West.

repeater   in digital transmission, equipment that receives a pulse train, amplifies it, retimes it, and then reconstructs the signal for retransmission. In IEEE 802 local area network (LAN) standards, a repeater is essentially two transceivers joined back-to-back and attached to two adjacent LAN segments. See *transceiver*.

routers   in IEEE 802 local area network (LAN) standards, devices that connect autonomous networks of like architecture at the network layer (layer 3). Unlike a bridge which operates transparently to communicating end-terminals at the logical link layer (layer 2), a router reacts only to packets addressed to it by either a terminal or another router. Routers perform packet (as opposed to frame) routing and forwarding functions; they can select one of many potential paths based on transit delay, network congestion or other criteria. How routers perform their functions is largely determined by the protocols implemented in the networks they interconnect.

satellite communications   entails microwave radio, line-of-sight propagation from a transmitting earth terminal (i.e., usually ground based but potentially ship or airborne) through the atmosphere and outer space media to a satellite, and back to earthbound receiving terminals. In essence, satellites are equivalent to orbiting microwave repeaters.

server   in a network, equipment that makes available file, database, printing, facsimile, communications or other services to client terminals/stations with access to the network. A gateway is a server which permits client terminal/station access to external communications networks and/or information systems.

service management system (SMS)   in virtual private networks a facility used to build and maintain a VPN database allowing customers to program specific functions for unique business applications. The SMS contains complete specifications of customer defined private network specifications including location data, numbering plan, features, screening actions, authorization codes, calling privileges, etc. This information is downloaded (transmitted) to network control points (NCPs) which implement its instructions on a customer-by-customer basis.

shielded twisted pair   twisted copper paired wire cable with an outer metallic sheath surrounding insulated conductors. See *unshielded twisted pair*.

signal   usually a time-dependent value attached to an energy propagating phenomenon used to convey information. For example, an audio or sound signal in which the data is characterized in terms of loudness and pitch.

signaling System No. 7, SS #7   international common channel signaling system recommendations established by the CCITT.

signaling   the process of generating and exchanging information between components of a telecommunications system to establish, monitor, or release connections (call handling functions) and to control related network and system operations (other functions).

simple network management protocol (SNMP)   the application protocol offering network management service in the Internet suite of protocols. A structure for formatting messages and transmitting information between reporting devices (agents) and data collection programs. Developed jointly by the Department of Defense, industry and the academic community as part of the TCP/IP protocol suite; ratified as an Internet standard in Request for Comment (RFC) 1098.

simplex   a transmission path capable of transmitting signals in only one direction.

single mode optical fiber(s)   have sufficiently small core diameters in relation to the wavelength (frequency) of operation that electromagnetic (light) wave is constrained to travel in only one transverse path from transmitter to receiver. This requires the utmost in angular alignment of light emitting devices at points where light enters the fiber and results in higher transmitter/termination costs than multimode fiber systems. See *multimode optical fiber(s)*.

space division switch   a switch which implements the switch matrix using a physical, electrical, spatial link. Where older space division switches used electro-mechanical mechanisms with metallic contacts, modern space-division switches are implemented electronically using integrated circuits. (Usually denoted by "S" in combined time and space division switches).

special services   any of a variety of LEC and IXC switched, non-switched, or special rate services that are either separate from public telephone service or contribute to certain aspects of public telephone service. Examples include PBX tie trunks, foreign exchange (FX) and private line services. These services are important to business telecommunication planners/users.

station equipment   a component of telecommunications systems such as a telephone or data terminal, generally located on the user's premises. Its function is to transmit and receive user information (traffic), and to exchange control information with the network to access communications services.

superframe format (SF) framing format (D3/D4—mode 3)   The most widely used T1 carrier framing format in which the bipolar bit stream is organized into superframes each consisting of 12 frames. To ensure timing, the signal must consist of at least one "1" bit in every 15 bits, and at least 3 "1" bits in every 24 bits. See *extended superframe*.

supervisory signals   signals used to indicate or control the states of circuits involved in a particular switched connection. A supervisory signal indicates to equipment, to an operator, or to a user that a particular state in the call has been reached and may simplify the need for action.

switch matrices   the mechanism that provides signal paths between its input and output terminations. Modern matrices are electronic and involve either time or space division switching. A time division switch employs a TDM process, in a time-slot interchange (TSI) arrangement. In space division, a physical, electrical, spatial link is established through the switch matrix. Whereas older space division switches used electro-mechanical mechanisms with metallic contacts, modern space-division switches are implemented electronically using integrated circuits.

switching   refers to the process of connecting appropriate lines and/or trunks to form a desired communications path between two station sets, or more generally, any two arbitrary points in a telecommunications network. Included are all kinds of related functions, such as signaling, monitoring the status of circuits, translating address to routing instructions, alternate routing, testing circuits for busy conditions, and detecting and recording troubles.

switching systems   interconnect transmission facilities at various network locations and route traffic through a network.

Systems Network Architechture (SNA)   IBM's proprietary description of the logical structure, formats, protocols, and operational sequences for transmitting information units through and controlling network configuration and operation.

T1 Carrier   a time-division multiplexed digital transmission facility capable of supporting 24 voice channels, (each encoded as a 64 kbps PCM DS0 signal), producing an aggregate multiplexer output signal at the 1.544 Mbps DS1 rate. Developed in the 1960s, T1 carrier, is designed to operate, full duplex over two pairs in unshielded twisted pair (UTP) cable.

TCP/IP   transmission control protocol/internet protocol. The transport layer and internet layer, respectively, of the Internet suite of protocols. TCP corresponds to layer 4 of the OSI protocol stack; IP performs some of the functions of layer 3. It is a connectionless protocol used primarily to connect dissimilar networks to each other.

tandem switching system   a broad functional category describing systems that connect trunks to trunks, and route traffic through a network.

tariff   a published rate for a specific telecommunications service, equipment, or facility that constitutes a public contract between the user and the telecommunications supplier (i.e., carrier); tariff services and rates are established by and for telecommunications common carriers in a formal process in which carriers submit filings for federal or state government regulatory review, public comments, possible amendment and approval.

telecommunications   any process that enables one or more users to pass to one or more other users information of any nature delivered in any usable form, by wire, radio, visual, or other electrical, electromagnetic, optical means. The word is derived from the Greek *tele*, "far off," and the Latin *communicare*, "to share."

telecommunications business applications   At the highest level, business applications are unique aggregations of telecommunications services that satisfy particular enterprise needs, for example medical/health care, hospitality, airline reservation, etc. Lower level business applications, for example, station-to-station calling within a premises, are enterprise independent. For these situations telecommunications services correspond directly to generic business applications.

telecommunications closet   in a premises distribution system, an area for connecting the horizontal and backbone wiring and for containing active or passive PDS equipment.

telecommunications network   a system of interconnected facilities designed to carry traffic from a variety of telecommunications services. The network has two different but related aspects. In terms of its physical components, it is a facilities network. In terms of the variety of telecommunications services that it provides, it can support a set of many traffic networks, each representing a particular interconnection of facilities.

telecommunications service   is a specified set of information transfer capabilities provided to a group of users by a telecommunications system.

terrestrial microwave radio   transmission systems consisting of at least two radio transmitter/receivers (transceivers) connected to high gain antennas (directional antennas which concentrate electromagnetic or radiowave energy in narrow beams) focused in pairs on each other. The operation is point-to-point, that is, communications are established between two and only two antennas (installations) with line-of-sight visibility. This can be contrasted to point-to-multipoint systems like broadcast radio or television.

time division multiplexing (TDM)   a transmission facility shared in time (rather than frequency), i.e., signals from several sources share a single channel or bus by using the channel or bus in successive time slots. A discrete time slot or interval is assigned to each signal source.

time division switch   a switch which implements the switch matrix using the TDM process, in a time-slot interchange (TSI) arrangement. (Usually denoted by T in combined time and space division switches).

token passing bus LAN (IEEE 802.4)   a LAN using a deterministic access mechanism and topology in which all stations actively attached to the bus "listen" for a broadcast token or supervisory frame. Stations wishing to transmit must receive the token before doing so; however the next logical station to transmit may not be the next physical station on the bus. Access is controlled by preassigned priority algorithms.

token passing ring LAN (IEEE 802.5)   a LAN using a deterministic access mechanism and topology, in which a supervisory frame (or token) is passed from station to adjacent station sequentially. Stations wishing to transmit must wait for the "free" token to arrive before transmitting data. In a token ring LAN the start and end points of the medium are physically connected, leading to a ring topology.

traffic   the flow of information within a telecommunications network.

transceiver   a generic term describing a device that can both transmit and receive. In IEEE 802 local area network (LAN) standards, a transceiver consists of a transmitter, receiver, power converter and for CSMA/CD LANs, collision detector and jabber detector capabilities. The transmitter receives signals from an attached terminal's network interface card (NIC) and transmits them to the coaxial cable or other LAN medium. The receiver receives signals from the medium and transmits them via the transceiver cable and NIC to the attached terminal. The jabber detector is a timer circuit which protects the LAN from a continuously transmitting terminal.

transfer mode   a generic term for switching and multiplexing aspects of broadband integrated services digital networks (BISDN), adopted by CCITT Study Group XVIII.

transmission control protocol/internet protocol (TCP/IP)   see *Internet suite of protocols.*

transmission facilities   provide the communication paths that carry user and network control information between nodes in a network. In general, transmission facilities consist of a medium (e.g., free space, the atmosphere, copper or fiber optic cable) and electronic equipment located at points along the medium. This equipment amplifies (analog systems) or re-generates (digital systems) signals, provides termination functions at points where transmission facilities connect to switching systems, and may provide the means to combine many separate sets of call information into a single "multiplexed" signal to enhance the transmission efficiency.

transmission impairments   degradation caused by practical limitations of channels (e.g., signal level loss due to attenuation, echo, various types of signal distortion, etc.), or interference induced from outside the channel (such as power-line hum or interference from heavy electrical machinery).

transmission medium   any material substance or "free space" (i.e., a vacuum) that can be, or is, used for the propagation of suitable signals, usually in the form of electromagnetic (including lightwaves), or acoustic waves, from one point to another; unguided in the case of free space or gaseous media, or guided by a boundary of material substance.

transport services   network switching, transmission and related services that support information transfer capabilities between originating and terminating access service facilities.

trunk   in a network a communication path connecting two switching systems used to establish end-to-end connections between customers.

twisted pair   see *unshielded twisted pair.*

twisted pair   the most common type of transmission medium, consisting of two insulated copper wires twisted together. The twists or lays are varied in length to reduce the potential for interference between pairs. In cables greater than 25 pair, the twisted pairs are grouped and bound together in a common cable sheath.

unguided media   any medium in which boundary effects between "free space" and material substances are absent. The "free space" medium may or may not include a gas or vapor. Unguided media including the earth's atmosphere and outer space support terrestrial and satellite radio and optical transmission.

unshielded twisted pair (UTP)   two wood pulp or plastic-insulated copper conductors (wires), twisted together into pairs, capable of propagating electromagnetic waves. The twists, or lays, are varied in length to reduce the potential for signal interference between pairs, in multipair cables. Wire sizes range from 26 to 19 gauge (i.e., 0.016 to 0.036 inch in diameter) and are typically manufactured in cables of from 2 to 3600 pairs. Shielded twisted pair cable is similar to UTP, but the twisted pairs are surrounded by a cylindrical metallic conductor which is clad with an insulating sheath. See *cable; conductor.*

very small aperture terminal (VSAT) earth terminals using small antennas (1.5-6 feet in diameter). This technology typically operates in the Ku band (11/14 GHz), and Ka band (20/30 GHz).

video conferencing the real-time, usually two-way transmission of voice and images between two or more locations. Today, both voice and video analog signals are digitized by video codecs before transmission which can involve wide bandwidths. To conserve bandwidth, some systems employ "freeze frame," where a television screen is only "repainted" every few seconds. Codecs for higher quality full motion video, attempt to minimize bandwidth requirements by taking advantage of intervals with relatively little motion (which require smaller bandwidths), and by trading-off smooth motion tracking and picture resolution.

virtual private networks (VPN) services using public network facilities augmented by network control point and service management system facilities wherein traffic is routed through the public network under computer control in a manner that makes VPN service indistinguishable from dedicated facilities-based private networks. Customers can define, change and control network resources with the same or more flexibility as afforded by facilities-based private networks.

waveform amplitude (magnitude) versus time representation of signals.

wide-area telecommunications services (WATS) a service permitting customers to make (OUTWATS) or receive (INWATS) long distance voice or data calls and to have them billed on a bulk rather than an individual call basis. The service is provided by means of special private access lines connected to WATS equipped central offices. A single access line permits inward or outward service but not both.

wink-start a supervisory signal that consists of an off-hook followed by an on-hook signal, exchanged between two switching systems. The wink-start signal is generated by the called switch to indicate to the calling switch that it is ready to receive address signal digits.

work area in a premises distribution system, an area containing stations and the connections between those stations and their telecommunications (information) outlets.

IXC Selected Tariffs

AT&T Tariffs

FCC #1	Long Distance MTS	
	Custom Network Services including:	
	Software Defined Network	
	MEGACOM WATS	
	MEGACOM 800	
FCC #2	WATS	
	WATS 800	
FCC #4	Switched Digital Services including:	
	ACCUNET Packet Service	

FCC #9	Private Line Service including:
	ACCUNET T1
	Dataphone Digital Service
	Voice Grade
	ACCUNET Spectrum of Digital Service
	Private Network Switching

FCC #10   Rate Center Directory
Points-of-Presence
Service Availability

FCC #11   Private Line Local Channels

FCC #12   Custom Designed Integrated Service

FCC #16   Competitive Government Service including:
FTS 2000
Defense Commercial Telecommunications Network

CT   Contract Tariff Service

MCI Tariffs

FCC #1   Customized Business Service including:
MTS
WATS
PRISM I, II, III
800 Service
Private Lines
VNET

FCC #7   Government Telecommunications Service

US Sprint Tariffs

FCC #1   Long Distance MTS

FCC #2   WATS Services

FCC #3   Analog Private Line Service

FCC #5   Virtual Private Network

FCC #6   Video Teleconferencing Service

FCC #7   Digital Private Line Service

FCC #8   Access Services (Local Channels)

Acronyms and Abbreviations

ACD	automatic call distribution
ACF	access coordination fee
AMI	alternate mark inversion
ANI	automatic number identification
ANSI	American National Standards Institute
ASB	asynchronous balanced mode
ASI	alternate space inversion
ATM	asynchronous transfer mode
AWG	American Wire Gauge
B8ZS	bipolar eight-zero substitution
BF	framing bit
BFt	terminal framing bit
BIPS	billion instructions per second
BISDN	broadband integrated services digital network
BOC	Bell-operating company
BRI	basic rate interface
CALC	customer access line charges
CAP	competitive access provider
CBR	continuous bit rate
CCIS	common-channel interoffice signaling
CCS	centi-call seconds
CCS	common-channel signaling
CCSA	common-control switching arrangement
CDMA	code division multiple access
CMIP	common management information protocol
CO	central office
COC	central office connections
CPU	central processing unit
CSDC	circuit-switched digital capability
CSMA/CD	carrier sense multiple access with collision detection
CSR	Centrex station rearrangement
CSU	channel service unit
DACS	digital access & cross-connect system
DARPA	Defense Advanced Research Projects Agency
DCE	data circuit terminating equipment
DCP	digital communications protocol
DCS	digital cross-connect system
DDD	direct distance dialing
DDN	Defense Data Network
DLCI	data link connection identifier
DOD	direct outward dialing
DQDS	distributed queue dual bus
DSP	digital signal processor
DSS	digital subscriber service

DSS/BLF	direct station selection/busy lamp field
DSU	data service unit
DTMF	dual tone multiple frequency
DTS	digital termination service
E-mail	electronic mail
ECSA	Exchange Carriers Standards Assoc.
EKTS	electronic key telephone system
EMI	electromagnetic interference
EO	end office
EPSCS	enhanced private switched communications service
ESF	extended superframe
ESN	electronic switched network
FDDI	fiber-distributed data interface
FDL	facility data link
FDM	frequency division multiplexing
FEP	front-end processor
FIPS	Federal Information Processing Standard
FSS	fully separated subsidiary
FTAM	file transfer access & management
FX	foreign exchange
GOSIP	Government Open Systems Interconnection Profile
HDLC	high-level data link control
HVAC	heating, ventilation and air conditioning
I-MAC	isonchronous media access controller
IDN	integrated digital network
IEC	International Electrotechnical Commission
IOC	interoffice channel
IP	internet protocol
IPX	internetwork packet exchange
ISDN	integrated services digital network
ISO	International Organization for Standardization
ITC	independent telephone company
ITU	International Telecommunications Union
IVR	integrated voice response
JPEG	Joint Photographic Experts Group
JTM	job transfer manipulation
KSU	key service unit
KTS	key telephone system
LAN	local area network
LATA	local access and transport area
LCR	least-cost routing
LDN	listed directory number
LEC	local exchange carrier
LSI	large-scale integrated circuit
MAAP	maintenance & administration panels
MAC	media access control

MAC	moves, adds, and changes
MACSTAR	multiple access customer station rearrangement
MAN	metropolitan area network
MCU	mobile control unit
MF	multiple frequency
MFJ	*Modification of Final Judgment*
MFOTS	Military Fiber-Optic Transmission System
MHS	message handling system
MIB	management information base
MIPS	million instructions per second
MPEG	Moving Pictures Experts Group
MSS	metropolitan switching system
MTS	message telecommunications service
MTSO	mobile telephone switching office
NBEC	non-Bell exchange carrier
NCP	network control point
NCTE	network channel terminating equipment
NI	network interface
NIC	network interface card
NIST	National Institute of Standards and Technology
NOC	network operations center
NOS	network operating system
NPA	numbering plan area
NSEP	National Security & Emergency Preparedness
NT	network termination
NTSC	National Television System Committee
OA&M	operation administration & maintenance
OCC	other common carriers
OPX	off-premises extension
OSC	operating system control
OSI	open systems interconnection
OSS	operations support systems
P-MAC	packet media access controller
PAD	packet assembler-disassembler
PCB	printed circuit board
PDS	premises distribution system
PHY	physical-layer protocol
PMD	physical-layer media-dependent
POP	point-of-presence
POT	point of termination
PPSN	public packet switched network
PRI	primary rate interface
PSN	packet switched network
PSPDN	packet switched public data network
PSTN	public switched telecommunications network
PTT	postal, telephone, and telegraph

PUC	public utility commission
PVC	permanent virtual circuit
RBHC	regional Bell holding company
RBS	robbed bit signaling
RF	radio frequency
RFC	request for comment
RFI	radio frequency interference
RFP	request for proposal
RSU	remote switching unit
SAFENET	survivable adaptable fiber-optic embedded network
SDH	synchronous digital hierarchy
SDN	software defined network
SF	single frequency
SMDR	station message detail record
SMDS	switched multimegabit data system
SMT	station management technology
SNA	Systems Network Architecture
SNI	subscriber network interface
SNMP	simple network management protocol
SONET	synchronous optical network
SS	signaling system
SSN	switched service network
STM	synchronous transfer mode
STP	shielded twisted pair
STP	signaling transfer/point
TA	terminal adapter
TCP	transmission control protocol
TDM	time division multiplexing
TDMA	time division multiple access
TIA	Telecommunications Industry Association
TP	transaction processing
UIS	Universal Information Services
UNI	user-network interface
UTP	unshielded twisted pair
VAD	value-added distributor
VAN	value-added network
VAR	value-added reseller
VBR	variable bit rate
VCI	virtual circuit identifier
VCS	virtual circuit switch
VPI	virtual path identifier
VPN	virtual private network
VRU	voice response unit
VSAT	very small aperture terminal
WAN	wide area network
WARC	World Administrative Radio Consortium
WATS	wide-area telecommunications services

Bibliography

1. Briere, Daniel D. *Long Distance Services A Buyer's Guide.* Massachusetts: Artech House, Inc., 1990.
2. Cavanagh, James P. *Applying the Frame Relay Interface to Private Networks.* New York: IEEE Communications, 1992.
3. Dicenet, G. *Design and Prospects for the ISDN.* Massachusetts: Artech House, Inc., 1987.
4. Freeman, Roger L. *Telecommunication System Engineering. 2nd ed.* New York: John Wiley & Sons, 1989.
5. Martin, James. *Telecommunications and the Computer. 3rd ed.* New Jersey: Prentice Hall, 1990.
6. Malamud, Carl. *STACKS—Interoperability in Today's Computer Networks.* New Jersey: Prentice-Hall, Inc., 1992.
7. Minzer, Steven E. *Broadband ISDN and Asynchronous Transfer Mode (ATM).* New York: IEEE Communications, 1992.
8. Noll, A. Michael. *Introduction to Telephones and Telephone Systems.* Massachusetts: Artech House, Inc., 1986.
9. Rey, R. F., ed. *Engineering and Operations in the Bell System. 2nd ed.* New Jersey: AT&T Bell Laboratories, 1984.
10. Rider, Michael J. *Protocols for ATM Access Networks.* New York: IEEE Network, January 1989.
11. Sapronov, Walter. *Telecommunications and the Law.* Maryland: Computer Science Press, 1988.
12. Scott Schauer. *Frame Relay: Designing Virtual Private Networks.* Illinois: Busines Communications Review, 1991.
13. Stallings, William. *ISDN: An Introduction.* New York: MacMillan Publishing Company, 1989.
14. Tanenbaum, Andrew S. *Computer Networks.* New Jersey: Prentice-Hall, Inc., 1981.
15. Tedesco, Eleanor Hollis. *Telecommunications for Business.* Massachusetts. PWS-Kent, 1990.
16. Weik, D., Sc., Martin H. *Communications Standard Dictionary.* New York: Van Nostrand Reinhold Company, 1983.
17. *The Buyer's Guide to Frame Relay Networking.* Virginia: Netrix Corporation, 1991.
18. *Definity and System 85 System Description.* Colorado: The AT&T Documentation Management Organization, 1990.
19. *IEEE Standards for Local Area Networks: Supplements to Carrier Sense Multiple Access with Collision Detection (CSMA/CD) Access Method and Physical Layer Specifications.* New York: Institute of Electrical and Electronics Engineers, Inc., 1987.
20. *IEEE Standards for Local Standards for Local Area Networks: Carrier Sense Multiple Access with Collision Detection (CSMA/CD) Access Method and Physical Layer Specifications.* New York: Institute of Electrical and Electronics Engineers, Inc., 1988.
21. *IEEE Standards for Local Area Networks: Token-Passing Bus Access Method and Physical Layer Specifications.* New York: Institute of Electrical and Electronics Engineers, Inc., 1988.
22. *IEEE Standards for Local Area Networks: Token-Passing Access Method and Physical Layer Specifications.* New York: Institute of Electrical and Electronics Engineers, Inc., 1988.
23. *IEEE Standards for Local Area Networks: Logical Link Control.* New York: Institute of Electrical and Electronics Engineers, Inc., 1988.

24. *Notes on the BOC Intra-LATA Networks—1986.* Technical Reference TR-NPL-000275 Issue 1, April 1986.
25. *Project 802—Local & Metropolitan Area Networks.* Institute of Electrical and Electronics Engieers, Inc. Unapproved Draft—Published for Comment Only.
26. *Telecommunications Transmission Engineering. 3rd ed. Vol. 1:* Principles. Bellcore, 1990.
27. *Telecommunications Transmission Engineering. 3rd ed. Vol. 2:* Facilities. Bellcore, 1990.
28. *Telecommunications Transmission Engineering. 3rd ed. Vol. 3:* Networks and Services, Bellcore, 1990.
29. *Guide to Networking Services*—CCMI, Inc., 11300 Rockville Pike, Rockville, MD 20852-3030, 1992.

Index

unshielded twisted pair (UTP), 30, 32, 107, 237, 239
user network interface (UNI), 280

V

variable bit rate (VBR), 255-256
very small aperture terminal (VSAT), 36
videoconferencing and imaging, 306-309
  market participants, 309
  standards, 308-309
  technology, 308
video dial tone services, 12
videophones, 306-309
virtual circuit identifier (VCI), 226
virtual circuit switch (VCS), 258
voice services, 141-205
  and networks, 143-152
  overview, 175-176
  selecting network, 175-194
  selecting network services, 153-173
  selecting premises services, 195-205
  taxonomy, 146
voice systems
  alternative coding techniques, 25-27

Centrex message processing, 77-78
KTS message processing, 89
PBX message processing, 64-65
signaling, 93-106
transmitting digital signals over voice networks, 24-25
VF, 36-37

W

waveforms, 17
wide-area network (WAN), 253-266
  data communication services, 258-266
  overview, 253
  requirements and environmental trends, 254-257
  selecting data network services, 267-276
wide-area telecommunications services (WATS), 144, 165
  toll charges, 159
Wireless In-Building Network (WIN), 306
wiring, (see also cable)
  horizontal, 110-112
  universal, 197
  wire-center coordinates, 119

About the Authors

Joseph A. Pecar is President of Joseph A. Pecar and Associates, Inc. He has 34 years of systems engineering experience in the design, development, implementation, and integration of command, control, communications, and intelligence systems for the government and private industry. **Roger J. O'Connor**, telecommunications director, Vertech, Inc., provides technical consulting services to government and commercial user-level organizations, consulting services to government and commercial user-level organizations, consulting services to government and commercial user-level organizations, consulting services to government and commercial user-level organizations, vendors, and manufacturers. **David A. Garbin** is a principal engineer in the advanced Information Systems Division at MITRE Corporation and has 20 years of technical and management experience in all facets of telecommunications networks.